Y0-BUP-165

GLOBAL CHALLENGE AND LOCAL RESPONSE

Initiatives for Economic Regeneration in Contemporary Europe

COOPERATION FOR DEVELOPMENT SERIES
Series Director, ***Ervin Laszlo***
EUROPEAN PERSPECTIVES VOLUME 2

This volume grew out of research undertaken as a part of the United Nations University's European Perspectives Project, a component of the University's Regional Perspectives Programme. The purpose of this programme has been to identify the distinctiveness of the various regions and to analyse their position within the world system and, further, to explore the possibilities of constructive links between them. Analysis at the regional level has been insufficiently realized, and, as a result, the specific regional dynamics often escape proper notice. Also, the region as a level of political decision-making has been underutilized, too much prominence being given to nation-states on the one hand and 'global negotiators' on the other. Thus, it is hoped the Regional Perspectives Programme will prove to be of both theoretical and practical relevance and value.

The particular focus in this book is the way in which local economies and communities in Europe have undertaken their own development and employment initiatives to mitigate problems resulting from declining global economic growth and the transformation of European industry.

The Project from which this book emanates was commissioned by the United Nations University to W. B. Stöhr, who developed the conceptual and operational framework, selected case-study authors, formulated the guidelines for individual contributors and coordinated their work.

GLOBAL CHALLENGE AND LOCAL RESPONSE

Initiatives for Economic Regeneration in Contemporary Europe

edited by
Walter B. Stöhr

THE UNITED NATIONS UNIVERSITY

MANSELL

LONDON AND NEW YORK

First published 1990 by
Mansell Publishing Limited, *A Cassell Imprint*
Villiers House, 41/47 Strand, London WC2N 5JE, England
125 East 23rd Street, Suite 300, New York 10010, USA

and

The United Nations University
Toho Seimei Building, 15-1, Shibuya 2-chome,
Shibuya-ku, Tokyo 150, Japan

British Library Cataloguing in Publication Data
Global challenge and local response: initiatives for
economic regeneration in contemporary Europe —
(Cooperation for development series. European perspectives
project; V.2).
1. Europe. Regional economic development
I. Stöhr, Walter B, *1928-* II. Series
330.94

ISBN 0-7201-2064-0

Library of Congress Cataloging-in-Publication Data
Global challenge and local response: initiatives for economic
regeneration in contemporary Europe / edited by Walter B. Stöhr.
p. cm. — (Cooperation for development series. European
perspectives; v. 2)
Includes bibliographical references.
ISBN 0-7201-2064-0
1. Europe—Economic conditions—1945- 2. Regional economics.
I. Stöhr, Walter B. II. Series.
HC240.G54 1990
338.94—dc20 89-29069
CIP

This book has been typeset by Colset Private Limited, Singapore and
printed and bound in Great Britain by Biddles Ltd, Guildford and King's Lynn.

Contents

The contributors

Walter B. Stöhr is Professor and Director of the Interdisciplinary Institute for Urban and Regional Studies (IIR) and Dean of the Faculty of Arts and Science at the University of Economics, Vienna, Austria. He was leader of the project on local development and employment initiatives within Europe that is reported in *Global Challenge and Local Response.*

Etele Baráth is a senior researcher at the Centre for Regional Studies of the Hungarian Academy of Sciences, Budapest, and Director of Construction Planning Company.

John Bryden is Programme Director of the Arkleton Trust, which studies new approaches to rural development and education.

Roberto Camagni is Professor of Economics at the University of Padua and also teaches at Bocconi University in Milan, Italy.

Roberta Capello is a researcher at Bocconi University in Milan, Italy.

Clive Collis is Associate Head of the Department of Economics at Coventry Polytechnic, United Kingdom.

Bohdan Gruchman is Professor and Director of the Institute of Planning at the Academy of Economics in Poznań, Poland.

Gerd Hennings is Professor of Local Economic Policy at the School of Spatial Planning, University of Dortmund, the Federal Republic of Germany.

Bengt Johannisson is Professor of Entrepreneurship and Business Development, Lund University, and Chairman of the Department of Management and Law at Växjö University, Sweden.

Klaus R. Kunzmann is Professor and Director of the Institut für Raumplanung (IRPUD) at the School of Spatial Planning, University of Dortmund, the Federal Republic of Germany.

Denis Maillat is Professor of Regional Economy and Director of the Institute of Economic and Regional Research (IRER) at the University of Neuchâtel, Switzerland.

Michael Marshall was formerly a member of a research and analysis team at the West Midlands Enterprise Board with responsibility for developing the Board's economic and industrial strategy for the region.

Deborah J. Moya is on the staff of Empirica, a Bonn- and London-based consultancy firm, where her main area of work is related to the use of training and new technology in the development of training programmes.

David Noon is Associate Head of the Department of Urban and Regional Planning at Coventry Polytechnic, the United Kingdom.

Andreas Novy is a graduate research collaborator at the Interdisciplinary Institute for Urban and Regional Studies (IIR) at the University of Economics, Vienna.

Peter Roberts is Professor and Head of the Department of Urban Planning at Leeds Polytechnic, the United Kingdom, and Chairman of the Regional Studies Association.

Günter Scheer is Manager of the Austrian Association for Indigenous Regional Development (ÖAR) in Vienna.

Ian Scott is a freelance rural community development writer, lecturer and consultant and co-ordinates the Rural Adult Education Project of a newly founded national rural charity in England (ACRE), which provides community development courses for rural communities.

Wolfgang J. Steinle is Managing Director of Empirica, a Bonn- and London-based consultancy firm.

Peter Szaló is a researcher at the Centre for Regional Studies of the Hungarian Academy of Sciences in Budapest.

Antonio Vazquez-Barquero teaches regional and urban economics in the Department of Applied Economics at the Autonomous University of Madrid.

Anneliese Zobl is a trainee in the marketing division of a mail order chain.

Synthesis

Walter B. Stöhr

Crisis is a productive state provided you do not associate it with catastrophe.
(Max Frisch)

Today it is commonplace to hear that a large multisite enterprise is to close down several of its plants and transfer production to other regions. Tens of thousands of workers in different parts of Europe lose their jobs. Small local enterprises are also affected and entire towns are in danger of losing their economic base.

Local economic crises are of concern to all parts of Europe today. Major European industrial regions, several decades ago still pioneers of worldwide capitalist modernization, bastions of power and motors of postwar reconstruction – regions such as the German Ruhr or the English West Midlands – have seen many of what were once their most prestigious plants closed down and lost large proportions of their jobs.

East European industrialization, until recently still the showpiece of socialist economic and social progress, could keep up full employment only by artificial pricing. Major industrial areas are burdened with severe environmental problems and a lack of functional flexibility and competitive capability on the world market. Here too, plants have recently been forced to cut employment, with popular upheaval being the near consequence.

In peripheral rural areas national governments have in the past tried to stem underdevelopment by attracting branch plants, most of which have failed once initial subsidies have run out.

Regions and localities have relied for their development on large-scale enterprise and central-government policies. Only recently has local action moved to the foreground as an alternative. This applies not only to 'old' industrial as well as rural areas, but also to large and intermediate-size cities. Selected examples across Europe are analysed in this book (cf. also Map 1.1 and Tables 1.1 and 1.2).

Has it failed then, this traditional, large-scale development paradigm –

whether driven by market forces or initiated by central government and oriented mainly towards industrial mass production and quantitative economic growth?

In contrast to earlier periods, the causes of the present crisis are international in origin and therefore barely tangible. An abstract process known as worldwide economic restructuring is held responsible for their symptoms and traditional government policies are unable to cope with them. This is true for Western as well as Eastern Europe.

The symptoms of crisis have been spreading rapidly since the middle of the 1970s from country to country and from location to location. Many territorial communities affected have shown no resistance to them. It is almost as though entire local and regional communities had succumbed to a societal 'acquired immune deficiency syndrome' as a result of the 'virus' of international economic restructuring. Could there be a parallel to the medical condition, acquired immune deficiency syndrome (AIDS), which deprives the body of resistance to infections from outside? Whereas the medical AIDS is propagated by the increased interaction and mobility of people, could local community 'AIDS' be propagated by the high mobility of capital and its almost complete severance from location?

It would appear that monocentric reliance on traditional large-scale market-driven, large-organization and central-government-initiated development processes have steadily weakened the capability of territorial communities to confront the challenges of worldwide economic restructuring by indigenous innovation and flexibility.

Central policies have frequently aggravated this. They may have been able to redistribute growth during growth-dominated periods but they have been unable to generate local innovative capacity and promote flexibility during periods dominated by restructuring needs.

Fortunately, this societal immune deficiency syndrome appears to be more easily reversible and less deterministic than the medical one. This assertion is supported by the fact that despite these widespread symptoms of crisis local and regional communities are successfully organizing themselves to meet the worldwide challenges through initiatives distinguished, whether large or small, by their innovation and flexibility.

The hypothesis underlying this book is that the basic prerequisites of development - initiative and entrepreneurship - were broadly latent in all populated areas. Not only were they present in dynamic towns, but also in peripheral areas where these qualities were continually nurtured by the daily natural hazards of survival. As societies become more highly organized and complex, this innovative and entrepreneurial capacity is either encouraged by societal reward (whether financial, religious, organizational or in some other form) or impeded by societal discouragement (central authority, large enterprises, dominance by strong status-quo-preserving power groups, external dominance, etc.). The latter monoculture can lead in many cases to the societal immune deficiency syndrome described above.

For entire local communities to escape this syndrome, however, innovative and entrepreneurial opportunity must not be restricted to individuals or classes, but must encompass the entire territorial community. What is needed is the 'social entrepreneur' described in Chapter 3. But, beyond this, societal mechanisms for the transfer and broad support of innovative and entrepreneurial initiatives, flexible institutional structures and a high degree of local interaction and cooperation are also required to overcome this societal immune deficiency syndrome.

The case studies in this book describe some efforts made in this direction and their results. In one of the oldest industrial areas, the English West Midlands, a broad upgrading of existing industries has been successfully initiated through concerted sector strategies and cooperation between regions with similar structural problems (Chapter 6). In the Ruhr region of Germany (Chapter 7), the Dortmund area has developed a three-phase restructuring strategy of consolidating the existing economic structure (strategy for 'today'), speeding up the innovation process of indigenous firms (strategy for 'tomorrow') and attracting new high-tech firms to the area (strategy for 'the day after tomorrow'). In a further case it seemed necessary to abandon completely the traditional industrial base and convert to an entirely new non-manufacturing base by replacing the formerly world-renowned metallurgical complex of the lower Swansea Valley in South Wales to a leisure, shopping and commercial centre (Chapter 5). Emphasis in these strategies has ranged from conversion within the existing structure, on the one hand, to the development of a completely new structure on the other. Another interesting example from a more recently industrialized area is the conversion in the Jura Arc in Switzerland of an integrated fine-industries complex (watchmaking, machine tools, precision engineering) through the transition from micro-mechanics to micro-electronics facilitated by the internal transfer and mobilization of skills (Chapter 8). In all these cases, technological and organizational changes at the local and regional level and in the cooperative structures within and between regions were required.

Case studies from COMECON countries (Chapters 9 and 10) reveal a similar problem but with different dimensions. Industrial towns formerly developed mainly through central planning mechanisms underwent an often dramatic process of quantitative growth but showed considerably less sustained potential for innovation and qualitative upgrading than neighbouring towns with a higher degree of local initiative and self-regulation.

In rural, frequently peripheral, areas, a dynamic individual can often trigger a sustained series of local development initiatives. The successful ones were oriented towards the broadening of local economic structures based on a more diversified utilization of local resources, combined with technological upgrading, training efforts, design and cultural activities, local financing and improved forms of local cooperation and information exchange. This not only helped to improve local employment and living conditions but also strengthened the local negotiating position with outside entrepreneurs, central government or international organizations. A unilateral dependence on central state funding, on the other hand, often led to extreme vulnerability and constrained local room for manoeuvre. This is in contrast to established mainstream economic development doctrine, which relies mainly on central macro-policy instruments and disregards the territorial societal and structural conditions within which the individual entrepreneur operates.

The advantage of local or regional action in the restructuring process is that it can identify, mobilize and combine diverse potential local resources much better than central policies. Central policies must concentrate on facilitating access to information at the local level and promote the capability to make use of it.

SOME CONCLUSIONS

We shall now summarize some key experiences of local development initiatives in different parts of Europe.

In almost all cases studied, initiative was started by a key individual. Very often, however, these key individuals were no longer synonymous with those groups traditionally considered determinants of local development – local administrators or large-scale entrepreneurs – but, instead, 'unconventional' initiators such as a local veterinarian, a curate or clergyman, a retired military officer, a crafts design specialist, a local intellectual or some other community activist. In short, there is a high proportion of 'informal' initiators. Where such key initiators were to be found in established institutions such as local government or a chamber of commerce, they were often 'new types' of officials, either newly elected (e.g. in Spain after the new 1978 constitution) or especially contracted for this purpose. In metropolitan or old industrial areas these initiators were predominantly local or indigenous, while in rural areas they were predominantly external, often return migrants.

The emergence of an individual initiator was an important but not a sufficient condition, however. In order to become effective this initiator (particularly an external one) had to have (or gain) access to existing local information through social and institutional networks. In other words, individual initiators had to plug into or work through local networks, which for this purpose sometimes had to be restructured or even newly created. The latter was necessary particularly in cases where different local factions were isolated or even antagonistic towards each other. In this case the individual initiator had to become what in Chapter 1, Fig. 1.1 is called a 'local development agent', interrelating different local actors and institutions with each other and motivating them to collaborate. This motivation was often facilitated by the awareness of external threat. A further prerequisite for successful initiatives was high local mobility of relevant information through formal channels (such as a local newspaper) or informal channels (such as intensive social interaction, even at street corners as in Mediterranean countries).

The main actors had to identify with the local community, but they did not necessarily need to be local residents or return migrants. Sometimes an external branch plant manager would act as initiator. The prerequisite seems to be that actors can expect more benefit from local than from external cooperation or interaction, and even branch plants of multisite enterprises increasingly become aware of the importance of the local milieu for their successful operation, be it in terms of the local labour market or of local subcontracting. For similar motives (and not only out of philanthropy) private enterpreneurs may become local 'social entrepreneurs' (Chapter 3).

LOCAL RESTRUCTURING IN THE NORTHERN AND WESTERN PERIPHERY OF EUROPE

About ten of the fifty case studies of local initiatives analysed in this book are located in the northern and north-western periphery of Europe (four Scandinavian countries and two 'Celtic fringe' countries, Scotland and Ireland). Again half of these are located in rural, sparsely settled areas, based primarily on agricultural (including livestock or fishing) activities which are subject to the general restructuring and overcapacity problems of these sectors. The other half of these case studies are located in better accessible areas of Sweden and Denmark, mainly characterized by industrial monostructure of traditional sectors such as glassworks, tobacco processing and shipbuilding. These locations are suffering from the worldwide industrial restructuring process in these specific sectors.

Local initiatives considered successful were found not only – as might have been expected – in the more accessible areas (southern Sweden and Denmark) but also in more remote areas such as Selbu in Norway, or in extremely remote ones such as Connemara in the west of Ireland and Pairc on the Scottish Western Isles (see Map 1.1). The local initiatives considered successful were mainly indigenously triggered and oriented towards mobilization of local entrepreneurial resources, economic diversification, the introduction of new products, the upgrading of skills and the introduction of new organizational forms for economic, cultural and training activities.

The characteristic traits of some of the local initiatives considered successful by the respective authors are as follows. For Maleras, a traditional glassworks area in the south-east of Sweden, Johannisson (Chapter 3) shows how a former glassworks designer acted as a 'social entrepreneur' in mobilizing local community action. This action focused on the stimulation of broadly based local initiatives for the mobilization of local entrepreneurial resources, the creation of locally interrelated manufacturing and service firms and the successful takeover of the glassworks by fifteen former employees and eighty other community members.

For Connemara in the west of Ireland, Scott (Chapter 4) shows how in an extremely remote and less developed rural area the activities of a local curate (similar to the much larger Basque Mondragón Cooperative Federation) led to a number of local development initiatives. These manifested themselves in the creation of a local credit union, a locally controlled farmers' cooperative which was triggered by the closing of a private agricultural supply store, and a local community development council. The main economic impact was achieved by the creation of a community company running self-catering holiday cottages, the surplus from which was invested in the creation of a local cultural and training centre. This centre became the basis for a local craft training programme and a community resource and education programme concentrating on woodworking and design skills, as well as for service support to small enterprises. These local activities were able to attract substantial external assistance, by the European Combat Poverty Programme, for example, which, in its second phase, was already primarily locally managed. These as well as other case studies in this volume show that once indigenous initiatives have been developed and are receiving broad local support, negotiations with outside actors (state agencies or private enterprises) can be undertaken more successfully in the interest of the local community.

For Pairc in the extremely remote Western Isles of Scotland, Bryden (Chapter 4) shows that even in such a very difficult location modest success can be achieved by field workers (contracted by the Scottish Highlands and Islands Development Board) in promoting local resource-based crafts (knitting of island wool, sheepskin rugs) by design training and improvement in commercial functions (project costing, marketing ideas). Under these conditions, however, local negotiations with outside actors on the control of local resources (a multinational company interested in fish farming) became much more difficult.

Less successful local development schemes on the other hand seem to be characterized by a heavy reliance on external (state) agencies, the concentration of efforts on intensifying – rather than diversifying – existing local activities and a lack of local entrepreneurial capacity.

For Pilgrimstad, Johannisson (Chapter 3) shows how, in spite of its relatively favourable location in central Sweden, efforts to compensate for substantial employment and economic loss were unsuccessful; these losses were caused by the transfer of a dominant fibreboard branch plant of a multi-locational company to a coastal location. This locality seems to have become accustomed to external dependence and hierarchical structure, thereby inhibiting local entrepreneurship. The willingness of the population to participate in local initiatives was small and the solutions envisaged by the major local actor, a local politician representing the trade unions (in Sweden very centralized), were mainly of a centralized and bureaucratic character.

In the 'Celtic fringe', local development efforts with higher state dependence (West Kerry) were also amongst the less successful ones. Bryden and Scott (Chapter 4) feel that '... too great a degree of dependence on state funding leads to vulnerability on the one hand and a constraint on the room of manoeuvre on the other'.

A key question, however, seems to be how in areas which lack entrepreneurial capacity and modern organizational and administrative skills (emphasized in the studies on the Scottish Ness Community Cooperative, Chapter 4, for example) self-sustaining local initiatives can be induced. Substantial 'management-assistance' grants like those offered by the Highland and Islands Development Board were considered counter-productive as they fed the habit of high overhead costs which could not be sustained in the long run (Chapter 4). Outside (such as state) agencies also proved little able to identify potential new production lines and new technologies which could be introduced successfully by local enterprises.

The attraction of entrepreneurial capacity from outside such as by the multinational fish-farming enterprises in the Scottish and Irish case studies also appeared to have had debatable success as it led to external control over local resources, thus reducing potential local benefits and local control, and at the same time reinforcing dependence on external agents.

In the cases where local development efforts have had little or no immediate material effect in terms of new economic activities or jobs, unquantifiable advances in terms of increased problem awareness and mobilization of the local population were reported as the major outcome (Chapter 4). It has to be seen whether in a later stage these intangible advances will in fact lead to the needed material benefit for the respective local communities.

RESTRUCTURING IN OLD INDUSTRIAL CORE AREAS OF EUROPE

Especially urgent restructuring problems are usually encountered in 'old' industrial areas. Their structural conditions mostly go back to the first industrialization wave based on mass production of standardized commodities and on what today is termed Taylorist or Fordist production technology, making maximum use of economies of scale. Consequently these areas were usually dominated by a small number of large firms producing commodities pertaining to late phases in the product cycle. The corresponding production processes seldom rely upon recent technological innovation and employ a work force with a low qualification profile.

Case studies analysed in this book are taken from early-industrialized areas in England, Wales and West Germany, but also from some more recently industrialized areas in northern Italy, Switzerland and southern France. Similar problems have emerged in enclaves of early industrialization in the Prato region in Italy and in the Basque country in northern Spain.

Some of the regions analysed, such as County Durham, England, and Swansea in Wales (Chapter 5) or the heavy-industry areas in the Basque country (Chapter 13), were already *persistent problem areas* in the crisis period of the 1930s. Others were amongst the most prosperous economic regions of their respective countries during the Second World War and the subsequent reconstruction period, and were only hit by drastic decline in the early 1970s. Examples of this kind include the English West Midlands (Chapter 6) and the German Ruhr (Chapter 7).

Decline in most of these regions was due not only to their sectoral concentration on traditional mass production industries (mining, iron and steel, shipbuilding, textiles etc.), however, but also due to the displacement of previously existing small and medium-sized firms (SME) – often craft based – by a few large firms which often assumed oligopolistic positions in the regional economy, its labour market and its local government structure. This reduced the probability of entry for new firms which might have contributed to greater diversification of the economic structure and the labour market. New entries were often also discouraged by militant trade union policies as is shown for the Birmingham region (Chapter 6). Problems were further aggravated when during crisis periods the few regionally dominant large firms tended to shift investment abroad. This happened in the Birmingham region of England during the 1930s (Chapter 6) and has been happening in the Ruhr in Germany since the 1970s (Chapter 7); similar disinvestment occurred in the old industrial areas of Spain studied in Chapter 13.

Restructuring was furthermore made difficult by the restricted range of labour skills and yet comparatively high labour costs in these areas, by the fact that environmental conditions had frequently deteriorated, by low levels of local entrepreneurship and by rigid, usually hierarchical, administrative and large-firm structures. All this, along with large-firm disinvestment, in most cases had led to a neglect of future-oriented research and development (R&D) and training activities and to low rates of product and process innovation. A typical case is the West Midlands (Chapter 6).

Crisis symptoms in the case study areas were accordingly severe: the West Midlands lost 29 per cent of manufacturing jobs within a five-year period (1979–84, cf. Chapter 6), the Greater Nottingham area lost 44 per cent of its textile manufacturing jobs in a ten-year period (1971–81) and in the Swansea area total employment declined by 18.5 per cent in a ten-year period (1976–86, cf. Chapter 5). In the Ruhr, employment in the coal industry had fallen by two-thirds within three decades (1956–84), that in the iron and steel industry by 38 per cent within a decade (1974–85) and in the entire manufacturing sector by 31 per cent in fifteen years (1970–85, cf. Chapter 7). Even in the Jura Arc in Switzerland, industrial employment fell by 30 per cent during the same period (1970–85, cf. Chapter 8).

In most of these regions the magnitude of crisis symptoms *surpassed central government capacity* for dealing with them. The gravity and speed of the decline, however, often triggered strong local initiatives and the emergence of new local leading figures (besides the traditional ones from public administration and large enterprises): local bank managers, owners of SME, scientists and research managers, often forming local discussion circles as shown in the case study on the Ruhr (Chapter 7).

In view of the magnitude of crisis symptoms, however, *'success' criteria* have to be seen in terms different from those applied to other, less affected, areas. Usually in the severely depressed areas, even a mitigation of decline must be considered a success, particularly if it is accompanied by structural and qualitative change leading to increased competitiveness and survival chances in future: by improvements in product quality, job qualification, new technologies and/or the introduction of new products in established or new firms. Frequently such success criteria have also been matched by social, educational, environmental and infrastructure improvements (Chapter 5).

Some of the quantitative *short-term results* of local restructuring reported on were: in the English West Midlands under the aegis of its enterprise board, the establishment of thirty-three new enterprises (of which only three failed), of over 2,600 newly created jobs and a multiplier of 1:4, every pound sterling of board investment triggering four pounds of total investment (Chapter 6). In the north-east of England the National Programme of Community Interest helped create 1,700 jobs in the region, and of the assisted firms 90 per cent were locally based. In the Nottingham area twelve fashion workshops were newly established, and in the dyeing and finishing sub-sectors some fifty jobs were newly created and 320 reported saved, while in the Swansea area 1,245 jobs were newly created in the Maritime Quarter (Chapter 5).

While the bare bones of these quantitative results appear very modest compared to the magnitude of the problem, behind them *important structural and qualitative transformations* have sometimes been initiated which might substantially improve the future auspices of development of these regions. Such transformations range from the modernization of dominant existing sectors (emphasized in the English West Midlands), through the insertion of high-technology components into existing structures (as in the Jura Arc in Switzerland) or a planned gradual three-phase shift from modernizing existing sectors to the insertion of new high-technology sectors (the Dortmund case) to the complete substitution of the past mining and manufacturing base by a new service base (Swansea).

The most pronounced emphasis on the *upgrading of existing sectors* is recounted for the English West Midlands in the motor-vehicle industry, the foundry industry and the clothing industry (Chapter 6). Entire industry groups there were the objects of regional restructuring programmes: in the motor-vehicle industry key attempts were to introduce innovation via vehicle design and transmit it through the vehicle producers to components suppliers; in the traditional steel-based sectors technological upgrading was attempted via the machine-tool industry, such as in the development of computer controlled numerical equipment. Workers (many of them of Asian origin) who in the process had become redundant and had taken refuge in the spontaneous creation of small, low-grade textile and clothing firms received systematic upgrading assistance. The intervention programme to upgrade the latter sector comprised the establishment of the West Midlands Clothing Resource Centre offering advisory assistance, training, computer design facilities, and so forth; its aim further was to reduce dependence of small firms on sub-contractors and sales agencies. A condition for using the centre's facilities was the commitment of firms to upgrade pay and working conditions.

The core objective of regional action in the West Midlands, however, was to *recover its major strength of intensive intra-regional industrial linkages* which had disintegrated during the recent crisis. In order for these intra-regional linkages to create positive multipliers, local intervention provided for long-term (equity) investment in fixed capital to overcome the lack of regionally based investment agencies, for investment in human skills to upgrade relevant qualifications and integrate disadvantaged groups, for the promotion of new product and process technologies and the mobilization of community energies and talents for strengthening the West Midlands' capacity for indigenous growth. This required mobilizing diversified interest groups around common projects in order to articulate broad regional support alliances as a territorially defined regional 'fourth interest', apart from the functionally defined interest groups of management, trade unions and national government. On such a basis of a territorially defined 'fourth interest', inter-regional cooperation was sought with other areas experiencing similar problems. These areas include other European car-producing regions and the UK-wide regional cooperation of 'Local Action for Clothing and Textiles' to provide information exchange services and collaborative training schemes, and articulate a common voice to the outside including issues of national and EC policy (Chapter 6).

A similar emphasis on restructuring existing sectors rather than escaping to new ones has been applied in the Nottingham area of the English East Midlands to a sector frequently discarded as non-competitive in highly industrialized countries: textiles and clothing (Chapter 5). Local action there concentrated on the creation of a strong research base for the structural problems of existing industry, on the promotion of collaborative marketing and production ventures in this formerly highly fractionated industry, on the retraining of workers to more promising subsectors such as design, dyeing and finishing, and on the establishment of decentralized fashion enterprise workshops. Local initiative started at the local level (city council) from where it spread upward to the county council level.

In the Ruhr, the Dortmund local action programme aspired to transform the city into a modern centre of technology and services and to improve its environmental and residential quality in a three-phase strategy for 'today', 'tomorrow' and 'the day after tomorrow'. This included the creation of a technology centre in Dortmund with an adjoining new-technology area in cooperation with local university, local government and local business (Chapter 7), similar to Japan's 'third-sector strategy' in their nationwide 'technopolis' policy. Here also initiative started from the Dortmund city council and subsequently mobilized state, federal government and EC support.

In the Prato region of Italy (Chapter 12), a traditional textile area, the splitting up of large firms increased flexibility and elasticity in production, while at the same time the extended family structure and intensive social relations promoted the local diffusion of innovation and skills. In the industrial and metropolitan areas of Turin and Milan, increased flexibility and innovation was achieved by the introduction of new managerial, organizational and technological models (such as flexible automation, 'just-in-time' production, remote-control systems, the introduction of modern fibre-optic networks, and the development of science and technology centres), while social and class integration reduced the political conflict potential.

Local action in most of the old industrial areas analysed in this book *operated through existing institutions* rather than creating entirely new ones. In some cases it involved a change of personnel in these institutions (as in the Nottingham and Dortmund cases, Chapters 5 and 7), incorporating a new generation of administrators with trans-departmental thinking and the willingness to collaborate with other agencies and persons. Frequently this also required the breaking up of traditional structures and the crossing of traditional ideological barriers such as between capital and labour. The drastic crisis symptoms, however, helped overcome such traditional barriers and made new initiatives imperative also in this respect. The acute crisis made it easier for local government, even in areas monopolized by traditional industry, to allocate substantial resources also to other economic sectors.

Organizationally in many cases the *establishment of independent commercial implementing agencies* (in the form of limited liability companies, for instance) were considered more efficient and more flexible than implementation directly through the public sector. In one English case study (Sedgefield, Chapter 5), the transformation of a formerly public-sector-dominated area to an 'enterprise economy', where the public sector created the climate for private-sector investment, was considered successful. While the trend in successful restructuring seems to move away from nationalized industry solutions which during the past decade have proved unsuccessful in most European countries, many of the cases analysed in this book also do not rely purely on private initiative, but rather on *public–private partnership*. Examples are the public-equity investment schemes in private firms (rather than outright public grants) and the 'planning and investment agreements' of local agencies with private firms in the English West Midlands, providing for the *promotion of private enterprise under local community control* (Chapter 6).

A primarily *knowledge-based restructuring strategy* was applied in the two case studies from French-speaking areas, the Swiss Jura Arc and the French Languedoc-Roussillon region. Although predominantly rural in spatial structure, both have substantial non-agricultural activities which, however, have been declining rapidly during the past decades. This applies particularly to the traditional watchmaking industry in the Jura Arc. The regional restructuring programme concentrated on the elimination of some of the traditional production stages in the watch industry, the introduction of new management and marketing strategies, on product modification and on the concentration of watchmaking enterprises (Chapter 8). The surplus in labour skills in the watchmaking industry was absorbed in the regional fine-mechanical industry while at the same time within the watchmaking industry potential electronic knowledge was mobilized (the quartz watch having been initially invented in Switzerland). This knowledge transfer and upgrading was supported by systematic training programmes, partly related to the newly created Swiss Electronics and Micro-Electronics Centre in the region at Neuchâtel.

The restructuring process in the Jura Arc was substantially aided by the existence of a closely integrated regional light industry complex (watchmaking, machine-tool and precision engineering industry) and by a concerted effort for the transfer of skills between these sectors and for their technological upgrading.

In the Languedoc-Roussillon region the previous decline of traditional small and medium enterprises was reversed in a rather unconventional way by the attraction of a large IBM branch plant which subsequently produced multiplier effects. It was attracted on the basis of regional conditions such as old-established research and training centres, a well qualified labour force and good environmental conditions. After a transition period, local SME increasingly utilized the subcontracting potential of the IBM plant, which turned out to be an innovation catalyst for other activities in the region as well (Chapter 8).

In summary, key strategy elements in the restructuring experience in old industrial areas are as follows. Instead of the traditional regional policy approach of attracting inward investment, which had proved to offer only a short-term and partial respite to the problems of old industrial area problems (Chapter 5), the promotion of the indigenous regional development potential was in most cases given preference. Local firms were usually found to have a higher degree of local loyalty and therefore tended to give higher priority to local restructuring (instead of shifting investment and activities to other regions) than multisite firms.

Apart from economic restructuring, a broader approach to area renewal and community development was in most cases considered necessary: most local efforts included environmental, social and political change and sought to rectify the fundamental weaknesses of an area's economy. Integrated packages of measures and initiatives linking economic policy to local action for social and physical regeneration, including urban development policy, seemed essential (Chapters 5 and 7).

Local action seemed particularly important in the definition of objectives, opportunities and programmes for local restructuring; in priming finance; in exerting pressure upon other participants; and in the creation of an 'integrated and coordinated local mechanism for design and implementation of policy' (Chapter 5). On these bases the negotiations with a wide range of other public and private actors, also external to the region, proved most successful.

Important *preconditions which improved the chances of restructuring* were:

- the existence of remnants of a (pre-industrial) crafts history was found important in the Italian case studies, in the Jura Arc and even in the English West Midlands, though in the latter case this history had existed only up to the 1870s;
- a high share of locally based enterprises;
- the recent existence of intensive intra-regional linkages forming an interrelated regional complex described as important in the Jura Arc and in the Ruhr, and to be re-established in the English West Midlands;
- physical proximity and cooperation between research, training, strategic planning, marketing and production units;
- the existence of intensive multiple communications and exchange networks between commodity and non-commodity sectors, formal and informal activities, which helped to produce, retain and diffuse specific territorial competences as well as the promotion of local solidarity;
- the existence/promotion of a local technical culture compatible with new technologies which can be applied to other factors of existing regional strength.

DIVERGENT LOCAL DEVELOPMENT PATTERNS IN EASTERN EUROPE – CASES FROM TWO COMECON COUNTRIES

The case studies from COMECON countries (Chapters 9 and 10) are drawn from Poland and Hungary. These two countries were among the first COMECON countries able to experiment effectively with and implement policies for the decentralization of responsibility and decision-making power, even before recent drastic reforms. Their experiences can therefore be considered among the most advanced ones regarding local development initiatives in the COMECON framework. It must be added that the case-study areas analysed are located in the western parts of the respective countries, not only located closest to the respective borders towards the West but historically, in relative terms, also most influenced from there (western Poland for instance was under Prussian rule for most of the last century). These areas also represent those parts of the respective countries with the highest density of urban centres able to serve as catalysts for local development initiatives. All in all, the conditions in these areas are likely to be among the most favourable for local initiatives within COMECON countries.

The case studies focus on comparing, within the specific regions, the performance of selected towns which since the Second World War were developed mainly by central-government initiatives with towns which developed to a higher degree on the basis of locally induced initiatives.

It should be mentioned that the (often implicit) objectives, and therefore also the 'success criteria', of local development initiatives in COMECON countries must be considered differently from those in more market-oriented countries. Regional full employment, for instance, was an important objective but could not be evaluated as a success criterion in COMECON countries, as at that time by definition there existed neither unemployment nor relevant statistics. Local employment was mainly determined by central government decisions. In many cases, however, the objective of local initiatives was to mitigate the local consequences of central-government-induced large industrial projects, such as bottlenecks of local infrastructure and services, environmental deterioration, etc. The 'success' of local development initiatives at that time must furthermore be evaluated using different standards as the legal manoeuvring space of local development initiatives in centrally planned countries generally has been much smaller than in other European countries.

The comparison of the development performance between towns of mainly centrally induced development with that of towns with more local initiatives yielded interesting results. Whereas centrally induced development frequently led to a rapid population increase in the affected locations, their medium- and long-term population growth rate was often lower than that of towns with more locally induced development, particularly in the Polish case studies. Local strains and supply deficits were also handled more easily and at relatively lower cost than in towns with predominantly centrally induced growth.

While in towns with centrally induced development technological innovation was often transferred in substantial magnitude from outside, its further development frequently met with substantial local bottlenecks (such as lack of trained personnel). This was particularly the case if the respective R&D facilities were not also transferred to the respective location. Particularly in the Polish case studies, it appeared that technological progress was more continually sustained in towns with a higher degree of locally induced development.

In towns with a higher degree of locally initiated development labour-market disequilibria were smaller (both regarding disparities between supply and demand, but also between different strata of qualification). Particularly in the Polish case studies, export dynamics were also greater there than in towns with predominantly centrally induced development (the latter strangely enough being oriented more to national or regional markets).

In towns with more locally initiated development the degree of cooperation between enterprises and with local authorities was usually higher. In many cases from both countries, this cooperation took place in local directors' clubs where leading personalities from enterprises, and from local administrative and political bodies met regularly. Centrally induced (mainly large-scale) development on the other hand frequently led to local defensive action and triggered the mobilization of local citizen groups to protect their environment, preserve their local heritage, and promote knowledge of local history, in local cultural, educational or 'friends of town' societies and the like.

Local initiatives historically have often been based on resistance against central government or outside interference. In western Poland (also called Greater Poland) a seedbed for local initiatives was formed by a traditionally strong movement of cooperatives both in urban crafts and rural activities, dating back to the resistance against the dominance of Prussian landowners (Junkers) during the last century.

The development of local communities in centrally planned economies at that time depended to a great extent on their ability to attract central-government funds and on their personal and political access to central decisions as is shown in Chapter 10. This is comparable to the wooing of local communities for multi-locational enterprise in market economies. These centrally allocated funds have usually had a low level of 'intellectual capital' and therefore contained relatively little self-sustained innovative capacity and flexibility. However, quality, innovative capacity and flexibility on a sustained basis in centrally planned economies also depended to a great extent on what Baráth and Szaló (Chapter 10) call the 'structural energy' of local territorially organized socio-cultural systems, as both the case studies for Poland and Hungary indicate. This fact has assumed growing importance as the development of centrally planned economies – like that of market economies – has during the last decade become increasingly dependent on innovation and flexibility rather than on pure quantitative growth.

LOCAL RESTRUCTURING IN CENTRAL AND SOUTHERN EUROPE

Case studies for this part of Europe were taken from Austria, Italy and Spain. We shall here concentrate on the experiences of rural areas and small towns, as those of some old industrial areas and large urban centres from these countries have already been referred to in the corresponding section above.

The peripheral rural areas dealt with – the Mühlviertel in Austria along the former 'Iron Curtain', and the Andalusian Lebrija region in southern Spain – were both among the least developed in their respective national contexts. Owing to a longstanding tradition of internal and external dependence and an orientation towards obedience and passivity rather than criticism and initiative, local initiatives had to rely mainly on external actors and externally induced political and institutional change. In the Mühlviertel (Chapter 11) these external actors frequently came from larger urban centres where they had received their training, while many of them originated from rural areas and were therefore more knowledgeable and personally concerned with problems and potentials of rural areas. In Andalusia, an area with a longstanding feudal tradition, the new Spanish constitution of 1978 provided for free municipal elections and the mayors, for the first time democratically elected, became important new leaders of local initiatives. These local initiatives led to the creation of local cooperatives of agricultural day-workers (*jornaleros*), a mechanics, ceramics and cotton production cooperative, and a local tomato-canning company which was financed by 160 local residents jointly with local government. Furthermore two industrial estates, a training and a cultural centre were initiated (Chapter 13).

In neither of these rural areas had there existed regional universities, research centres or a strong regional entrepreneurial tradition. In both cases, however, increasing economic difficulties and the gradual withdrawal of central government from local development tasks because of fiscal bottlenecks (in Spain accompanied by political and economic decentralization following the new 1978 constitution) triggered a higher degree of local action. In the Austrian Mühlviertel, this was promoted by the proximity of the adjoining provincial capital, industrial and university city of Linz, to which surplus labour could commute and from which social and economic change radiated (though much of it in a passive form by inward investment in residential land). Still, within Austria, the case-study author considers the Mühlviertel as the cradle for the implementation of a self-reliant regional development strategy for which Austria has been a precursor within Europe.

In Italy (Chapter 12), both in the rural Marche area and in the Bassano (a medium-sized town) area important preconditions for local initiatives were considered to be low labour costs, high information mobility and strong political cohesion as well as a high degree of local inter-sectoral economic integration between agriculture and industry and between local industry and finance. Particularly for the 'Third Italy' experience, frequently considered a success story of 'diffused innovation', the following conditions were considered important: the organization of production in 'system areas' characterized by sectoral specialization, the physical proximity of firms and the nature of the environment, which was non-metropolitan or mainly rural. The homogeneity of such a productive structure in restricted geographical areas, according to the case-study authors, promoted the achievement of high rates of technological innovation and upgrading of labour skill, at the same time creating benefits from economies of scale at the district level and from productive flexibility. Other important factors were the creation of new synergies between production, research, strategic planning and marketing, both at the enterprise and communal level. This was accompanied by the emergence of new local solidarities between different social strata and formerly conflicting classes.

EUROPEAN APPROACHES TO LOCAL DEVELOPMENT INITIATIVES

The great relevance of local development initiatives in meeting the challenges of global economic restructuring has led to broad support not only by almost all national governments but also by OECD and the Commission of the European Communities. Details of their activities are given in Chapter 2.

The case study in Chapter 14 describes the experiences of a specific European programme promoting local development strategies by local training programmes sponsored by the European Social Fund. These training programmes concentrate on the pre-operational phase of job creation. They aim at creating flexible training structures by matching perceived development potentials with available human resources in the context of specific community objectives, rather than to follow traditional training practices which were usually oriented towards globally perceived demand for certain skills. These local training programmes, located in different countries and different area types, aim at matching local labour supply and demand by intensive (often informal) information rather than mainly via the anonymous market mechanism. They apply a collaborative teaching approach encompassing the local labour market as a holistic entity, and the orientation of these local training programmes depends on specific local history, structure and problems, using private non-profit organizations as instruments. Course contents include vocational and skills training as well as management and creation of new businesses.

LOCAL PRECONDITIONS FOR INNOVATION

The case studies assembled in this book as well as experience manifested elsewhere show that a number of factors represent important preconditions for a broad distribution of innovation within local communities:

- crisis conditions such as those resulting from changes in the international division of labour represent a strong potential trigger for innovation and entrepreneurship in the sense of Schumpeter's creative destruction and are roughly comparable to the challenges presented by natural conditions and catastrophes in pre-industrial societies;
- societal incentives and rewards must be offered (1) for individual initiative and entrepreneurship and (2) for their orientation toward broader benefits to local society. Financial and also social rewards such as esteem and recognition can serve this purpose;
- the institutionalized transfer of information, innovation and entrepreneurial initiative from outside and within the local community are further key prerequisites. From the outside, this may take the form of a rotation of personnel between local and external sources of innovation (e.g. with external universities, research centres, but also between local branch plants and external research units attached to multisite firms);
- within the locality or region this transfer can be implemented by promoting joint research between firms, thus helping to reduce research 'introversion', and by rotating personnel between research and production units to improve the transfer of knowledge between research and application. Synergistic local interaction networks for the exchange of

information, commodities and services as bearers of innovation and cooperation have proved to be worthy of promotion as important vehicles for these transfers;

- the promotion of local entrepreneurial cooperation as a framework for individual initiative and the orientation of its benefits. This is usually facilitated if the marginal advantage to individuals or firms from internal cooperation is kept higher than that from external interaction and cooperation. At the same time such cooperation promotes the formation of common objectives among local actors;
- broad democratic decision-making processes are usually an important prerequisite for the broad local distribution of benefits. They can also, however, lead to inefficient resource allocation and rigid local structures;
- the formation of rigid local hierarchies that limit incentives for innovation and the broad diffusion of their benefits is to be avoided. Hierarchies in the form of differentiated responsibility and decision-making power should be tied to specific functions, rather than specific individuals or groups. In this way the barriers preventing individuals from taking up those functions best suited to their abilities can be removed. The periodic rotation of key functions helps to reduce such rigidities.

EXTERNAL PRECONDITIONS FOR LOCAL DEVELOPMENT INITIATIVES

Specific external conditions seem to be as important as internal ones for the emergence of local development initiatives. This applies particularly to modified and (sometimes) to new roles played by central government. Traditional central government instruments for local and regional growth, oriented mainly towards incentives for capital to increase the quantity of jobs, need to be redirected towards more qualitative measures to improve the quality of labour and the flexibility of the capital stock, technology and organization.

The following functions fulfilled by central government are therefore to be emphasized:

- the facilitating of access to information on marketing opportunities (national and worldwide), new technologies, new organizational and management forms, and the learning experiences of other local development initiatives;
- the co-financing of regional training and research development centres;
- the co-financing of local development organizations; and
- the promotion of persons and groups with innovative potential at the local level.

Experience presented in this book shows, however, that direct external intervention by central government can also distort local action processes and weaken local feedback mechanisms.

Other important external conditions for local development initiatives appear to be:

- the allocation of substantial action, decision-making and financial scope to local agencies, 'allowing for more effective operation of local networks' (Chapter 4);
- the reinforcement of local and regional feedback mechanisms between decisions on economic, social, political and environmental matters and their respective results. These feedback mechanisms appear to be important prerequisites for indigenous adaptation and innovative capability; and finally
- the promotion of flexible institutional structures, both at the local level and for external purposes.

CHAPTER 1

Introduction

Walter B. Stöhr

The intention of this book is to present the richness and diversity of local and regional action undertaken - under varying conditions in different parts of Europe - with a view to confronting the challenges of the international economic restructuring process.

This chapter, by way of introduction, describes the problem situation caused by recent global economic restructuring frequently considered 'deterministic' in comparison with earlier economic crises, and shows the conditions for 'voluntaristic' room for manoeuvre still available to local communities. It then introduces the UN University project on which this book is based.

Chapter 2 goes on to deal with the theory and practice of local development in Europe, analysing the factors which have led to the present (re)emergence of local development initiatives and the substantive and institutional factors behind them.

The following case studies are grouped in five sections proceeding roughly from the north to the south of Europe. In Part I case studies are assembled from the northern and western periphery of Europe and are taken mainly, though not exclusively, from rural areas dominating there. Part II contains restructuring experiences from the early industrialized core areas of Europe, most of these experiences being concerned with the local and regional restructuring of 'old' industrial areas dating back to the last century. Part III brings together case studies from two COMECON countries, namely Poland and Hungary, focusing mainly on the effects of former central planning practice on the innovative and adaptive capacity of local communities and their lack of resilience to the effects of international economic restructuring. These analyses have in a way anticipated the reform process which meanwhile has taken place so abruptly in these countries. At the same time they demonstrate the latent potential for indigenous restructuring that existed in these countries even before the recent reforms. Part IV is concerned with local restructuring in the generally less industrialized central and southern European countries; case studies therefore are taken mainly from rural areas, but also from some older industrial enclaves typical of these internally very heterogenous countries. Part V deals with international programmes for promoting local employment initiatives on the part of the European Social Fund, OECD and the Commission of the European Communities. Finally, a select annotated bibliography guides the reader to further sources on the theory and practice of local development initiatives. The manuscript of this book was delivered to the UN University in spring 1988 and therefore predates recent drastic reforms in Eastern Europe.

Since the mid-1970s in particular there has re-emerged a rich experience of local development action in many parts of Europe. Transnational experience shows that long-lasting and self-sustained development needs to be flexible and locally rooted, based on indigenous rather than inward investment, while simultaneously, however, related to open information systems and cooperation in learning.

Let us hope that this compilation of knowledge and practical experience will influence all those interested in local development and contribute to broader action in a field we consider essential for dignified, self-determined and at the same time peaceful human progress in all parts of our world: to act locally and think globally.

1.1 THE BACKGROUND

The process of international economic restructuring has speeded up considerably since the middle of the 1970s. This was the result of declining aggregate economic growth rates, the exhaustion of markets for traditional products and the accelerated emergence of new technologies and products. At first repercussions at the national level were the main focus of concern regarding aggregate phenomena such as increasing disequilibria in the national balance of payments or of trade. Only more recently have the repercussions at sub-national levels – at the level of the regional or local community – also been given broader attention (Muegge and Stöhr, 1987; Henderson and Castells, 1987). It is here, after all, that individual human action is concentrated and here that repercussions are therefore generally most intense.

The reason for these local repercussions is that comparative advantage has shifted with increasing speed between local economies as a result of the exhaustion of local resources, changes in worldwide demand, the emergence of new competitors producing at lower cost, the development of new technologies and organizational forms, and the poor ability of specific locations to provide sufficient leadership in these change processes or adapt to them fast enough. Absolute advantage, traditionally dependent on locational access to localized resources and markets, has recently become determined primarily by organizational access to the control of capital and innovations. A key question has therefore arisen about how this control was shared between global functional units (such as transnational firms) and territorial units (such as regional or local communities).

The ever-deepening crisis of state finances has made it impossible for central governments to cope with the local and regional problems of economic restructuring. At the same time, the traditional state instruments of regional and local policy, developed during earlier growth periods and implemented through localized capital incentives and infrastructure investment, have proved ineffective for problems of structural change (Stöhr, 1985, 1987).

As recently as the early 1980s, when the alternative concept of local and regional development 'from below' was proposed (Stöhr and Taylor, 1981), a widely heard reaction was that a more indigenously induced local and regional development of this nature was hardly feasible under present international conditions. Three types of reason were most frequently given for these doubts. (1) *Economic* doubts were that most local economies were too small and therefore commanded too few resources and scale economies to be able to stem the tide of the international economic restructuring process and the deterministic changes in the worldwide division of labour. (2) *Political* doubts were that power vested at local or regional levels was too small to confront the dominance of large multinational enterprises, international financial institutions or even national government. Local and regional communities – like small and medium-sized enterprises – were assumed to be condemned to the role of 'history takers' (Muegge and Stöhr, 1987) *vis-à-vis* the power of international firms and central institutions considered as 'history makers'. (3) *Information* was lacking on the existence and especially on the success of effective local or regional development initiatives during recent decades. This applied not only to developing countries (Stöhr and Taylor, 1981), but also to highly developed countries such as the USA, where in recent decades the role of individual states and other sub-national governments in economic development had been receding even further behind that of central government (Schmandt, 1987).

The absence of compiled evidence on the existence, failure or success of local and regional development initiatives appears as a major reason for the persistent belief that they were unfeasible. This lack of compiled evidence, however, seems largely due to the fact that such local initiatives were not sufficiently 'newsworthy' for the media or the mainstream scientific community. For the news media they were too small to promise a sufficient market for their transmission to be profitable, while for the mainstream scientific community they seemed too varied and therefore not sufficiently amenable to generalizable conclusions to be considered worthy of scientific endeavour.

One objective of this book, therefore, is to help make such 'small-scale' local experiences 'newsworthy' again – just as small firms have again become newsworthy since Birch's (1979) study on the job-creating capacity of small enterprise in the USA. In Europe this was expressed by the European

Parliament's declaration of 1983 as the 'European Year of Small and Medium Sized Enterprises' (ECOSOC, 1982). A second objective is to introduce in local development an inductive process of learning from practical experience instead of the deductive one of learning from generalized laws and models, as development policy and most of the established disciplines have done during recent decades. A third objective of this book is to show, on the basis of systematic analysis of the available evidence, that while being subject to the 'deterministic' forces of the worldwide economic restructuring process, local and regional communities have substantial 'voluntaristic' room for manoeuvre in confronting global change and 'making' their own history. In this context the role of specific actors has proved important. In some cases local development is triggered by external actors (central institutions, multilocational firms), in others by local actors or institutions (cf. Fig. 1.1).

A key prerequisite, however, for broader local development proved in all cases to be intensive interaction, information exchange and cooperation between local actors. Where this did not exist, some form of a local development agent needed to act as a catalyst for bringing about this cooperation.

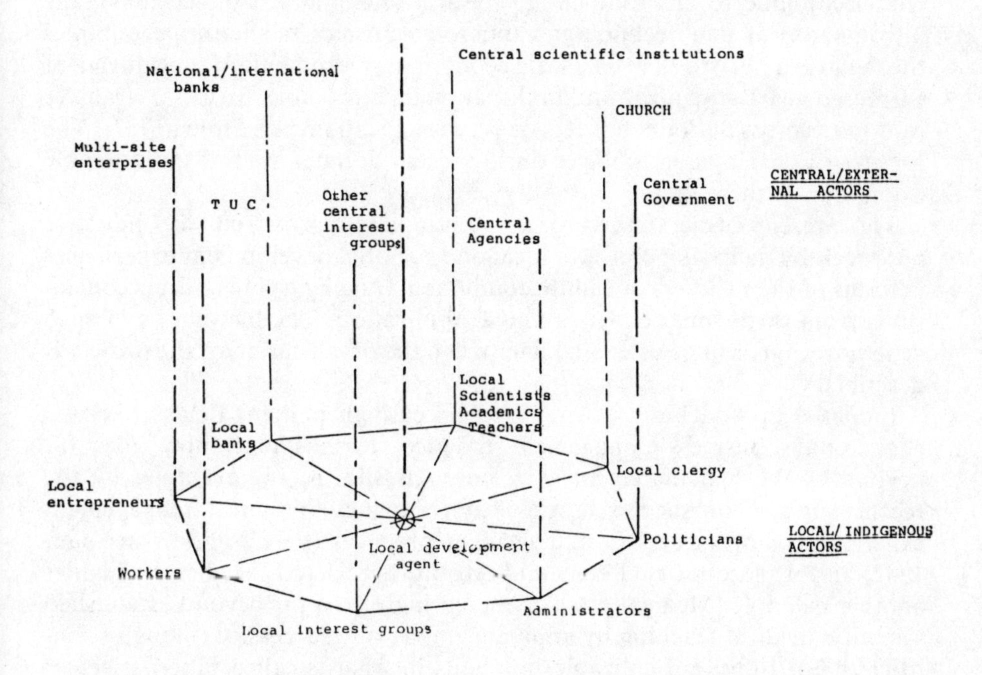

Figure 1.1 Major actors and interaction patterns for local development. - - - -, horizontal interactions; —·—·—, vertical interactions.

Sometimes such agents emerged on their own initiative, sometimes they were contracted by local institutions or the community and sometimes by outside organizations.

1.2 THE UNDERLYING CONCEPTIONS

1.2.1 INDUCTIVE LEARNING FROM PRACTICAL EXPERIENCE

Development policy has in the decades following the Second World War been largely based on established scientific theory such as (sociological) modernization theory, (economic) sector theory and export base theory, (regional economic) growth centre and diffusion theory and concepts such as scale and agglomeration economies. By and large, development policy applied these theories in a deductive way to practical situations, although many of the theories were either insufficiently tested empirically or falsified by empirical tests (as in the case of growth centre theory). They also failed to take account of national, regional or local differences in institutional, political and cultural conditions, which often had a decisive influence on the application of these theories. Many of their implicit assumptions have furthermore lost validity as a result of changes in the perception of development (broadening from economic to environmental, cultural and political dimensions), the introduction of new technologies (micro-electronics facilitating economies through flexibility rather than scale economies in production), new forms of entrepreneurial organization (multinational firms able easily to transfer resources across national borders or between locations; cf. Stöhr, 1987). The failures of development policies during recent decades are to a considerable extent due to these facts.

The present project, therefore, intentionally starts from an inductive approach by analysing concrete local and regional development experiences in terms of their success or failure components and by trying to draw conclusions from them for broader practical application. The limitations of such generalizations will be discussed later when the organization of the project is described.

Inductive approaches are being used increasingly in many fields of knowledge today. Business-management practice, for instance, now goes far beyond established management science, as shown, for example, by the recent analysis of success stories from Japanese practical management experience or of successful management practices (Peters and Waterman, 1982) and its successors (Peters and Austin, 1985, etc.). Medicine provides another example. Medical practice increasingly goes far beyond established scientific medical teaching by applying empirically successful (though scientifically not fully rationalizable) methods in what is called 'alternative' or 'holistic' medicine. Excellence or success, however, are not mere 'snapshots'; they cannot be measured in terms of single achievements. 'Best practice', as it is known, is a continually evolving body of experience.

While established theories are usually static and aim at the establishment of absolute knowledge, alternative approaches tend to be based on the creation of constant monitoring processes in the form of appreciative or learning systems, enabling adaptation to changing external conditions.

The main characteristic of these 'alternative' approaches is that they are based on practical success of a broader and more sustained nature (although often not yet fully rationalizable and consistent in scientific terms) rather than mainly applying the results of consistent and rationalizable partial theories despite their frequently observed inability to deal with real situations in all their complexity. Traditional established theories and concepts are rational in the sense that they are based on 'reason as the chief source and test of knowledge and on self-consistency as apprehended by the intellect' (Blanshard, 1978) within specific disciplines, while they frequently disregard cross-disciplinary interdependencies often contained in sense experience. Over the last two decades these inadequacies have led to a paradigm shift in many fields of science.

1.2.2 DISCOVERING THE 'MANOEUVRING SPACE' OF LOCAL DEVELOPMENT ACTION

Local development theory and policy have so far been mainly framed in deterministic terms. The development of localities was considered a function of their accessibility to resources, markets and technologies, their activity structure and the magnitude of scale or agglomeration economies. In this deterministic framework three main streams of thinking had emerged: a 'defeatist' one, related to the dependency school, inclining towards the hypothesis that existing spatial disparities of development are tenaciously stable and tend to reproduce themselves rather than to change. A second, 'automatist' stream was based on the neoclassical hypothesis of an inbuilt equilibrium tendency: provided the mobility of factors and commodities is increased sufficiently, free entrepreneurial choice will progressively eliminate spatial disparities and allow all the localities to approach a similar level of development with different sectoral specialization. A third, 'structuralist' theoretical stream, particularly since the accelerated worldwide economic restructuring process during the last decade, postulates the hypothesis of spatially rotating disparities in the form of either cycles – approximating to the product cycle – or more complex socio-political cycles such as those described by Marshall (1987), who explains the 'long waves' of the development of different regions in Great Britain by the role they have played in the (capitalist) world economy at different periods of the British Empire and by their internal class structure.

In an analysis situated between the last two theoretical streams, Camagni and Capello (Chapter 12) have empirically shown for Italy how comparative advantage has shifted through different regions during the past three decades. By relating composite indicators of regional productivity to those of regional labour cost they show that relative comparative advantage has shifted from the Mezzogiorno (1961) to the central region of Italy (1971) and finally to the north-eastern region or so-called 'Third Italy' (1981). On the basis of this (deterministic) analysis, however, they postulate, like Marshall, that local and regional communities possess a certain 'manoeuvring space' to deviate from such a deterministic trend and exert their own influence on the international division of labour. They derive this local manoeuvring space mainly from the proximity (and consequent 'reduction of market transaction costs') between spatially concentrated specialized small firms and the intensive synergies between economic and social actors due to common cultural and political values (Chapter 12). Marshall (1987, p. 237) sees it in changes of local political-economic structures and in new modes of planning and political accountability. Both explanations imply increased local control over innovation and capital.

Attempts to generalize from such inductive approaches to local innovation have been made before (Andersson, 1985; Stöhr, 1986; Törnquist, 1987; Storper and Scott, 1989, and others) and are summarized in Chapter 2. They identify as major prerequisites for local innovation such factors as societal instability, local competence, intensive information exchange, local synergy and cooperation, cultural diversity, organizational flexibility and common local objectives.

The present project is concerned mainly with the processes whereby local or regional communities can deviate or 'break out' from an externally determined development trajectory. A further objective is to arrive at some generalizable conclusions about the processes by which local communities can successfully confront ongoing changes in the international division of labour. A frequently heard argument is that each local development initiative is a unique case and therefore not transferable. At first sight, it does indeed seem contradictory to emphasize the individuality of each local development initiative while at the same time searching for generalizable results which can serve as learning experiences for other cases. Despite the considerable differences between the cases with regard to conditions and goals, however, it turned out that certain aspects of successful social processes could be generalized. The explanation may be that certain aspects of human behaviour can be generalized (at least within Europe, to which this project refers), that global challenges such as those of economic restructuring become increasingly similar for all areas (in Eastern or Western Europe), that the strategies of transnational firms become increasingly worldwide in their reach and that the existence of one world as a common fate and responsibility is increasingly accepted as an unavoidable fact, not only in economic but also in environmental, political and other terms.

The experiences contained in the case studies presented in this volume are intended to stimulate new approaches in similar efforts undertaken by other local or regional communities. Similar transfers of learning experiences have already been made, for instance by applying some of the elements of Irish local initiatives to the Scottish isles as Bryden and Scott show (Chapter 4), or by the international exchange of experiences between old industrial areas in the frame of RETI (Communauté de travail des Régions Européennes de Tradition Industrielle).

1.3 THE PROJECT

The project presented in this book focuses on the social processes which have led to innovation and structural adjustment at the local or regional level. Systematic empirical studies of these processes have hardly been done so far for Europe at this 'meso-level' between the individual enterprise and the national level with the exception of that of Sutton (1987), who focuses, however, on the more narrow aspect of employment initiatives and small-firm creation.

This meso-level is important, on the one hand, because the innovative capacity of individual enterprises depends to a high degree on the innovative conditions of the local or regional milieu. This has been shown for Europe particularly by the work of GREMI (Aydalot, 1984, 1986), the European Research Group on Innovative Milieux founded by Philippe Aydalot and since his premature death headed by Jean-Claude Perrin and Roberto Camagni. On the other hand the total rate of innovation at the national level also seems in the long run to depend substantially on the innovative capacity of regional milieux (Nijkamp and Stöhr, 1988).

The United Nations University, Tokyo, commissioned this project in the framework of its European Perspectives Programme with the relatively short time horizon of one year. It was therefore necessary to draw on existing or on-going research, as the commissioning of original empirical research was hardly possible within this time-frame. It is hoped, however, to carry out original empirical research in succeeding projects.

A team of authors was assembled to present experiences of local initiatives from all major parts of Europe. Close to fifty primary case studies were prepared by these authors (see Map 1.1 and Tables 1.1 and 1.2). Furthermore, secondary information on several hundred local initiatives from OECD and EC reviews was assembled in Chapter 15. Authors were selected according to the criteria that they should be action-oriented, have direct working knowledge of the areas they write on and command sufficient scientific rigour for a systematic study based on common guidelines.

Case studies are taken from parts of Europe ranging from Scandinavia in the north to Italy in the south, and from Spain in the west to Poland and Hungary in the east. It was the explicit intention of this coordinator to include not only case studies from Western European market economies, in many of which local initiatives are already common practice, but also from centrally planned East European COMECON countries, in some of which local development initiatives have emerged in recent years and have gradually been permitted to become more effective, thanks partly to ongoing

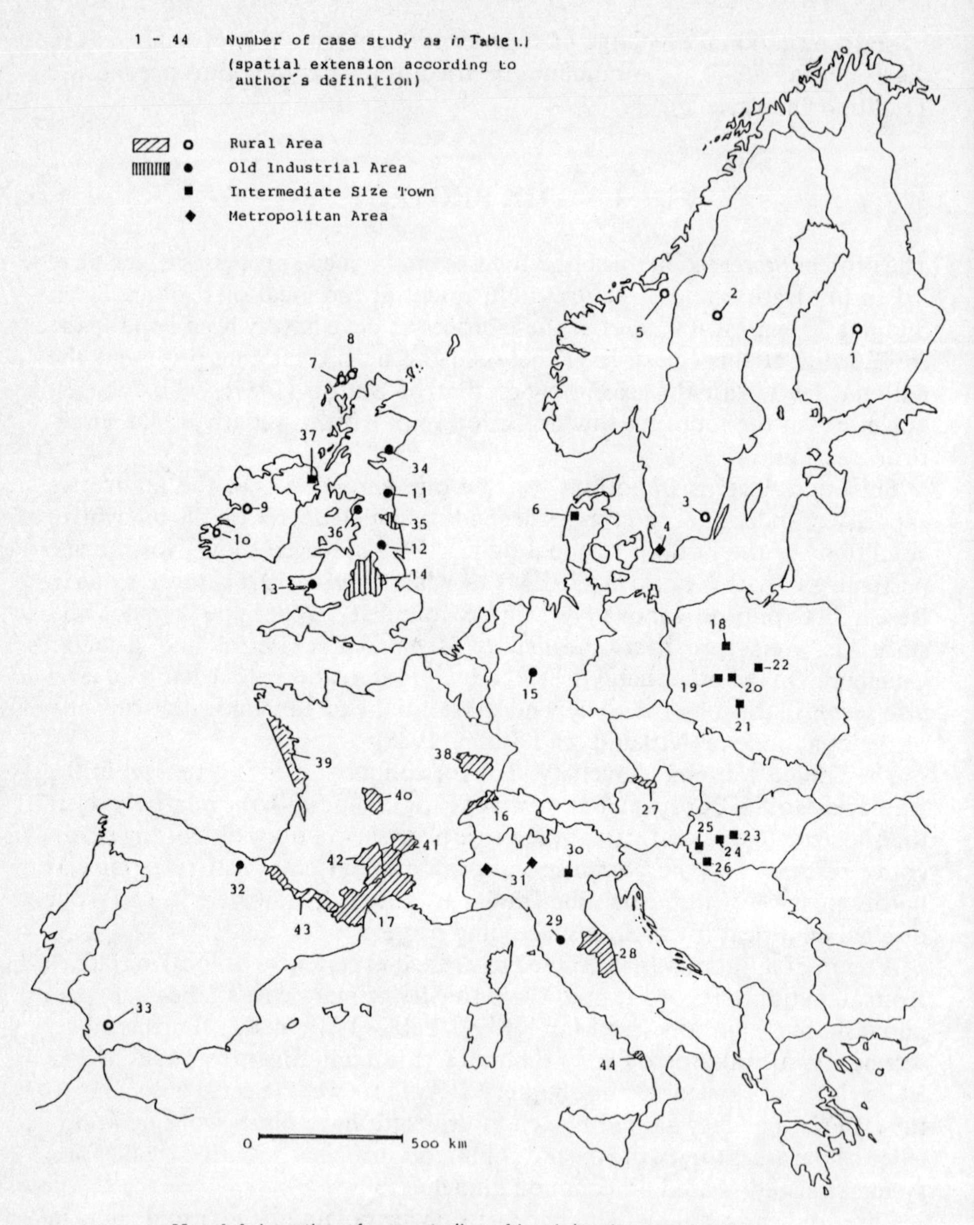

Map 1.1 Location of case studies of local development initiatives.

Table 1.1 List of case studies of local development initiatives

No. on map	*Local development initiative*	*Author(s) of case study*
1	Lannevesi (Finland)	Johannisson
2	Pilgrimstad (Sweden)	Johannisson
3	Maleras (Sweden)	Johannisson
4	Fosieby (Sweden)	Johannisson
5	Selbu (Norway)	Johannisson
6	Holstebro (Denmark)	Johannisson
7	Pairc (Western Isles of Scotland)	Bryden/Scott
8	Ness (Western Isles of Scotland)	Bryden/Scott
9	Letterfrack (Ireland)	Bryden/Scott
10	West Kerry (Ireland)	Bryden/Scott
11	Sedgefield (England)	Roberts/Collis/Noon
12	Nottingham (England)	Roberts/Collis/Noon
13	Swansea (Wales)	Roberts/Collis/Noon
14	West Midlands (England)	Marshall
15	Dortmund (West Germany)	Hennings/Kunzmann
16	Jura Arc (Switzerland)	Maillat
17	Languedoc-Roussillon (France)	Maillat
18	Oborniki (Poland)	Gruchman
19	Koscian (Poland)	Gruchman
20	Srem (Poland)	Gruchman
21	Ostrzeszów (Poland)	Gruchman
22	Kolo (Poland)	Gruchman
23	Székesfehérvár (Hungary)	Baráth/Szaló
24	Veszprém (Hungary)	Baráth/Szaló
25	Zalaegerszeg (Hungary)	Baráth/Szaló
26	Nagykanizsa (Hungary)	Baráth/Szaló
27	Mühlviertel (Austria)	Scheer
28	Marche (Italy)	Camagni/Capello
29	Prato (Italy)	Camagni/Capello
30	Bassano (Italy)	Camagni/Capello
31	Milan and Turin (Italy)	Camagni/Capello
32	Onate (Spain)	Vazquez-Barquero
33	Lebrija (Spain)	Vazquez-Barquero
34	Dundee (Scotland)	Steinle/Moya
35	Newcastle upon Tyne (England)	Steinle/Moya
36	Wigan (England)	Steinle/Moya
37	Carrickfergus (Northern Ireland)	Steinle/Moya
38	Vosges (France)	Steinle/Moya
39	Iles du Ponant (France)	Steinle/Moya
40	Creuse (France)	Steinle/Moya
41	Haute-Loire (France)	Steinle/Moya
42	Tarn and Aveyron (France)	Steinle/Moya
43	Pyrenees (France)	Steinle/Moya
44	Cilento (Italy)	Steinle/Moya

Table 1.2 Local development initiatives - distribution by countries and area types

	Rural areas	*Old industrial areas*	*Intermediate size towns*	*Metropolitan areas*
Finland (1), Norway (1) and Denmark (1)	2		1	
Sweden	2			1
Ireland (2) and Scotland (2)	4			
England (3) and Wales (1)		4		
Federal Republic of Germany		1		
France (1) and Switzerland (1)	2			
Poland			5	
Hungary			4	
Austria	1			
Italy	1	1	1	2
Spain	1	1		
European Social Fund Projects: France (6), UK (4), Italy (1)	7	3	1	
OECD and EC reviews on ILEs	x	x		
Total (except OECD and EC reviews)	20	10	12	3

economic reform stimulated by the worldwide industrial restructuring process and internal popular pressure. Yugoslavia was not included because of the special complexity of its local and regional development situation, which would have required a broader study than was possible in this context. Of the COMECON countries only Poland and Hungary are represented because local initiatives appeared to be most active there, and also because competent contributors were readily available from these countries. The personal availability of contributors also influenced the selection of case studies for Western Europe: Philippe Aydalot's sudden death deprived the project of an inspiring collaborator and a contributor of case studies on French-speaking parts of Europe; Denis Maillat then kindly undertook to contribute two case studies on this area. Selected contributors from Greece and Portugal were unable to complete their case studies for personal and health reasons. Both were burdened by the multiple tasks of combining motherhood with professional obligations in social environments where this is still rare.

Common guidelines were elaborated for case studies in a meeting held in Vienna in November 1986. Agreement was reached that 'development' was to be defined not only in quantitative terms of economic growth such as local product and employment, but also in qualitative and structural terms such as changes in the quality of employment; migration patterns; skills levels; sectoral, size and technological characteristics of plant openings and closures, including their ownership, control and organizational characteristics; availability of mechanisms for social conflict resolution and of participatory structures; changes in environmental quality etc. (see Stöhr, 1984). Case studies were to focus on the actors and processes which produced local innovation and restructuring, distinguishing between local, regional and external actors and their interrelations.

In order to do justice to the great variety of local development experiences in different parts of Europe and still retain a maximum degree of comparability between them, the external 'macro-conditions' (degree of political/administrative decentralization, historical, economic, cultural, and other constraints and facilitators) of case studies were to be described as well as the type of area (rural, old industrial, intermediate-size town, metropolitan areas). Not only the preconditions but also the social processes and their results were expected to depend on these external conditions and on the internal structure of the respective areas. This could be brought out in different case studies to varying degrees. The distribution of case studies between countries and area types is shown in Map 1.1 and Tables 1.1 and 1.2.

Individual case studies define the major actor(s) and the social processes which have been initiated inside or outside the respective local areas and try to identify the effect this has had on local employment, plant stability, social stability, political organization, environmental quality, and so forth. Furthermore, they analyse how local initiatives have been related to outside actors or institutions as sources of innovation, expertise, finance and power, for example. They also try to show how local communities have organized themselves internally to be able to control the economic, social, environmental and political consequences of, and possible dependence on, such external inputs.

1.4 WHAT ARE LOCAL DEVELOPMENT INITIATIVES?

At first sight this question seemed as simple to answer as the one about 'What is a camel? You see it and you know it is one.' With diversity of practical experiences, however, it appeared much more difficult to define a local development initiative (LDI) than one might expect.

As any development requires an initiative and takes place in a locality, is

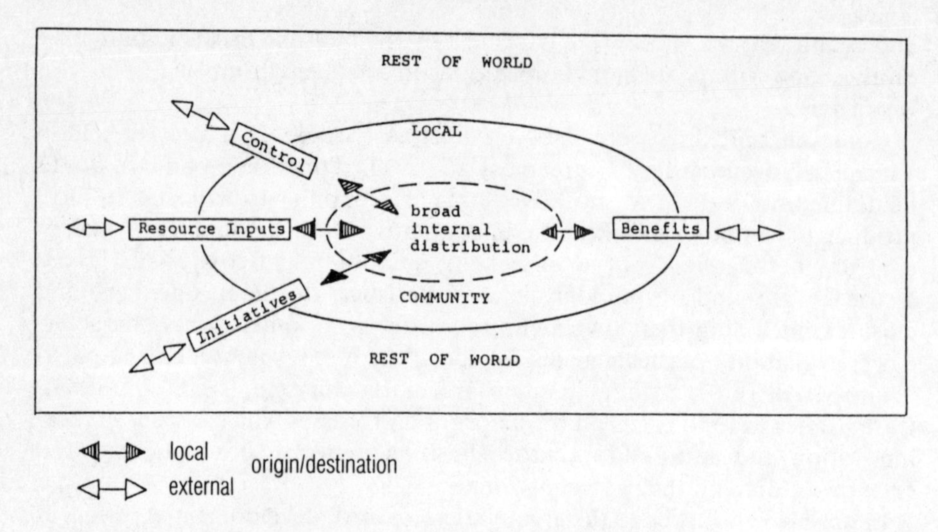

Figure 1.2 Definition of local development initiatives.

any development a local development initiative (LDI)? Development initiatives can be local in origin, in application, or in both. In this connection, we need to consider four aspects of local development: the origin of the initiative, of the resource inputs and of the control mechanisms, and the destination of the benefits (cf. Fig. 1.2).

Take a few practical cases: (1) A local entrepreneur takes the initiative to exploit a local resource (such as a mine) in a highly mechanized way, creating hardly any local employment, exporting the mineral and paying licence fees to central government under whose jurisdiction mineral exploitation falls. In this case the initiative and the inputs are local, but the benefits and control are mainly external. (2) Central government or a multinational firm takes the initiative to invest in a locality by establishing a plant generating local employment, income and local taxes in the municipality. This is almost the reverse case: the initiative, resource inputs and control are external, but the benefits mainly local. (3) A local municipality succeeds in attracting the branch plant of a multinational company by heavy subsidies, tax concessions and deregulation of the labour market and of environmental conditions; the plant offers predominantly unqualified jobs: the initiative is local, but external resources have been 'bought' by the commitment (or possibly over-commitment) of local budgetary means; the local benefits remain small (unionization and environmental protection are sacrificed) and control over local labour and the local environment remains mainly external. Which of these cases should we consider a 'local development initiative'?

The origin and destination of the four elements mentioned before therefore seem important for characterizing local initiatives: the origin of the initiative, of the resource inputs, of the benefits (output) and of control. Ideally a majority of these factors should be predominantly local: a local initiative using mainly local resources under local control for predominantly local benefit; in other words, local development by local forces and for local benefit. But practice always falls short of the ideal. It seems particularly important that benefits and control (Fig. 1.2) should be predominantly local, as under these conditions external resources and initiative can, at least temporarily, be harnessed in the local interest until these factors can also be mobilized internally to a higher degree. A further important factor is that the benefits should be distributed broadly within the locality. The cases analysed in this book correspond to these criteria in varying degrees.

Case study areas also vary in size. Rather than aiming for a uniform formal or administrative definition, studies comprise areas of intensive common action for restructuring at the local or regional level. In some cases the size of the area chosen depended on the spatial extent of specific restructuring problems, in other cases on the level at which common action could best be mobilized.

ACKNOWLEDGEMENTS

In writing this chapter I have benefited considerably from discussions with faculty and students at various US universities where I was invited to present this project. Thanks are also due to Mike Luger, Peter Roberts and Dale Whittington for their comments on an earlier draft. Responsibility remains, as always, with the author.

REFERENCES

Andersson, A. E. (1985), 'Creativity and Regional Development', *Papers of the Regional Science Association*, vol. 56, pp. 5–20.

Aydalot, P. (1984), *Crise et Espace*, Paris: Economica.

Aydalot, P. (1986), *Milieux Innovateurs en Europe*, Paris: GREMI.

Birch, D. (1979), *The Job Generation Process*, Cambridge, Mass.: MIT Program on Neighbourhood and Regional Change.

Blanshard, Brand (1978), 'Rationalism', *Encyclopaedia Britannica*, Vol. 15, pp. 527ff.

ECOSOC (1982), *The Promotion of the SMEs in the European Community*, Brussels: European Parliament.

Henderson, J., and Castells, M. (1987), *Global Restructuring and Territorial Development*, London: Sage.

Marshall, M. (1987), *The Long Waves of Regional Development*, London: Macmillan.

Muegge, H., and Stöhr, W. B. (1987), *International Economic Restructuring and the Regional Community*, Aldershot: Gower.

Nijkamp, Peter, and Stöhr, W. B. (1988), 'Technology policy at the crossroads of economic policy and physical planning', *Environment and Planning C, Government and Policy*, vol. 6, no. 4, pp. 371–4.

Peters, T., and Austin, N. (1985), *A Passion for Excellence*, New York: Warner.

Peters, T., and Watermann, R. (1982), *In Search of Excellence*, New York: Warner.

Schmandt, J. (1987), 'Regional roles in governance and the scientific state', paper presented at the meeting of the Association for Public Policy Analysis and Management, Baltimore, Md, 30 October.

Stöhr, W. B. (1984), 'Quantitative, qualitative and structural variables in the evaluation of regional development policies in Western Europe', in G. Demko (ed.), *Regional Development and Policies in Eastern and Western Europe*, Beckenham: Croom Helm.

Stöhr, W. B. (1985), 'Changing external conditions and a paradigm shift in regional development strategies', in Stefan A. Musto and Carl F. Pinkele (eds), *Europe at the Crossroads (Agendas of the Crisis)*, New York: Praeger.

Stöhr, W. B. (1986), 'Regional innovation complexes', *Papers of the Regional Science Association*, vol. 59, pp. 29–44.

Stöhr, W. B. (1987), 'Regional economic development and the world economic crisis', *International Social Science Journal*, vol. 112, pp. 187–98.

Stöhr, W. B., and Taylor, D. R. F. (1981), *Development from Above or Below?*, Chichester and New York: Wiley.

Storper, M., and Scott, A. J. (1989), 'The geographical foundations and social regulation of flexible production complexes', in J. Wolch and M. Dear (eds), *The Power of Geography: How Territory Shapes Social Life*, Boston, London: Unwin Hyman.

Sutton, A. S. (ed.) (1987), *Local Initiatives: Alternative Path for Development*, Maastricht: Presses Interuniversitaires Européennes.

Törnquist, G. (1987), 'Creativity: a geographical perspective', paper presented at the European Science Foundation Workshop on the World Economy and the Spatial Organisation of Power, Jerusalem, 1–4 September 1987.

CHAPTER 2

On the theory and practice of local development in Europe

Walter B. Stöhr

2.1 LOCAL DEVELOPMENT IN AN EXTENDED TIME PERSPECTIVE

Ever since people became sedentary they have organized themselves in territorial communities, the wellbeing of which has depended primarily on their own initiative, creativity, material resources, and ability to organize themselves to make best use of their resources. Local communities have intermittently attempted to expand the territorial range from which they could organize resources, in order to increase their wellbeing. According to Rokkan (1973), this took place along three dimensions of differentiation and retrenchment: economic, cultural and military. There has been a fluctuating but gradually increasing trend of spatial interaction and dominance leading up to the emergence of our present world economy (Laszlo, 1974; Stöhr, 1981; Rokkan *et al.*, 1987).

Important stages in this process of the spatial expansion of dominance and interaction, which were frequently followed by subsequent periods of contraction (Stöhr, 1981; Rokkan *et al.*, 1987), were the formation of nation states particularly since the eighteenth century, followed by the expansion into colonial empires and, after the Second World War, the integration of multinational markets such as those of the European Community and COMECON. Intermittent contractions of these radii had also occurred temporarily, for example during the physiocratic period (second half of the eighteenth century), during the second Kondratieff downswing (about 1873–95), and during the third Kondratieff downswing (about 1926 to the Second World War: cf. Stöhr, 1981, pp. 50ff.). Since the Second World War, after the dissolution of most of the colonial empires, a continuous internationalization of local economies has taken place which has made them increasingly reliant on resource allocations from a central government or from multilocational firms. Local communities increasingly felt relegated to the passive role of 'history takers' and subject to a feeling of helplessness.

Traditional mass-production-oriented 'Taylorist' industrialization has not only increased the number of individuals dependent on work being supplied to them – rather than creating it themselves – but has also made entire local communities dependent for their employment on decision-making centres elsewhere, thus depriving them of economic control over important aspects of their life. Many local communities have become 'dependent on money coming from outside and going straight out again, not circulating locally and thereby supporting local work' (Robertson, 1987, p. 59).

Local levels of development can essentially be increased by the spatial extension of local dominance, the attraction of outside resources (from central government or multilocational firms, for example) or local mobilization and a better organization of indigenous local resources and technological upgrading. The first two strategies have dominated while the latter has been widely neglected until very recently. Local and regional development policies since the Second World War were primarily mobility-oriented and – to use Hirschman's (1970) terms – guided mainly by 'exit' strategies (of capital, people, environmental damage and so forth) to other territories, rather than the promotion of the 'voice' strategies of local initiators of development within their own territorial systems.

Since the last century, the industrial revolution (and particularly the introduction of Taylorist mass production) has promoted strategic alliances between the state and large-scale industry. Industry relied on the state to expand and secure its market, while in many capitalist countries industry helped the state to maintain power. This was accompanied by the reinforcement of hierarchical, vertical organizational and decision-making structures in both the state and the industrial corporate domains.

These vertical organizational and decision-making structures could be maintained and proved efficient during periods of economic growth and market expansion in which scale economies of industrial mass production played an overriding role. The central state tried to alleviate the most extreme social (poverty) and spatial (lack of urban infrastructure) repercussions caused by the industrialization process in general and its cyclical recessions in particular.

These vertical organizational and decision-making structures, however, proved less and less suited for making use of the opportunities offered, and for solving the problems caused by the accelerated economic restructuring process since the early 1970s. In particular, they could not create the conditions needed for a local entrepreneurial climate or directly solve local problems of unemployment and lack of innovation.

Structures far more flexible, decentralized and horizontal were required to deal with these problems. The technological possibilities opened up by microelectronics, such as flexible automation, and the increasing diversification of demand contributed to the development of such decentralized structures. This stimulated a resurgence of interest in small and medium-sized enterprises related to local milieux and of local and regional initiatives.

From the early 1970s onwards, therefore, a rapidly increasing awareness of 'indigenous' local development efforts can be observed (Stöhr, 1985). This has manifested itself in 'an emphasis on indigenous business creation which replaced smokestack chasing in most areas' (Naisbitt, 1985, p. 11, for the USA, quoting the president of a consulting group based in Washington, DC). But large numbers of local development initiatives are emerging not only in the USA (Bergman, 1986), where they had precursors in the community development corporations after the 1960s (Ford Foundation, 1973), but also in Europe, Latin America and other continents (cf. Commission of the European Communities, 1985, 1986; Bassand *et al.*, 1986; Marshall, 1987, p. 237; Sutton, 1987; Galtung, O'Brien and Preiswerk, 1980; Max-Neef, 1982; Hirschman, 1984; IFDA, 1987; Mayer, 1988; as well as the Select Bibliography at the end of this book).

Some authors feel that present local development strategies can benefit particularly from the experiences of the two contraction/crisis periods of the 1870s and 1930s (the second and third Kondratieff downswings) when a great number of local self-help schemes were started (Novy, 1986). This proposition, however, requires thorough scrutiny as the two earlier crises were considered to be generally of a cyclical nature of demand constraint. Attempts were made to overcome these constraints during the first crisis of the 1870s by means of aggressive colonial world market expansion and during the second one of the 1930s with Keynesian policy instruments of demand management. In contrast the present 'crisis' is characterized primarily by an international and spatial restructuring process (Muegge and Stöhr, 1987). Whereas general cyclical depressions were mainly dealt with by means of central gevernment quantitative expansionary measures (external market expansion, deficit spending, public works projects, and so on), the present international restructuring crisis requires mainly decentralized qualitative measures geared to increasing innovative capacity and flexibility (Stöhr, 1986).

Comparable to these earlier periods, however, is the re-emergence of a broad discussion on unorthodox new policy instruments including local employment and development initiatives (now considered 'new' only in a short-term perspective following the previous period of central Keynesian policies). Their apparent 'novelty', however, was also due to their small scale (though great number) which reduced their appeal somewhat for the mass media (Stöhr, 1985) and lent them insufficient 'dignity' for science (Novy, 1986, p. 365). Marshall (1987) suggests that such locally based movements will emerge at 'turning points' in social, economic and political development and that the present wave of local initiatives is by no means 'new', but basically represents a 'reconstruction of the historic tradition of municipal enterprise and civic works' (p. 239). What may be new in many current local initiatives, however, is their 'grassroots' character. In comparison, many of the municipal enterprises and civic projects of earlier periods were often dominated by small and powerful local minorities. The intervening centralization period in some cases may have had the side effect of breaking up these local 'fiefdoms' (J. Bryden, personal communication).

2.2 LOCAL DEVELOPMENT AND THE PRESENT INTERNATIONAL ECONOMIC RESTRUCTURING PROCESS

In contrast to the above-mentioned Kondratieff downswings, during which high unemployment was mainly caused by aggregate imbalances between investment and effective demand, the present phase of unemployment is primarily accompanied by a spatial restructuring process of economic activities on a global scale due particularly to differences in the rate of innovation between sectors and regions. At present, unemployment and other crisis symptoms are therefore concentrated in certain areas (particularly 'old' industrial areas); their spatial incidence is also changing much more rapidly than before. In the formerly centrally planned COMECON countries this latent restructuring requirement has in the past been disguised by protection from the world market and by central resource allocation which has frequently preserved pockets of low productivity and labour surplus on the one hand and areas of overemployment on the other. Ongoing reforms in various COMECON countries, triggered by the need to open up to world markets, are delegating more responsibilities to individual enterprises and will also speed up this restructuring process and some of its concomitant problems, such as local unemployment.

The important characteristics of the present international restructuring process for local development are as follows (Stöhr, 1985; Muegge and Stöhr, 1987):

- new production and communications technologies permit the spatial segmentation of discrete production and distribution processes which before had been spatially unified; individual localities are therefore no longer the seat of entire enterprises but only segments thereof;
- the recent expansion of multisite enterprises and new forms of polycentric entrepreneurial organization have favoured the worldwide distribution of specific entrepreneurial functions according to their specific locational advantages; many locations have therefore been stripped of their former entrepreneurial key functions and have often been left with low routine functions;
- the increasing integration of international finance and capital markets has promoted high capital mobility and a disengagement between capital and location; this has facilitated often massive local disinvestment;

- the mobilization of a potential reservoir of industrial workers in practically all parts of the world not yet industrialized has led to rapid shifts, particularly of standardized 'old' products, to low-wage areas, often in the Third World;
- 'war' between regions and localities for production and distribution activities (particularly with high-technology content) has set in, localities bidding against each other for transnational companies;
- reduced aggregate growth rates have meant that local development can rely less and less on expanding markets but must be achieved by gaining a greater share of existing markets through higher productivity, the creation of new products, and the application of new technologies. Localities and regions are increasingly trying to capture as many new technologies and products as possible to enable them to stay at the 'young' edge of the product cycle;
- the requirement of innovation and flexibility deriving from the above conditions have made traditional regional and local development policies widely ineffective and have necessitated new approaches to stimulating development with increasing emphasis on the local level.

The above characteristics are mainly related to the increased worldwide mobility of capital which calls for new strategies for local and regional development. Many countries are already taking this into account by changing state policies.

2.3 ALTERNATIVE LOCAL-DEVELOPMENT POLICY APPROACHES

Basically, one can distinguish three groups of local-development policy approaches, which are not, however, necessarily mutually exclusive:

- a *centrally initiated* localized development policy approach ('from above');
- a *private enterprise* approach to local development based mainly on the operation of market mechanisms; and
- a broad *locally/regionally initiated* local development process ('from below').

The *centrally initiated* localized development policy approach has been practised most since the Second World War. It basically followed a central redistributive strategy through the spatial allocation of public infrastructure investment and the spatial differentiation of incentives for private activities (in primarily market-oriented economies, cf. Vanhove and Klaassen, 1987), or through the spatial allocation of nationalized activities (in economies oriented towards central planning). In another context I have referred to such development which was mainly redistributive and equity-oriented as 'regional development from above' (Stöhr and Taylor, 1981). While this strategy was reasonably successful during the period of relatively high aggregate growth rates (up to the early 1970s), when it seemed still feasible to influence the spatial distribution of this growth, it lost its effectiveness widely in the 1980s, when aggregate growth rates declined and regional restructuring and innovation became the main concern of local and regional development (Stöhr, 1985; Wadley, 1986).

In West European market economies these new requirements of local/regional development were already evident in the mid-1970s. Some of the East European centrally planned COMECON countries did not become aware of the problem until the mid-1980s, partly because they had been sheltered from the world market longer.

The key reason why such central redistributive policies are ineffective is that they cannot influence the entrepreneurial climate and innovative capacity of structurally weak regional communities (Premus, 1986; Dyckman and Swyngedouw, 1987). Similar conclusions were drawn for traditional labour-market and employment policies (Commission of the European Communities, 1983; Maier and Wollmann, 1986). In more concrete terms, the reasons why predominant central allocation policies have not been able to produce satisfactory results for local development are as follows:

- Central allocation policies have traditionally been organized along sectoral lines, which in most cases produced segmented results at the local level. While the coordination of sectoral policies at the local/regional level has been the traditional task of centrally steered regional and local development policy, it has frequently represented only the extended arm of central power, at best able to provide technical coordination of (mainly central) projects, but not able to mobilize and coördinate local resources. Self-sustained local and regional development requires the predominance of local actors, of local democratic decision-making, of local control of resources and innovation, and of local benefits (Stöhr, 1981). Recently, central agencies at the national and even international level have tried to implement directly integrated local and regional development programmes such as the 'Integrated Mediterranean Programmes' of the European Community, but convincing results are not yet available.

- While central allocation policies during growth periods were relatively successful in cushioning some of the negative social and spatial side-effects of growth (through welfare, housing and urban/regional policies), in the recent period of economic restructuring such policies have become widely ineffective. One reason for this is that they have stifled rather than motivated local and individual initiative, innovation and entrepreneurship.

- Central allocation policies have not been able effectively to overcome the problems caused by the functional economic and labour-market segmentation which the process of international economic restructuring characterized above has produced at local and regional levels (Kamann, 1986; Storper and Scott, 1986).
- Central allocation policies are seldom effective in inducing local innovation in places where it would not happen anyway. They can only work through local innovative structures or try to promote these.
- While central allocation policies are effective in influencing aggregate supply and demand, it appears that the creation of new local supply and demand structures requires additional policies at the local level (Maier and Wollmann, 1986, pp. 27, 295ff.).
- Structurally weak regions are usually handicapped by standardized technological and organizational structures. Central allocation policies are by their very nature standardized and therefore ill-suited to solve their problems.
- Central allocation policies in most market and even in some centrally planned economies also have not been able effectively to reduce social disparities within localities and regions on a sustained basis: again one of the reasons was that they were unable to change local structural conditions.

For several decades it has been maintained that in the present internationalized economy, development efforts 'from below' are hardly effective; but recent local initiatives are showing that economic initiatives and popular involvement at local and regional levels are an essential ingredient of any national programme for economic regeneration which must accommodate and respond to a diversity of local problems and uneven opportunities for resolving them. This does not mean, however, that they do not need the state to facilitate them (Friedmann and Weaver, 1979; Marshall, 1987).

It is frequently maintained that strong central government policies will stifle local initiative, a hypothesis which is applied to both local and regional development policy (Premus, 1986; Wadley, 1986) as well as employment policy (Maier and Wollmann, 1986). It is asserted that withdrawing central power from these fields will restore greater scope to local and regional initiatives, although the weakest areas may have the least potential for carrying through such intitiatives, so that disparities might increase yet further. State support for local initiatives, particularly in the weaker areas, will therefore still be needed.

A *private-enterprise approach* to local development, based mainly on the operation of the market mechanism, is frequently proposed in the light of the ineffectiveness of established local and regional development policies, which were mainly of the central redistributive type. Two main categories of argument have been used to support political criticism of traditional local and regional policy (Wadley, 1986, p. 71): the first, propounded mainly by the extreme Right, state that local and regional policy instruments have proved ineffective and inefficient and that action should be left to private enterprise through deregulation and the market mechanism; the second category of arguments originated mainly from the extreme Left and maintained that the regional problem formed part of the existing (capitalist) socio-economic system, the abolition of which was the precondition for solving regional problems.

Meanwhile, it has been shown that regional problems could be effectively solved neither by the free enterprise/deregulation model (experimented with mainly in the US and Britain) nor by the centrally guided socialist systems operated partly under some Western socialist governments but mainly in COMECON countries (Blazyca, 1983). Both these models are unable to reduce spatial disparities in innovative and adaptive capacity and overcome the helplessness of regional communities in the face of the challenges set by the process of international economic restructuring. They have found themselves caught between the Right's assertion that 'There is no alternative' and the Left's accusation of 'the perverse and remorseless machinery of global capital accumulation' (Marshall, 1987, p. xiv).

The private-enterprise approach is usually based on 'micro' and 'macro'-level arguments: The *micro-level* argument asserts that the private entrepreneur is in the best position to identify new market opportunities, choose relevant new products and technological innovations and invest resources accordingly. The advantages of this unique ability of the private entrepreneur are stressed even more strongly in view of the recent rapid changes in the international division of labour.

The *macro-level* argument is that as soon as the mobility of commodities and services (free trade) and of production factors (capital, labour and technical knowledge) is increased sufficiently, spatial differences in prices and levels of development will tend to equalize in all localities and spatial disparities will disappear. This is essentially the argument maintained by neoclassical theory, as mentioned in Chapter 1. We shall not discuss here the extensive literature defending or refuting both these micro- and macro-hypotheses. A great number of studies, however, have shown that both these hypotheses disregard the real processes operating between these micro- and macro-levels, namely at the local and regional *meso-level*:

As regards the micro-approach, it has been shown that the ability of the individual entrepreneur to define new markets, introduce innovations and gain access to capital (risk capital in particular) depends largely on the respective support structures available in their vicinity, i.e. on the meso-level of their local or regional milieu. It has also been shown that technological innovation must be organized on a territorial basis (Nijkamp and Stöhr, 1988;

Perrin, 1988) and necessitates the careful coordination of local training, community development, education, research, tax and regulatory policies at the local and regional level (Premus, 1986). In an earlier paper I have referred to this as a 'regional innovation complex' (Stöhr, 1986a). It requires an effective interaction and support network of actors at the local/regional level cooperating in various aspects and to varying degrees with actors and institutions outside at the national or international level (see also the diagram of actors in Chapter 1 – Figure 1.1). These local interaction and support networks are particularly important for innovation in small and medium-sized enterprises which are increasingly considered the key elements of sustained local development and innovation (OECD 1982; Rothwell and Zegveld 1982; Wadley, 1986, pp. 76ff.). These studies show that innovation and entrepreneurship are not autonomous micro-processes but depend considerably on the meso-conditions of the local and regional milieu.

As regards the macro-approach referred to above, these meso-conditions would have to be introduced in the neoclassical model as differences in external economies which explicitly distort the functioning of the market mechanism and therefore impede the levelling of development disparities between localities and regions.

Neither the central policies of the previously described allocation-type nor micro-action by individual entrepreneurs has proved effective in changing these meso-conditions. Only broader actions by local communities can be effective in changing the respective local or regional milieu. This is the 'manoeuvring space' of local communities referred to in Chapter 1, concerned mainly with improving conditions for entrepreneurial and innovative activities. We shall deal with this question in greater detail in the locally/regionally initiated policy option described below.

In order to guide private entrepreneurial decisions more in the direction of broader local community objectives some local authorities have made, as a compromise solution, 'contractual arrangements' with private business. In Britain, for example, some local authorities make support (frequently in the form of equities for new or existing plants) contingent upon contracts stipulating that firms protect the interests of their workforce and the wider community with respect to working (see Chapters 5 and 6) or environmental conditions. In the USA local administration frequently contracts 'linkage arrangements' whereby local authorization or support of new investment and activities is made contingent upon their orientation towards local needs (local service provision, employment, and so forth) or their contribution towards upgrading the local infrastructure in general. Similar agreements are also made in some COMECON countries between local authorities and local firms.

The problem is that, generally speaking, such exigencies can only be made by local communities which are already attractive to investors, whereas the remaining communities might continue to be forced to beg for outside investments, or to overcommit local budgetary resources by means of tax concessions.

In many West and recently in some East European countries a voluntary sector has increasingly taken charge of these problems. Its activities have focused on providing community services and creating community industry (particularly in Western Europe, e.g. the British community industry programmes), as well as on cultural and expertise-oriented activities (predominant in most East European voluntary sectors: see for example the local friends of town societies described in the Polish case study in Chapter 9). In general, it appears that once basic infrastructure needs are fulfilled the sustained impact of the development of local expertise in structurally weak areas is greater than that of purely material inputs.

The relationship between business and local communities is manifested in two ways: the benefits created by enterprise for a local community (which the contractual arrangements try to maximize) and the milieu offered to enterprise by the local community. Particularly in less attractive local communities, both aspects will need a more broadly based locally/regionally initiated development policy as described below and on which many case studies in this book focus. Essentially, in centrally planned states very similar mechanisms operate, except that there local communities attempt to woo central authorities (and only very recently private foreign multilocational firms too) for the allocation of plants from which they subsequently try to gain financial support for local infrastructure.

A broad *locally or regionally initiated* development policy is a third approach which has re-emerged recently, particularly for structurally weak local economies for which neither of the two approaches earlier mentioned has proved successful in coping with the problems and possibilities of worldwide industrial restructuring. Central (redistributive) policies proved too inflexible to cope with local problems and incapable of creating an entrepreneurial and innovative local milieu. The private enterprise approach, though in many cases dynamic, proved too evasive to be harnessed for local communal objectives. It became increasingly apparent that local development could be left entirely to neither central government nor private enterprise. Local communities would have to take greater charge than before. Since the 'crisis of the state' of the 1970s, therefore, broad local or grassroots initiatives have multiplied rapidly, in Europe too, although only a small proportion have as yet been analysed systematically (Stöhr, 1981 and 1985; Senghaas, 1982; Musto and Pinkele, 1985; Bassand *et al.*, 1986; Maier and Wollmann, 1986; cf. also the Select Bibliography at the end of this book).

One interpretation for local initiatives is that the generation of 'regional crises' caused by the international economic restructuring process, particularly in market economies, has aroused new forms of local social unrest and increased the potential for local mobilization (Marshall, 1987, pp. 13, 235ff.). In other words local mobilization is – in market and also in formerly centrally planned economies – the product of the institutional decision-making vacuum in which local communities are trapped between the 'anonymous' and usually inflexible role of the state on the one hand and, on the other, dynamic worldwide economic restructuring. In centrally planned economies efforts towards local mobilization have often also been related to local environmental or quality-of-life problems caused by centrally planned industrial projects (see, for example, the Polish case studies in Chapter 9).

Broad-based local development initiatives tend to emerge when more than just a narrow stratum of the local population is affected by the symptoms of crisis. In the case of plant closures this is so when directly affected plants represent a substantial share of local employment (particularly if its labour is well organized) or if other local sectors are indirectly affected, for example, as suppliers or customers.

Recently an acceleration of local initiatives has been triggered by the growing inability of the state to deal with local unemployment and restructuring problems via the traditional centrally initiated localized development policies mentioned above. They have been further supported by the increasing confidence spread by a number of successful local development and restructuring efforts and by the growing awareness of the need to mobilize additional local resources to solve these problems. To encourage even further this confidence and mutual learning process is a major objective of the present project.

2.4 CENTRAL SUPPORT FOR LOCAL DEVELOPMENT INITIATIVES

Once the sum of local problems substantially exceeds the capacity of central agencies to deal with them, they too will be increasingly prepared to support local development initiatives.

Since the latter part of the 1970s, therefore, an increasing number of national governments have officially supported local employment and development initiatives. Even before this, in the 1960s, the US federal government supported community action agencies which initially focused on social service provision to poor urban minorities and later, under the designation of community development corporations (CDCs), broadened their scope to economic development projects (Ford Foundation, 1973; Garn, 1976). In Europe, Britain was the first country in which local restructuring problems became extremely pressing, and as early as the mid-1970s the Home Office was sponsoring community development projects (Kraushaar and Loney, 1980; Marshall, 1987, p. 213). It was followed by most other West European countries.

More recently, even international organizations have become active in this field. The OECD, for instance, initiated a 'local initiatives for employment creation' (ILE) programme in 1982, while the Commission of the European Communities in the same year introduced a similar programme and in 1984 officially provided for measures to 'promote the endogenous development potential of the regions' through the European Regional Development Fund (Decree of the Council of the European Community, No. 1787/84, 19 June 1984). In 1986 it launched its local employment development action programme (LEDA) in order to complement and reinforce the effectiveness of Community aid given through its structural funds (European Social Fund, European Regional Development Fund, European Agricultural Guidance and Guarantee Fund). Some of the experiences to be drawn from OECD and EC Reviews of these initiatives are presented in Chapter 15. In 1984, furthermore, an information network for local initiatives, ELISE, was created in collaboration with OECD and the Commission of the European Communities and financed by the latter. The Council of Europe has also dealt extensively with this question in recent years (e.g. Council of Europe, 1987), as has the International Labour Office (Mayer, 1988).

These programmes to promote local employment initiatives at the national and international level are, however, a response to previous action at the local level. While the general academic and political discussion about the most appropriate macro-economic measures to combat unemployment was still in progress, individuals and groups at the local level had anticipated the result of this discussion by undertaking various efforts aimed at creating jobs locally (Commission of the European Communities, 1985, Annex 1). According to this report, these local employment initiatives were distinguished from traditional employment and development policy in four ways: (1) they were not triggered externally but developed within the local community as a form of economic self-help; (2) they were based on a partnership between the different local groups; (3) they were usually oriented towards a mix of economic and social objectives; and (4) their activities were primarily geared to the interests of the local community.

The OECD programme on local initiatives for employment creation (ILE) involves nineteen European countries and the USA. It is a non-statutory cooperative programme financed separately by the participating countries. Major groups of activities are: (1) information exchange and dissemination (mainly through a 'liaison letter', the *ILE Notebooks* and *Feedback ILE*, later published as *Innovation and Employment*); (2) industrial diversification and employment generation in local economies, in particular the contribution made by social partners, public authorities and firms to diversification, especially in old industrial areas; (3) innovation and development for job creation in less developed regions and countries, concentrating on small and medium-sized firms and untapped local resources in agricultural and potential tourist areas; (4) new roles for local economies in stimulating economic development and jobs, concentrating on the role of local government; (5) local job creation: the need for innovation, information and suitable technology, concentrating on the factors which create a favourable climate for process and product innovation to support employment growth, including science parks and small business incubators; (6) education, training and support needs of new entrepreneurs and of local employment initiatives, concentrating on the role of educational, training, finance, marketing, legal and other services for the creation of new firms and jobs (OECD, n.d.).

The Commission of the European Communities has also, initially in the frame of its research and action programme for the development of the labour market, commissioned since 1982 a number of local consultations in different European countries on local employment initiatives. Reports on these local consultations, the first for the period 1981–3, a second one for 1984–5, were elaborated by the Centre for Employment Initiatives in London. Subsequently, continuous fieldwork was being sponsored by the LEDA programme of the EC in twenty-four pilot areas chosen for their high unemployment. The objective of the initial local consultations was to evaluate the results of the respective local employment initiatives, to identify the obstacles and difficulties that emerged and to suggest possible solutions. The consultations were concerned with local employment initiatives varying widely in ideology, motivation and activity and ranging from the creation of new private enterprises, cooperatives, rural training initiatives to social self-help and environmental protection groups and initiatives for social tourism in mountain areas (Commission of the European Communities, 1985). The second series of the above-mentioned consultations put special emphasis on maritime and peripheral regions and on areas in industrial recession requiring restructured employment and focused on different forms of local support structure for the launching and maintainance of local employment initiatives and on the relationship between local job creation and overall local and regional development. It further concentrated on the function of improved animation and cooperation for local employment initiatives and the identification of intermediate support structures which might qualify for support from community funds (Commission of the European Communities, 1986). The report finds that, in general, advances have been made in most European countries in instituting new support structures and programmes to benefit local employment initiatives, but that the knowledge and understanding of them is still for the most part limited to a narrow group of administrators and technical specialists. The authors of the report feel that some 'popular' and easily available publication together with other information for 'the general public including elected decision-makers at local, regional and national levels, educators, trade unionists and young people' would be important (p. 44). To the outside observer this European situation contrasts markedly with, for example, the broad publicity which has been given in practically all the media in Japan to the new local development and technopolis policies (Stöhr, 1986b; Kawashima and Stöhr, 1988).

2.5 LOCAL DEVELOPMENT INITIATIVES: ALTERNATIVE EFFECTS, APPROACHES AND LOCAL SUPPORT STRUCTURES

Initially, most local initiatives were conceived of as employment initiatives. Although the *employment effect* of local intitiatives remains relatively small compared to overall unemployment (Maier and Wollman, 1986; Commission of the European Communities, 1986b, p. 225), the social, economic and institutional significance for development, however, is very often considered of equal or even greater importance.

In terms of *social significance*, three approaches may be distinguished (Commission of the European Communities, 1986a, p. 46, 1986c, p. 10f.). The first relates to *'alternative' objectives* in the sense that an individual's working hours should not be wholly devoted to economic production and services. The second concerns initiatives to reach established objectives by means of *'alternative' instruments* aimed at assisting people with physical, mental, social and cultural handicaps to become integrated into society. This is frequently done by offering them a sheltered work environment which should progressively be made as 'businesslike' as possible to enable them to earn an increasing proportion of their income by marketing their products (possibly in cooperative form) and give them ever greater independence from welfare and subvention. A third type of social significance often consists in *satisfying concrete local needs* not fulfilled through the market mechanism such as local environmental improvement.

In *economic* terms the principal effects of the first approach are to redistribute formal work and complement it with informal types of work. The second approach serves mainly to reduce the public funding required to assist handicapped groups, aiming as it does to increase their self-reliance in psychological and economic terms. The third approach is directed towards creating non-market mechanisms for satisfying important local requirements.

The *institutional significance* very often goes much beyond this, however, in the sense that new forms of cooperation between individuals, social groups, and also enterprises and institutions emerge in connection with local initiatives. These are important prerequisites for the sustained impact and reproduction of local initiatives.

Local support structures in this respect assume a key role for local development initiatives. There exists a dichotomy regarding the support structures required for these different local initiative approaches (Commission of the European Communities, 1986, p. 46) in that some support structures represent predominantly traditional entrepreneurial values, while others represent primarily 'alternative' self-managed principles. In some cases, however, it was possible to combine these different approaches under one umbrella organization (p. 47). In these cases of local 'synergy' local solidarity seems to have become more important than particular sectoral, ideological or group interests where the group seeks to retain for itself the greatest possible benefit from development initiatives. This is a question of functional (frequently also group or class) solidarity as against territorial (local, regional) solidarity.

The actual provision of support structures for local initiatives will vary accordingly. Where mutual ideological distrust outweighs mutual understanding and appreciation, very often separate support structures between cooperative 'alternative' initiatives and those for more traditional 'small enterprise' initiatives will predominate, while in other cases a single 'one-stop-shop' support structure for all local initiatives will become feasible (p. 47). The report suggests that even if these support services were kept separate institutionally they should at least be located near to each other geographically (like shops in a department store or a shopping centre) in order to increase their general accessibility and enhance understanding and appreciation of each other's work.

Local employment or development initiatives oriented mainly towards traditional small and medium-sized enterprises will usually emphasize firm creation, technological upgrading and economic efficiency, while local employment initiatives aimed principally at marginal social groups will focus on employment and service provision, relegating the criterion of economic efficiency, at least initially, to a secondary role (p. 52). For sustained local development it will be important, however, that the objectives of both types of local initiative approach and complement each other. Here, the creation of new firms and the introduction of new technologies would be accompanied by increased employment, whereas social initiatives (with the exception of permanently handicapped groups) should increasingly endeavour to improve economic feasibility. The convergence of these two groups of local initiatives could be encouraged by increasing their mutual functional support through appropriate local community mechanisms so that new enterprises might increasingly play a social role, while at the same time social initiatives might be increasingly guided by economic objectives and criteria, thereby mutually sustaining each other.

The importance of close interaction between social/cultural and economic initiatives for self-sustained local development, as well as their mutually reinforcing effect, is illustrated especially in Chapters 3 and 10. In some cases local social or cultural networks and projects were also the basis for economic initiatives, as shown particularly in Chapters 4 and 9.

2.6 FACTORS OF LOCAL INNOVATION

In contrast to the deterministic macro-theories of local development referred to in Chapter 1, a number of recent studies have also been concerned with the micro-factors relevant to local innovation. Andersson (1985) postulated a combination of local competence, local synergy and societal instability as important factors for local innovation, basing his thesis to a great extent on the experience of turn-of-the-century Vienna. Törnqvist (1987) adds to these conditions the presence of key personages as catalysts of communication between individuals and areas of competence, cultural diversity, economic stagnation (similar to Schumpeter's creative destruction), organizational flexibility (or the sidestepping of formal organization structures) and networks between 'cultural circles' at regional, national and global levels. His examples are taken from historical studies of innovation in Renaissance cities, early industrial cities such as Manchester around 1840 and turn-of-the-century Vienna, of contemporary cultural, artistic and economic renewal in large metropolitan areas of London, Paris, Los Angeles and New York, and of business innovation in small towns and recent high-tech and science parks in various countries. Stöhr (1986a) has analysed the major components of 'regional innovation complexes' and their interaction in different societal forms such as a cooperative, private and 'third sector' (local public–private–university partnership) model.

Storper and Scott (forthcoming) relate innovation and future competitiveness closely to the introduction of technological and institutional forms of 'flexible production'. They consider as preconditions for the introduction of flexible specialization: local cooperation, association and coordination (including local consensus between labour unions, employers and local government); the existence of local information networks and the socialization of useful local knowledge; local input–output relations; and teamwork and cooperation on the labour market. They maintain that these conditions are usually not given in areas of traditional mass-production-oriented 'Fordist' industrialization and that flexible specialization therefore emerges mainly in the suburban peripheries of major metropolitan centres, in traditional craft communities or in previously unindustrialized communities. This approach also contains deterministic traits in that certain (for instance, 'old industrial') areas are all but excluded from opportunities for developing innovation and flexible specialization. In contrast, a number of case studies in the present volume show that in fact such innovation has also been possible in 'old industrial' areas and that even these areas can combat the societal 'immune deficiency syndrome' referred to in the Synthesis.

What then are the prerequisites for avoiding or combating this societal immune deficiency syndrome towards international economic restructuring?

The results to be drawn from the case studies in this book happen to coincide widely with those to be derived from the Japanese and other innovation experiences (NZZ, 1988; Stöhr, 1986b; Nijkamp and Stöhr, 1988). They are mainly concerned with the emergence of innovation and their transfer towards a broad distribution between and within local communities. The major local factors for this to happen as well as important external preconditions are listed at the end of the Synthesis.

ACKNOWLEDGEMENTS

Valuable comments on an earlier draft of this chapter were received from J. Bryden, H. Goldstein, B. Gruchman, B. Johannisson, M. Marshall, A. Novy, P. O. Pedersen and D. Whittington. Any remaining shortcomings are as ever the responsibility of the author.

REFERENCES

Andersson, A. E. (1985), 'Creativity and Regional Development', *Papers of the Regional Science Association*, vol. 56, pp. 5–20.

Andersson, A. E., and Johansson, B. (1984), *Knowledge Intensity and Product Cycles in Metropolitan Regions*, WP-84–13, Laxenburg, Austria: International Institute for Applied Systems Analysis.

Bassand, Michel, Brugger, E. A., Bryden, J. M., Friedmann, J., and Stuckey, B. (eds) (1986), *Self-Reliant Development in Europe. Theory, Problems, Actions*, Aldershot: Gower.

Bergman, E. (1986), 'Policy realities and development potentials', in E. Bergman (ed.), *Local Economies in Transition*, Durham: Duke University Press.

Blazyca, G. (1983), *Planning is Good for You: The Case for Popular Control*, London: Pluto.

Commission of the European Communities (1983), 'Gemeinschaftsaktion zur Bekämpfung der Arbeitslosigkeit', *Beitrag der örtlichen Beschäftigungsinitiativen*, Brussels, KOM (83) 662 endg.

Commission of the European Communities (1985), *Research and Action Program for the Development of Local Labour Markets – Local Employment Initiatives. Report on a Series of Local Consultations in European Countries 1982–1983*, Luxembourg: Centre for Employment Initiatives, London, Office for Official Publications of the European Communities (ISBN 92–825–5575–5).

Commission of the European Communities (1986a), *Cooperation in the Field of Employment – Local Employment Initiatives. Report on a Second Series of Local Consultations held in European Countries 1984–1985. Final Report*, Luxembourg: Centre for Employment Intitiatives, London, Office for Official Publications of the European Communities (ISBN 92–825–6086–4).

Commission of the European Communities (1986b), *Initiatives Locales pour l'Emploi. Relevé des expériences de création d'emplois non conventionelles*, Luxembourg: Programme de Recherche et d'Actions sur l'Evolution du Marché du Travail, Office des publications officielles des Communautés européennes (ISBN 92–825–6085–6).

Commission of the European Communities (1986c), *Analyse der Rolle und des Ausbildungsbedarfes von Entwicklungsberatern im Rahmen Lokaler Arbeitsplatzbeschaffung*, Luxembourg: Forschungs- und Aktionsprogramm zur Entwicklung des Arbeitsmarktes, Office for Official Publications of the European Communities (ISBN 92–825–6087–2).

Council of Europe (1987), Permanent conference of Local and Regional Authorities of Europe (CPL/AM/REG/87) 4, Strasbourg.

Dyckman, John W., and Swyngedouw, E. A. (1988), 'Public and private technological innovation strategies in a spatial context: the case of France', *Environment and Planning C*, pp. 401–15.

Ford Foundation (1973), *Community Development Corporations, A Strategy for Depressed Urban and Rural Areas*', New York: Ford Foundation.

Galtung, J., O'Brien, P., and Preiswerk, R. (1980), *Self-Reliance – A Strategy for Development*, London: Bogle-L'Overture.

Garn, Harvey *et al.* (1976), *Evaluating Community Development Corporations – A Summary Report*, Washington DC: Urban Institute.

Hirschman, Albert O. (1970), *Exit, Voice and Loyalty*, Cambridge, Mass.: Harvard University Press.

Hirschman, Albert O. (1984), *Getting Ahead Collectively: Grass Roots Experiences in Latin America*, Oxford: Pergamon.

IFDA (1987), *Urban Self-Reliance Directory*, Nyon: International Foundation for Development Alternatives.

Kamann, Dirk-Jan F. (1986), 'Industrial organisation, innovation and employment', in P. Nijkamp (ed.), *Technological Change, Employment and Spatial Dynamics*, Berlin: Springer-Verlag.

Kawashima, Tatsu-hiko, and Stöhr, W. B. (1988), 'Decentralized technology policy – the case of Japan', *Environment & Planning C, Government and Policy*, vol. 6, pp. 427–41.

Kraushaar, R., and Loney, M. (1980), 'Requiem for planned innovation: the case of the Community Development Project', in M. Brown and S. Baldwin (eds), *The Yearbook of Social Policy in Britain 1978*, London: Routledge & Kegan Paul.

Laszlo, E. (1974), *A Strategy for the Future*, New York: Braziller.

Leontief, W. *et al.* (1953), *Studies in the Structure of American Economy*, London: Oxford University Press.

Maier, H. E., and Wollmann, H. (eds) (1986), *Lokale Beschäftigungspolitik*, Basel, Boston and Stuttgart: Birkhäuser.

Malecki, Edward, J., and Nijkamp, Peter (1988), 'Technology and regional development: some thoughts on policy', *Environment and Planning C*, vol. 6, no. 4, pp. 383–401.

Marshall, Michael (1987), *Long Waves of Regional Development*, London: Macmillan.

Max-Neef, Manfred A. (1982), *From the Outside Looking In: Experiences in 'Barefoot Economics'*, Uppsala: Dag Hammarskjöld Foundation.

Mayer, Jean (ed.) (1988), *Bringing Jobs to People, Employment Promotion at Regional and Local Levels*, Geneva: International Labour Office.

Muegge, Herman, and Stöhr, W. B., with Hesp, P., and Stuckey, B. (eds) (1987), *International Economic Restructuring and the Regional Community*, Aldershot: Gower.

Musto, S. A., and Pinkele, C. F. (eds) (1985), *Europe at the Crossroads – Agendas of the Crisis*, New York: Praeger.

Naisbitt, John, and the Naisbitt Group (1985), *The Year Ahead, 1986, Ten Powerful Trends Shaping Your Future*, New York: Warner.

Nijkamp, Peter, and Stöhr, W. B. (1988), 'Technology policy at the crossroads of economic policy and physical planning', *Environment and Planning C, Government and Policy*, vol. 6, no. 4, pp. 371–4.

Novy, Klaus (1986), 'Aspekte einer Theorie der Arbeitsbeschaffung auf der Basis lokaler Selbsthilfe – historisch gewonnen', in Hans E. Maier and H. Wollmann (eds), *Lokale Beschäftigungspolitik*, Basel: Birkhäuser.

NZZ (1988), 'Mit Computern und Robotern die Welt umarmen, Wissenschaft und Technik in Japan III' (Embracing the World with Computers and Robots, Science and Technology in Japan III), *Neue Zürcher Zeitung, Fernausgabe*, no. 32, p. 39.

OECD (1982), *Innovation in Small and Medium Firms*, Paris: Organization for Economic Co-operation and Development.

OECD (n.d.), '*The OECD's Cooperative Action Programme on Initiatives for Local Employment Creation (ILE)*, Paris: Organization of Economic Co-operation and Development.

Perrin, Jean-Claude (1988), 'A deconcentrated technology policy – lessons from the Sophia-Antipolis experience', *Environment and Planning C, Government and Policy*, vol. 6, no. 4, pp. 415–26.

Premus, Robert (1986), 'Federal technology policies and their regional effects', paper presented to the Appalachian Regional Commission, 25 September.

Reuss, Karl (1986), 'Die klassische Gewerbeförderung vor neuen Aufgaben – Gewerbeförderung in Baden-Württemberg und ihre lokale Verankerung', in Maier and Wollmann, op. cit., pp. 148–78.

Robertson, James (1987), 'Work, money and the local economy: some directions for the future', in Sutton, op. cit., pp. 57–70.

Rokkan, Stein (1967), 'Geography, religion and social class', in S. M. Lipset and S. Rokkan (eds), *Party Systems and Voter Alignments*, New York: Free Press.

Rokkan, Stein (1973), 'Cities, states and nations: a dimensional model for the contrasts in development', in S. N. Eisenstadt and S. Rokkan (eds), *Building States and Nations*, vol. 1, Beverley Hills and London: Sage.

Rokkan, Stein, Urwin, Derek, Aarebrot, Frank H., Malaba, Pamela, and Sande, Terje (1987), *Centre–Periphery Structures in Europe*, Frankfurt and New York: Campus.

Rothwell, R., and Zegveld, W. (1982), *Innovation and the Small and Medium-Sized Firm*, London: Frances Pinter.

Senghaas, D. (1982), *Von Europa lernen – Entwicklungsgeschichtliche Betrachtungen*, Frankfurt: Suhrkamp.

Stöhr, W. B., (1981a), 'Development from below: the bottom-up and periphery-inward development paradigm' in Stöhr and Taylor, op. cit.

Stöhr, W. B. (1981b), 'Towards "another" regional development? In search of a strategy of truly "integrated" regional development', R. P. Misra and M. Honjo (eds), *Changing*

Perception of Development Problems, Singapore: United Nations Center for Regional Development.

Stöhr, W. B. (1985), 'Changing external conditions and a paradigm shift in regional development strategies', in Musto and Pinkele, op. cit.

Stöhr, W. B (1986a), 'Regional innovation complexes', *Papers of the Regional Science Association*, vol. 59, pp. 29–44.

Stöhr, W. B. (1986b), 'La politique japonaise des technopoles: innovation technologique et institutionelle', in Jacques Federwisch and Henry Zoller (eds), *Technologie nouvelle et ruptures régionales*, Paris: Economia.

Stöhr, W. B., and Taylor, D. R. (1981), *Development from Above or Below?*, Chichester and New York: Wiley.

Storper, Michael, and Scott, A. (eds) (1986), *Production, Work, Territory: The Geographical Anatomy of Industrial Capitalism*, Boston and London: Allen & Unwin.

Storper, Michael, and Scott, A. (forthcoming), 'The geographical foundations and social regulation of flexible production complexes', in J. Wolch and M. Dear (eds), *Territory and Social Reproduction*, London and Boston: Allen & Unwin.

Sutton, A. S. (ed.) (1987), *Local Initiatives: Alternative Path for Development*, Maastricht: Presses Interuniversitaires Européennes.

Törnquist, G. (1987), 'Creativity: a geographical perspective', paper presented at the European Science Foundation Workshop on the World Economy and the Spatial Organisation of Power, Jerusalem, 1–4 September.

Vanhove, Norbert, and Klaassen, Leo H. (1987), *Regional Policy, A European Perspective*, 2nd edn, Aldershot: Gower.

Wadley, David (1986), *Restructuring the Regions*, Paris: OECD.

Zegveld, Walter (1988), 'Technology policy and changing socio-economic conditions', *Environment and Planning C*, vol. 6, no. 4, pp. 415–27.

PART I

LOCAL RESTRUCTURING IN THE NORTHERN AND WESTERN PERIPHERY OF EUROPE

CHAPTER 3

The Nordic perspective: self-reliant local development in four Scandinavian countries

Bengt Johannisson

3.1 INTRODUCTION[1]

The Nordic countries are pluralistic corporate welfare states. During the late 1970s and in the 1980s these countries have run into the same structural problems as most other European countries, problems stemming mainly from the process of international economic restructuring. It is evident that the policy measures which once created wealth have not produced effective resistance to recent global challenges. The main question, then, is whether local initiatives with little respect for, or help from, national policies can not only survive these socio-economic challenges but, more important, turn them into opportunities. The aim of this chapter is to assess in what way the socio-political fabric of the Nordic welfare states counteracts or supports local economic initiatives which are taken expressly to cope with structural changes at the national and international levels. The study covers four Nordic countries - Denmark, Finland, Norway and Sweden, and omits Iceland. In these four countries the spatial distribution of social and economic life has been a key political issue during recent decades. The urbanization process, which was rapid in the 1960s, came to a standstill in the 1970s and then accelerated once again in the 1980s. This is especially true of Norway and Sweden. In the interest of national economic development - so the argument in Sweden goes - expansion in the metropolitan areas must be encouraged, or at least accepted. Traditional agglomeration arguments have given way to the acceptance of 'dynamic' urban advantages such as technological expertise, global communication networks and creativity. Table 3.1 presents some basic facts about the Nordic countries.

The Nordic countries can be considered to be a rather integrated area in geographical, historical, cultural and economic terms. They are also interrelated socially. They have formed a joint labour market with no passport constraints and their social policies are mutually compatible. With Finland's very rapid progress during the postwar period all Nordic countries have now

Table 3.1 The Nordic countries - some socio-economic indicators

Country	*Population 1984 (millions)*	*Land area (sq. km.)*	*Density*	*Urbanization (%)*[2]	*Unemployment 1986 (%)*[3]
Denmark[1]	5.1	43,000	119	84	7.9
Finland	4.9	340,000	14	72	6.9
Norway	4.1	320,000	13	70	2.1
Sweden	8.3	450,000	18	83	2.2

[1] The Danish figures exclude the partially self-governed territories of Greenland and the Faroe Islands.
[2] Percentage of population in urban units with 200 inhabitants or more.
[3] According to ILO (annual averages).

reached a welfare-state status. Table 3.1 indicates that there are, however, considerable differences in population density between Denmark and the other three countries. Because of this, and because of its location as well as its membership of the European Common Market (since 1972), Denmark's regional challenges have more in common with those of other continental European countries. To be more precise this applies mainly to Copenhagen and the eastern part of Denmark, the areas generally dominating economic and social life. The western part of the country, especially northern Jutland, has had many problems similar to those of any peripheral Nordic area. However, in the 1970s and 1980s growth in the manufacturing sector has mainly taken place in this part of the country. (The Danish territories, Greenland and the Faroe Islands, have of course special economic development problems, which will not be dealt with here.)

Denmark and Sweden were the first among the Nordic nations to embark on postwar industrialization. For obvious reasons, Norway, and then Finland, followed suit considerably later. In Finland, where this industrialization lag was some fifteen years, employment in the primary sector (agriculture, forestry and fisheries) dropped over the 1960–80 period from about 35 per cent of total employment to less than 15 per cent. In Denmark, although the industry was mainly engaged in the development of the dominant agricultural sector, this sector (including forestry and fishing) accounted in 1981 for the same low share (8 per cent) of the workforce as in Norway (NU, 1986; no. 4). In all countries except Denmark and in spite of heavy subsidies to the agricultural sector, this development caused major regional imbalances. The main force counteracting regional concentration has been a very strong public sector, in 1980 accounting for 33 per cent of total employment in Sweden, 26 per cent in Finland, 28 per cent in Norway and 40 per cent in Denmark (NU, 1986, no. 4).

The public sector administers substantial government support in terms of transfers with regional implications: general sectoral support, tax equalization grants, labour market measures, regional support measures and special measures such as cases of acute crises in the many one-company towns. Thanks mainly to this transfer system, the manufacturing sector has expanded in the peripheral and diminished in the metropolitan areas. Consequently, the regional income differences within the Nordic countries today are surprisingly small. A supplementary force also contributing to this equalization is provided by the strong organizations on the labour market.

The very fabric of the Nordic welfare states tends to limit both the need for and the potential of traditional regional and industrial policies. According to a study of Swedish public programmes with regional implications, only a small fraction in 1985 referred to measures emanating from regional and industrial policies (Ds I, 1987, no. 6). None the less, in the light of political pressure for a general reduction of the public sector and a concentration of immaterial resources on centres of national competence, these policies are very important to areas outside the metropolitan towns. The ideology behind and actual measures resulting from present regional and industrial policies need to be scrutinized in order to identify the terms for local economic development. The main characteristics of the different Nordic regional and industrial policies are therefore summarized in section 3.2. In section 3.3 a conceptual framework for the case studies is presented, elaborating on common Nordic characteristics believed to be relevant to the emergence of local initiatives. In addition, a tentative model for describing and interpreting the presented cases with respect to local economic development is introduced. The cases themselves are presented in section 3.4. In section 3.5 loyalty to the territorial strategy is scrutinized in a comparative analysis of the cases. In the final section lessons to be learned from the cases are interpreted in terms of the prospects for local economic development, focusing on certain process characteristics. These lessons form a basis for recommendations for future regional and economic policies in the Nordic countries.

3.2 CHARACTERISTICS OF THE NATIONAL REGIONAL POLICIES IN THE NORDIC COUNTRIES

Nationwide regional planning was introduced in Denmark, Norway and Sweden during the 1960s and in Finland at the beginning of the 1970s. In accordance with the regional policies in the Nordic countries, the *zones* designated for localized support cover the western part of Jutland in Denmark and, essentially, the northern parts of the other three countries. This means that in Norway about 80 per cent of the land area (with 35 per cent of the population) is eligible for support, in Finland about 70 per cent (40 per cent of the population) and in Sweden about 60 per cent (a mere 13 per cent of the population). In Denmark 20 per cent of the population live in areas eligible for regional policy measures.

Different levels of and terms for the aid are specified within the general support areas. Finland has the most detailed system, with eight levels in all, Norway has seven levels and Sweden three. The lack of specification in the Danish system is due partly to the small size of the country and partly to generally much more restricted government intervention in industry than in her fellow Nordic countries.

The Nordic countries, with the exception of Denmark, have placed the main responsibility for both *regional planning and policy implementation* at the regional level. In Norway, Sweden and Denmark there are publicly elected county councils. The Norwegian councils have particularly large discretionary powers. In addition, there is in Norway a special regionally structured organization which administers loans and grants within the regional policy (DU) and another organization which supplies consultancy services (INKO). In Sweden county councils finance jointly with the Swedish Board of Industry regional development funds and appoint their boards. The Swedish county administrative boards, the decentralized state administration, have recently been given a wider mandate to design and implement regional aid programmes. In Norway these boards are quite irrelevant in regional planning itself and in its implementation. In Finland regional planning is dealt with largely by a regional state organization (KERA). Finland differs again in having many of its municipal associations take on such responsibilities as physical planning at the regional level. In Denmark physical planning and regional development programmes are integrated at the national level.

A large variety of measures are taken in all Nordic countries; only a few of them will be commented upon here. Location grants and loans are especially common in Finland where in 1984 they amounted to some US$100 million; Sweden and Norway reached only half of that sum each and in Denmark such measures amounted to approximately US$20 million in 1986. These grants and loans are earmarked for investments in machinery and buildings, basically for manufacturing industry. However, as a result of recent changes, development grants are now also available for investment in 'software' in all four countries. Employment subsidies are granted in Sweden and Finland. This means that in Sweden, for example, if a firm hires new personnel in order to expand the business, the wage cost can be reduced by about 20 per cent for three to seven years. Subsidies for long-distance goods transport (mainly *out* of the region) for companies located in the northern part of Finland, Norway and Sweden are substantial; in Sweden similar subsidies have recently been introduced to encourage personal business contacts. Finally, several of schemes supporting service- and job-creation projects in very sparsely populated areas have been launched in these three Nordic countries. Similar schemes are now also being operated in Denmark.

To sum up, the Nordic countries (Denmark, possibly, to a lesser extent) have very elaborate regional policies with respect to geographical coverage, organizational structure and number of concrete measures. This would seem to provide an ideal context for economic development focusing on territorial units as prime objects. However, this impression is illusory. First, although implementation of policies is mainly regionalized, the programme mix is basically centralized. Secondly, the discretion of the local public administration within economic policy is very limited. In many other sectors the municipalities have wide responsibilities and powers. Thirdly, almost all measures within the regional and industrial policies are aimed at individual firms which are, moreover, approached as though they were 'footloose' actors on international markets. The role of individual firms within local industrial complexes is not at all recognized. Recent research findings concerning industrial development indicate that, for example, establishing and maintaining linkages between firms, small as well as large and independent of their location, are more vital to firm and industry development than the improved management of owner-controlled resources (cf. e.g. Johannisson, 1987). Fourthly, the public support system, being a bureaucracy, tends to undervalue the importance of entrepreneurs as leaders of the industrial change process and to overvalue the role of formal plans.

Considering these obstacles it is very difficult to integrate economic policy with other sectoral policies at the local level, the only level where sectoral interdependence can be truly realized. Initiatives taken by the local municipalities have, however, been welcomed by the central authorities inasmuch as they reduce the latter's responsibility and financial costs for the support of disadvantaged areas. In order to make such a laissez-faire policy work efficiently, the municipalities must be invested with the necessary authority and professional competence.

3.3 LOCAL ECONOMIC DEVELOPMENT – CONDITIONS AND PROCESS

All human endeavour manifests itself locally. Even autonomous enterprise, the nucleus of capitalism, has ties to place. The founder's identification with a certain territory is an important factor in the creation of a new firm, and interaction with the local environment is thus fundamental for the development of the firm. Locally, the economic sector is strongly linked to other behavioural sectors. Because of the internationalization and professionalization of economic activities, however, these insights have been eclipsed by business ideologies that disregard, or even contradict, values and rationales that represent the linkage between economic activity and other dimensions of societal life, including those associated with place. These two contrasting world views are reflected in, respectively, a *territorial paradigm* and a *functional paradigm* for economic development (cf. Friedman and Weaver, 1979; Johannisson and Spilling, 1983). In Table 3.2 some characteristics of the two strategies for local economic development are presented.

The concept of paradigm will here be used interchangeably with 'perspective' and 'strategy'. All three concepts indicate that the manner in

Table 3.2 Contrasting strategies for local economic development

Strategy elements	*Territorial strategy*	*Functional strategy*
Actors and goals		
Acting subject	The local community	The firm
Main objective	Sustainability	Profitability
Efficiency ideal	Economies of scope	Economies of scale
Development process		
Initiation	Crisis awareness	Strategic choice
Change agent	Social entrepreneur	Autonomous entrepreneurs
Pivotal competence	Cultural community	Professional experts
Dynamics	Dialectic process	Means/end analysis and implementation
Resources and organization		
Primary resources	Local identity and creativity	Financial and material resources
Organizing	Social networks	Production networks
Role of immediate environment	Active co-actors	Suppliers and rule providers

which economic problems are approached and solved is mainly a matter of choice, for the local actors as well as for the policy makers and researchers. The concept 'strategy' is introduced in Table 3.2, which addresses the issue: how can concrete economic problems be consciously dealt with?

In line with the theme of this project, the comments will be given *from the point of view of the territorial strategy*. The local community as a collective actor differs from the firm in many respects: absence of a uniform control structure, coordination through networks instead of hierarchy, and involvement of actors as *complete* human beings (not as role takers). This point is fundamental to an understanding of what will make the collective take action – that is, develop expectations, generate motivation and commit its members (Brunsson, 1985).

Territories stay put but people move, firms move, even centres of creativity move. The main objective of the territorial strategy, then, is to build a *sustainable community* integrating these three elements. This necessitates for example local variety in terms of economic activity as well as a high degree of self-sufficiency in non-material resources such as willpower, solidarity and open-mindedness. Individual local firms, however great their contemporary success, will always be vulnerable.

The phrase to remember, therefore, in aiming for sustainability is '*economies of scope*' – an awareness of their joint resources widely distributed among the actors within the territory (cf. Johannisson, 1987). Such insight represents a significant individual and collective problem-solving capacity. In a changing environment such flexibility is far more vital to overall effectiveness than the technical efficiency provided by 'economies of scale'.

Within a functional paradigm measures follow strategic proactive choices;

within a territorial strategy collective action is initiated when the basic values of the community are threatened. The necessary *crisis awareness*, then, is either the outcome of an evolutionary change created by complex interaction with the environment or the product of a 're-action' to sudden and radical change. Both individuals and collectives usually build crisis awareness in the latter manner. Only persons able to create visions of a better future for themselves and others will have the abilities needed to anticipate crises and turn them into opportunities. Within the functional paradigm those who have a vision of advances in knowledge are the autonomous entrepreneurs. Within a territorial strategy a similar role is played by 'community' or *'social' entrepreneurs*. Table 3.3 summarizes some basic differences between autonomous and social entrepreneurs (cf. Johannisson, 1986a).

The re-active change mode in a territorial strategy derives from the fact that *the cultural community* – the established way of life and its preservation – is the focal interest. While the community members will only mobilize themselves when the community's foundation is threatened, the individual firm, adopting the functional paradigm, will rely upon external experts, consultants and other professionals with a cosmopolitan outlook to diagnose the need for change.

Table 3.3 Social and autonomous entrepreneurs - a taxonomic outline

The social entrepreneur . . .	*The autonomous entrepreneur . . .*
considers the development of the community as a primary personal goal	considers the community as a means to personal goals
discovers and helps to build the abilities of other community members and increase their self-respect	enhances his/her own self-reliance and competence
endeavours to replace him/herself as a community manager	strives for personal development in the role of founder-manager
regards him/herself as a coordinating body within a federative structure	puts him/herself at the top of an (autocratic) organization to combine production factors
inspires others to start their own businesses	mobilizes material, financial and human resources within his/her own enterprise
approaches authorities and external parties as potential supporters	sees authorities and other interest groups in society as hindrances and threats if they are not subservient to his/her own objectives
builds local arenas and links different personal networks	exploits arenas and builds his/her own personal network
seeks projects which will reduce the socio-economic risks for the community.	seeks situations where hazardous activities can be organized as independent projects.

The *change process* itself varies between the two suggested perspectives. Within a functional strategy traditional planning models are applied: goals are formulated, alternative solutions/programmes defined, choices made and programmes implemented. Within the territorial paradigm the revitalization process is assisted by lessons from organizational change where learning of a new behavioural pattern must be preceded by the unlearning of the previous approach to problems. Further, the new world view must be institutionalized and legitimized. For these phases the metaphor 'unfreeze–change–freeze' has been used, especially in intra-organizational settings. The emerging crisis awareness creates the climate necessary for an 'unfrozen' position. The change itself, guided by the vision of the social entrepreneur, appears as a trial-and-error process integrating various sectors of the community successively in an *ad hoc* way (cf. Johannisson 1986a). The 'freeze' phase occurs when a territorial perspective has been accepted both locally and contextually.

Action research in local economic development in Norway and Sweden indicates that change itself will not necessarily lead to a permanent takeoff but sometimes merely results in a temporary vitalization followed by an emotional hangover. However, if this crisis is overcome, perhaps with the help of an emergent social entrepreneur or support from peer communities, the change process may evolve into a dialectic self-organizing procedure.

The primary resource that the social entrepreneur mobilizes and organizes is made up of *non-material assets inherent in the local identity and creativity* of the community members. These assets may correspond to those usually associated with autonomous entrepreneurship, such as existential motivation, willingness to work long hours, ingenuity and a positive attitude to calculated risks. A strong local identity will moreover produce the overview required for access to 'economies of scope'. Every member will not only mobilize his/her own personal abilities but will also offer the resources inherent in personal local and contextual networks to fellow community members. Local and global exchange will maintain the creative potential. This exchange, manifested in *social networks*, is based on instrumental commitment as well as affective and moral commitment (Johannisson, 1987). Ties based on kinship, friendship and neighbourhood seem to be especially capable of coordinating non-material resources. Networks with such complex ties may very well give indirect access to material and financial resources, central within a functional strategy, where economic activity is coordinated by production networks based on primarily instrumental commitment.

People, territories and firms are identified by the networks they build. The autonomous firm, according to the functional paradigm, should nurture its relationships on the market and perceive the environment solely as a rule setter and potential provider of unilateral resources. In contrast the territorial strategy, regarding as it does the local community as a microcosm of society at large, suggests that *external parties should be approached as active co-actors* regardless of whether they operate on the private or public market.

The latter perspective (considering the complete socio-economic environment as a potential resource-bank) is especially relevant in a Nordic setting. These complex welfare states are contexts where any initiative – a new business venture or a local economic-development effort – is circumscribed by an 'iron square' of strong institutions within the political system, the public bureaucracy, powerful interest groups (such as the trade unions) and the market. According to a thorough Norwegian study, negotiations between these parties tend to strengthen the power of the already powerful (NOU, 1982, no. 3). Those not enjoying access to centres of influence may feel they are being strait-jacketed inside a society. These, then, are the characteristics of an 'organization economy'.

The highly concentrated power structure in the Nordic countries means essentially that it is difficult to react to external change. The prevailing system is well suited to maintain stability or, possibly, change within a given structure. The prevailing functional strategy for economic development accepts that power is centralized within each party and distributed between the parties making up the 'iron square'. None the less, global economic development has shown that there is a need for systems which can promote change, not preserve the status quo. Consequently, the functionally oriented regional and industrial policies presented in section 3.2 have been supplemented by both a general decentralization policy and a recognition of the need for local initiatives (cf. also Figure 3.1). The magnitude and direction of these changes varies between the Nordic countries.

The mixed and negotiated context of any economic activity does have some advantages – at least from a territorial perspective. As the various power centres intermingle extensively, any entrance into the system may, through clever networking, give access to any other party. Thus, although the local municipalities in the Nordic countries are quite restricted as regards direct involvement in economic activities (cf. section 3.2), they are fairly strong from an international viewpoint, both financially and in terms of competence in other fields, such as the social sector. This strength can be used in acting as a broker between, on one hand, the overall 'organization economy' (of which they themselves are parts), and, on the other, the individual firms or projects for local economic development. The need for political influence and the ability to lobby vary according to national differences in respect of municipal discretion. In Norway municipalities can in fact own industrial firms while municipal economic policy in Denmark is restricted to the distribution of land available for building and the renting of premises. The general tendency, however, is that, in response to the inability of the state to provide viable general solutions, many municipalities have taken on an increasingly larger responsibility for solving their own economic problems.

In Figure 3.1 some concrete changes that seem to loosen the grip of the strait-jacket imposed by the 'organization economy' are indicated, on the basis of experiences from the Swedish scene (cf. Johannisson, 1986b).

Within the *political system*, many local politicians have used the absence of

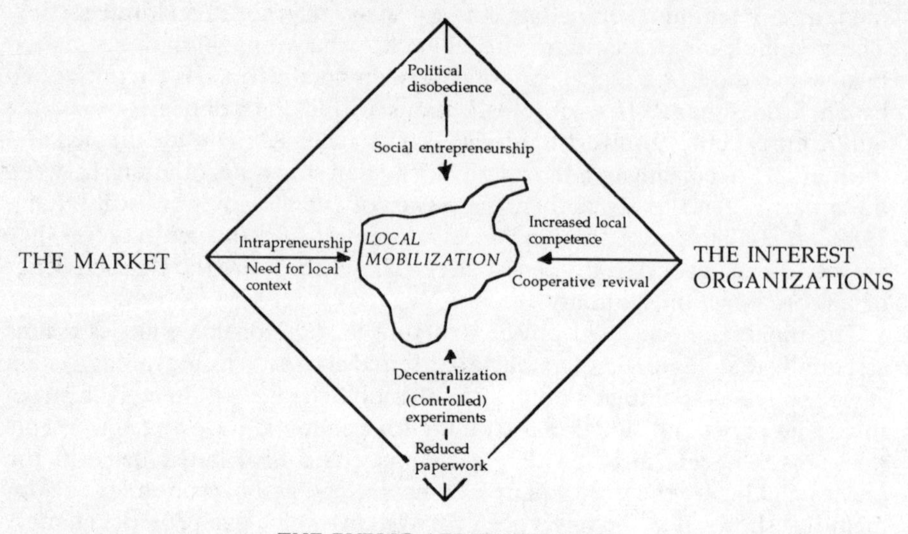

Figure 3.1 Creative dynamics in a stable state - emergent Swedish local initiatives.

an efficient regional policy to justify their own initiatives far beyond their formal mandate. In some cases the leading politician has personally assumed the role of social entrepreneur.

Within the *public administration*, decentralization (including quasivariants such as relocation of government offices outside the metropolitan areas) has been significant during the past ten years. At the same time the paperwork required by the public authorities for statistical purposes has been reduced – a fact which is especially appreciated by small and medium-sized businesses. At both the regional and local public level, freedom of action has increased and active experimentation has been encouraged.

The influential *interest groups* have also reconsidered their role. The labour movement has increased its general economic competencies and adopted a more supportive attitude to localized economic development in general and employee-owned companies in particular. The cooperative sector has started up cooperative resource centres and encouraged the creation of youth cooperatives jointly with municipal authorities.

Finally, the dominating parties on the *market*, the large corporations, have realized the need for building an internal (community) culture in order to promote 'intrapreneurship' – that is, corporate entrepreneurship. This, in combination with the need for a more varied local labour market, has increased the corporations' interest in supporting indigenous entrepreneurship (as well as the functionally oriented shared image of all the plants of the corporation wherever they are located).

It is very important to point out that a territorial strategy based on self-reliance does not mean isolation from the wider environment and restriction to internal resources and local markets. What it does imply is an acknowledgement of the importance of the social context of any economic activity, whether considered at the level of the individual firm or the (local) community. First it means looking for possibilities, not only potential conflicts, in the intersection between different societal sectors. Secondly, within the economic sector, the territorial strategy suggests looking for local solutions before external supply or product markets are penetrated. Identification with the local community will facilitate such local integration. However, the territorial strategy does not exclude external or 'global' operations of individual persons and firms. Such linkages are vital in any community, however self-reliant. Strong self-esteem will not only produce social support for external networking; local identity also supplies a mechanism whereby global contacts are established and maintained in the interest of the local community (cf. Johannisson, 1987). In practical terms this means that any experience gained in a certain context will be turned into a collective asset simply by being linked into the local network. In order to cope with changing external conditions the sustainable community needs members with both a strong local identity and a global outlook.

3.4 NORDIC EXPERIENCES IN LOCAL ECONOMIC DEVELOPMENT - SIX CASES

3.4.1 METHODOLOGY AND OVERVIEW

The cases are specifically chosen to represent all four Nordic countries and to cover differences with respect to the community sector in which the mobilization process was initiated (the economic/occupational or the cultural/social, for example), the duration of the process, the role of the municipal administration and the success of the local development effort. With respect to the last of these sampling criteria, there was initially a tendency to select successful local initiatives. Obviously, the choice of a relevant indicator of 'success' is a matter for debate. In view of the fact that an explicit objective of most support programmes within the industrial and regional policies in the Nordic countries is to create (local) employment, the success of Lannevesi and Pilgrimstad is dubious.

The case reports referred to here are all based on primary data and secondary statistics and documentation. The research process was rationalized partly by drawing on previous projects and partly by using members of the Scandinavian research society as liaison persons. Both strategies offered privileged access to the communities under scrutiny, which has probably considerably enhanced the quality of the findings. The Swedish cases include both a genuine action research study (Måleràs), a project encompassing some action elements (Pilgrimstad) and a traditional study where local exchange systems of entrepreneurs have been studied in particular detail (Fosieby). The three other cases are all based on personal interviews with, in each case, a dozen or so local actors. The cases have been presented more elaborately elsewhere and made available to the focal interviewees for comment.

The cases will be described individually and impressionistically. Where a conscious mobilization strategy has been adopted it is reported chronologically. Moreover, the reports describe what has been considered unique by the local actors, as noted in previous studies of the community or considered relevant in my own empirical research. In Table 3.4 a survey of the cases is given.

It may seem paradoxical to generalize about local economic development when one of the main arguments proposed here is that local conditions should determine both the goal and course of the mobilization process. The presentation of the cases and the following analysis shall therefore be restricted to illustrations of the different conditions for local economic development as outlined above and – possibly – the provision of a base for a future, more thorough, investigation aimed at establishing an understanding of success and failure in local economic development in advanced mixed

Table 3.4 Nordic cases of Local Economic Development Initiatives (LEDI)

FINLAND, NORWAY, DENMARK			
	Lannevesi	*Selbu*	*Holstebro*
Country	Finland	Norway	Denmark
Population 1987	770	4,200	38,000
Institutional status	None	Municipality	Municipality
Regional context	Rural area	Rural area	Rural area
Initiation of the mobilization	1960	1974	1961
Key projects of LEDI	Social networks Pluriactivity	Municipal economic strategy	Aggressive cultural policy
SWEDEN			
	Måleriås	*Pilgrimstad*	*Fosieby*
Population	280	500	—[1]
Institutional status	None	Village committee	None
Regional context	Rural area	Sparsely populated area	Metropolitan area
Initiation of the mobilization	1979	1986	1973
Key projects of LEDI	Local entrepreneurship	Job creation	Joint ventures

[1] The 'place' is an industrial park.

economies. The present material describing the cases, therefore, documents differences between various mobilization projects while at the same time offering a platform for the pursuit of a more general kind of knowledge.

In order to meet the two goals of the field research (to provide (1) illustrations within the suggested framework and (2) a base for future research) they are presented one by one in section 3.4.2. Explicit references to the suggested conceptual framework are deliberately omitted in favour of providing an image of uniqueness.

The section concludes with a comparative summary where the case reports are analysed according to the model of the Nordic national contexts as 'organization economies' (cf. Figure 3.1). In section 3.4.3, comments are made for each case with respect to how consistently the territorial strategy has been adopted.

3.4.2 THE CASE HISTORIES, SUGGESTING HOW THE FUTURE CAN BE CREATED

Lannevesi (Finland) – the potential of a strong community spirit

Lannevesi cannot be said to represent a success story in local economic development if the main development criterion adopted is the increase in the number of new firms and jobs created in manufacturing or service industries. During the period 1970–87 the population actually decreased from 1,080 to 770. In spite of its location in the centre of Finland, where structural change has dealt a severe blow to the dominating primary industry (agriculture and forestry), Lannevesi has demonstrated considerable resistance and has so far avoided the more severe population and economic decline of its neighbouring villages. How Lannevesi has organized its resistance is the focus of this case history.

Agriculture and forestry have always dominated economic and social life in Lannevesi. The first wave of activities to expand the economic base of the village was initiated by a veterinarian, now deceased. He encouraged farmers to supplement their agricultural activities with aquaculture, by breeding roe fish. Although Nordic aquaculture has recently run into difficulties because of international competition, the roe fish business has remained profitable. Such supplementary aquaculture operations now involve some twenty farmers. Other farmers have diversified their activities by starting berry farming and renting weekend houses to tourists. Although neither the salaried jobs nor the financial income generated by these activities are significant, they mean that some thirty farmers operate on markets outside conventional agriculture. This represents a considerable collective business experience.

Since Lannevesi lacks industrial traditions, the manufacturing industry has always been problematic. A factory for prefabricated houses had to cease operations in the late 1970s in spite of government support. Its founder has managed to continue his innovative activities on a small scale. A couple of years after the closedown, a sawmill, also heavily subsidized by the state, was established in the community. However, the sawmill burned down only a few years later and has still not been rebuilt. Two returnees - people who emigrated earlier basically because of the lack of jobs - have established small sawmills, each providing a small number of jobs. One of the sawmills is itinerant; moving from farm to farm it brings both new production and market technology right into traditional primary industry.

In the 1960s and 1970s economic decline forced many community members to commute to jobs - for example, to the municipal centre Saarijärvi (20 kilometres away with approximately 15,000 inhabitants) or to emigrate to expanding labour markets, primarily in Sweden. During the 1980s, however, quite a few small new firms, such as the sawmills, a car repair shop and an engineering firm, have been established. While the entrepreneurs include several returnees, what is perhaps more important in the longer term is that people with roots elsewhere have settled in Lannevesi while retaining their jobs in the municipal and regional centres. It is thanks to this that such basic community functions as the local primary school can be maintained. In addition, the municipal authorities have seen the necessity for considerable investments in infrastructure like the sewerage system etc. The stabilized population base also offers the local service structure - including a local bank, a post office, two food shops and a cafeteria - a safe base for future operations.

Lannevesi has its own cooperative bank, which, thanks to a dynamic manager over the past twenty-five years, has developed into an administrative centre in the village. Besides having practically all village members as its customers, the bank organizes a variety of local activities and excursions such as bus trips to cultural events elsewhere. Because of his background in the agricultural sector, the manager has been widely accepted locally. The character of the bank and its manager also explains why most new economic activities are in turn connected with the agricultural sector and are initiated by local people.

There seem to be two main reasons why people move to Lannevesi: the beautiful landscape and the Lannevesi spirit - a general concern for the village and its inhabitants. This spirit is kept alive through strong social networks, partially organized by local associations. These, as well as the industrial and entrepreneurial activities, are influenced by values and actors from the agricultural sector. The material symbol for Lannevesi's intense social life is the 'Youth Society' in the village centre - a centre for meetings, study circles and the local amateur dramatics association. Its renown reaches far beyond the village. Thus social and cultural life in Lannevesi integrates the native population, returnees, immigrants and temporary visitors. However subtle, the spirit of the village obviously strengthens, and is strengthened by, people's commitment to place. Out of such commitment motivation may very well emerge for local ventures, including projects in the economic sector in the future.

Selbu (Norway) – Local identity and municipal brokerage

Selbu municipality, located in the middle of Norway, has a very solid base in agriculture and forestry. Only fifteen years ago about 40 per cent of its population was employed within the primary sector; today the figure is less then 25 per cent. However, industry and trade also have deep roots in the community; for more than 500 years farmers supplemented their income from agriculture through seasonal work in the mountains, producing millstones for the national market. Copper ore was also mined in the municipality for more than 150 years. When the millstone market ceased and the ore was exhausted, people in Selbu, both men and women, turned to knitting to supplement their income from farming and forestry. Their Selbu design brought this handicraft national renown and the craftsmen the status of purveyor to His Majesty the King.

Although only fifty kilometres or so from the regional centre (Trondheim), Selbu is difficult to reach because of the area's mountainous landscape. Always considered an isolated area, Selbu's natural conditions have given rise to a strong local identity, characterized by, for example, its special dialect. The municipality is in turn split into several villages which have each retained their own character. This implies that there are many community houses and small schools. The strong ties to place have also meant that many ex-inhabitants have returned after an external career to take up influential positions in the community. These people include at present the mayor and the manager of the main local bank.

For almost a hundred years up to the mid-1970s, Selbu's population was in decline. In spite of elaborate support systems within regional policy, few local businessmen were interested in expanding their operations and few new entrepreneurs were forthcoming. Either they could not provide the securities needed for loans or they could not supply equity capital. However, the municipality then began to operate as a broker between the regional support system and the individual firms. The local authority received grants and borrowed money on more privileged terms. The municipality then rented the premises to the companies. Selbu was the first municipality in Norway to implement such a local economic policy. The system was adopted by both local firms and relocated branches within national or transnational companies, especially in the electronics/computer industry. When the leasing period had elapsed the companies were offered the premises at a price equivalent to the outstanding loans.

At the beginning of the 1980s municipal efforts to stimulate economic development increased dramatically. The most senior civil servant retired from that position to take formal responsibility for municipal business services as a liaison officer on a half-time basis, a function he had carried out informally until then. The reorganization aimed at turning the liaison function into a competitive advantage. The business liaison officer is in turn a member of the board of the local cooperative savings bank, where almost all community members and businesses are customers. The municipality has, in addition, established an advisory business board and set up special funds with seed-money for new firms or for temporary support to existing firms which have run into acute cash problems. The financial resources for this fund are supplied through fees paid by the non-local hydro-electric corporations which exploit the community's natural water-power resources.

Today industry in Selbu can be divided into three categories: first, farming activities, where, however, most owner-managers supplement their revenue with income from salaried work; second, the traditional manufacturing industry based on local resources, such as a sawmill, a firm producing prefabricated houses (partly supplied by the sawmill) and a dairy; third, half a dozen firms within the electronics/computer industry. Some of these moved to Selbu in the 1970s but most were relocated to the municipality in the 1980s. They are run by people attached to either transnational companies or the technical university in Trondheim. The secondary sector in Selbu supplies about 360 jobs divided equally between the traditional manufacturing industry and the new high-tech industry (which naturally includes some simple assembly work).

The main challenge for the municipality is to balance the interests of the indigenous and the relocated businesses. These have different preferences for such public investments as continuing education. Although the businessmen have their own local trade association this cannot deal with all the problems arising, especially as the organization is dominated by people belonging to traditional industry. However, many of its leaders do pay special attention to these problems and the mayor has been co-opted to the board of the trade association. The tensions between the different industries may have been sufficiently relieved in the many arenas to permit informal coordination offered by the intense social life, such as in the many sports associations and a busy music scene.

The Selbu handicraft industry is a province where local historical values and practices are concretely linked with global markets. Subsidized by the public support system, knitted handicrafts are at present being 'modernized' by well-known designers and distributed worldwide through international fairs.

Holstebro (Denmark) – a cultural strategy for renewal

Holstebro, a commercial centre since the thirteenth century, is located in north-western Jutland, a peripheral region from the Danish point of view. At the end of the 1950s Holstebro's industry was dominated by a large locally owned tobacco company. Even back then the municipal leaders were worried about the vulnerability implied by this dominance. Although prominent in the municipality, the town lacked both educational and recreational facilities and the young were leaving Holstebro to seek higher qualifications in the university cities. Very little was done to lure them back after their exams. Pressure to develop a more stimulating local environment was openly exerted by the well-educated employees at the military regiment located in Holstebro since 1954.

More stable municipal finances, derived, for example, by levying taxes on the regimental staff, supplied the municipal authorities with the resources needed for an aggressive development policy. A restructuring process in the Danish tobacco industry in the early 1960s caused a 90 per cent reduction in jobs in the space of just a few years in the Holstebro firm. Jobs *per se* were not, however, the main problem – the boom of the 1960s quickly absorbed unemployment. A more robust community structure seemed to be a basic requirement. The municipality decided, therefore, that their community building efforts should not focus directly on the manufacturing sector.

Headed by a visionary mayor and a resourceful chief administrator, the municipality linked planned measures to the principal economic activity in Holstebro: retail trade. They decided on a scheme (offering, incidentally, fascinating analogies with Vienna a hundred years earlier) to restructure the town plan on the one hand, and, on the other, to take unique and radical measures to increase the attractiveness of Holstebro's shopping area. The result of the renewed environmental planning was that two 'rings' were created, one leading bypass traffic around the city, and the other bringing visitors to (free) parking lots very close to the commercial centre of the town. Considerable investments were made in sculptures to line the central streets and the artists had other works exhibited in the several local museums that were also established in the mid-1960s. Official buildings were decorated with paintings and textiles exhibited by the museums on a collection basis.

The cultural profile of Holstebro was raised further by providing venues for guest performances of theatre groups from elsewhere. Today several of these have settled in Holstebro and the local theatre association has more than 4,000 members. From the local music school – the largest in Denmark with almost 2,000 students (mainly part-time) – some graduates have gone on to play for international orchestras. The business community both supports and exploits the town's musical life, for promotion purposes at international fairs, for instance.

Vocational training schools in the commercial and technological field have also been established in Holstebro. Denmark's 'open university' has now extensive operations in the town, which is aiming to institutionalize them. Generally, local organizations receive strong support; the municipality has invested, for example, in a large sports centre.

The strategy adopted by the municipality, therefore, has been to build a broad societal base for various economic activities and to promote the city as a complete milieu. The efforts to support the relocation of firms to Holstebro have focused on small and medium-sized firms (between 1958 and 1979 Holstebro belonged to a region eligible for regional industrial support). However, these efforts have been closely linked to the general promotion of the city and have only marginally affected the business community.

The secondary industry in Holstebro is rather 'footloose'. Even those entrepreneurs who have been running their businesses in the municipality for a long time point out that their location is more or less incidental. The industry is diversified, or, to be more precise, fragmented, since the local commercial ties between the companies are very few and weak. The lack of a strong local identity and the fact that both supply and product markets are regional and national, if not global, also contribute to the seemingly disunited industrial picture. The existence of a stimulating socio-cultural milieu seems, then, to be the vital ingredient that keeps entrepreneurs and their qualified employees in Holstebro. This local social life supplies arenas for informal information exchange. Its full potential, however, has not yet been realized.

The municipal authorities have realized recently that more systematic efforts are needed to develop the local business structure. In 1986 a corporation for commercial development was established, jointly financed by the community, local business and its organizations. It has a liaison officer and a secretary who are expected to supply services in accordance with business demands and also to follow up their own ideas in developing cooperative projects and educational programmes for businessmen, etc. The corporation's board of directors includes representatives from the political and business community and from the trade union movement.

The Holstebro development strategy has clearly had a long-term orientation, and appears to have been successful in creating general attractiveness. Over the period 1970–83 the number of inhabitants in the municipality increased by 5,000. The number employed also increased, as did the number of firms, although proportionately not as much as the municipality's population. The challenge for future development appears to be to inject into the business community the same kind of dynamics that have characterized the creation of Holstebro's socio-cultural milieu.

Målerås (Sweden) – social entrepreneurship and firm cooperation

In 1950 Målerås was a small prosperous community with a population of 700 and a thriving glassworks with some 200 employees. During the following thirty years, however, the population shrank by two-thirds and the glassworks' existence was threatened several times by business failures. By the 1960s the plant was reduced to a simple remote-controlled production unit. A closedown was averted only because the glassworks' very successful designer was also a dedicated member of the Målerås community. He was prepared to continue his professional career in Målerås only as long as the glassworks were kept in operation.

Målerås is located in the 'Kingdom of Crystal' in southern Sweden, about thirty kilometres from the municipal centre (Nybro) and sixty kilometres from the regional centre (Kalmar). At the end of the 1970s the trade union movement initiated study circles at the twenty small single-company towns in the region. Their ambition was first and foremost to strengthen local identity by reviewing, and documenting, the history of the community and its glassworks. The next step would be to develop a strategy for revitalizing each community. Målerås was one of the few places where such future-oriented projects were started. Two universities gave their support but the prime mover was the designer mentioned above; he adopted the role of a social entrepreneur, personifying the vision of a prosperous local community, encouraging others to start their own businesses and supplying them and already existing businesses with product ideas.

The revitalization work was organized by a project group which met every month during the first two years. Subgroups were set up for special tasks, such as finding premises for business operations, securing the few public services that were still left and, where possible, expanding them. The group members included both local and external participants who jointly represented significant network resources. The actual project meetings meant that established and new entrepreneurs came together, building local information and production networks. Thus a new engineering shop was able to start up a business supplying the existing small firm engaged in building truck platforms, and the small casting shop was furnished with product ideas by the designer. A relocated accounting business soon had most local businesses as customers. The ambitious promotion of the project in the mass media brought other entrepreneurs to the village. A few of them turned out to be less talented and ceased operations after a few years.

Although the glassworks was indisputably of practical importance in the local economy – in 1980 it was still the largest employer – its symbolic role was even more significant. Well aware of this, the designer and others made it their ambition to take over the glassworks. In 1981 their efforts succeeded and the factory was bought by fifteen employees, eighty other community members and a sales corporation from an adjacent village. The premises were rented from the municipality. The glassworks expanded operations progressively and for the last few years it has been the most profitable in its field in Sweden. In 1987 it has about seventy employees. The premises are to be taken over by the glassworks, as well as the equity once held by the external sales corporation.

Besides the various production networks, there are other collective activities involving the local businesses, such as a joint permanent exhibition of the local products. Together with the project organization, the businessmen have also operated as a pressure group on local and other external bodies for basic social services. The municipal administration (initially rather sceptical about the project) was persuaded to build some residential flats, and the co-operative movement (which is very centralized in Sweden) to retain the only food store in the village and, subsequently, to reopen the petrol station. The project group, now recognized by local and regional authorities, has encouraged a young couple to reopen the local cafeteria. The project did not, however, manage to stop the closing down of the school or the removal of the local post office.

In 1984 the population of Målerås began to increase again. Today a significant number of workers are also commuting to Målerås from adjacent villages – to the glassworks or to the other local firms. A leather industry, under new management, has taken over its main competitor in the suitcase industry. There is constant pressure on the municipality to provide more housing and soon the school may have to reopen. Once more or less abandoned by local and regional authorities, Målerås has managed to create a sustainable community, relying basically on local entrepreneurial resources.

Pilgrimstad (Sweden) – overprotection in the organization economy

Pilgrimstad is located in the geographical centre of Sweden, some 500 kilometres north of Stockholm and about thirty kilometres from the regional centre, Östersund, with approximately 35,000 inhabitants. In 1986 Pilgrimstad was a typical single-company town, completely dependent on an externally controlled fibreboard plant built in the early 1950s. At the beginning of 1986 the plant owners decided to concentrate production in another unit located on the Baltic coast. This illustrates the general process whereby the Swedish forest industry is becoming increasingly concentrated in terms of both ownership and location of production units.

Bräcke municipality, where the village of Pilgrimstad is located, is one of the best-known in this sparsely populated region. Its chief political leader, the mayor, is the chairman of a special delegation for sparsely populated areas set up within the Department of Industry. The municipality was one of the first in Sweden to introduce municipal district boards, i.e. to decentralize political power to district boards (although the political membership of the district boards reflects that of the municipal board). In addition, village committees have also been established in different areas, including Pilgrimstad. Bräcke municipality is also involved in two supplementary experiments which both aim at breaking down the strong sectoral boundaries within the public administration. Thus, the local public arena in Bräcke, injected with a spirit of change, has proved to be very adaptable.

One of the most significant political leaders in Bräcke was until 1985 also chairman of the municipal board for the district including Pilgrimstad. He was also a union representative on the company board. In terms of competence and influence he was therefore a very resourceful person. When the company announced in January 1986 that it was going to close down the factory in six months' time, he organized a project group to cope with the situation. The group's ambition was either to rescue the factory or to find alternative production. For this mission the leading actor mobilized his extensive network within the political sector and public bureaucracy. In spite of his enormous commitment he did not succeed in preventing the factory from being closed down. So far (1987) neither he nor anybody else has managed to encourage any local person to start his or her own business; neither regional nor national groups have been seriously interested in starting up any alternative production.

When the ninety employees at the plant were laid off only about one-third were interested in participating in local efforts to create new jobs locally. One-third of the previous workforce were granted early retirement and the youngest third preferred to move immediately to jobs in adjacent villages or to Östersund. For the remaining twenty or thirty people an education programme was organized. This programme aimed partly at consolidating their basic knowledge in mathematics and languages and partly at providing an introduction to the skills necessary for running a business. Within the framework of the programme researchers affiliated to the regional university helped the participants to carry out a survey of the local population's expectations of the future. Every community member above the age of 12 was asked to complete a questionnaire. The most significant result was a desire for better local services, especially leisure amenities.

The mobilization process was dominated from the very beginning by the former chairman of the municipal district council. There was no reason to question his competence, nor his centrality, whatever mobilization strategy was chosen. However, he alone decided which measures were to be deemed appropriate. His background led to a bias for established solutions taken from the political system, the public bureaucracy and the trade union movement. He seemed convinced that any solution to Pilgrimstad's problems would be found outside the community and that it would have to be worked out in cooperation with other bodies, such as the municipal industrial development company. He saw few dangers in letting the community members find jobs in other communities while he was working for local solutions. The lessons to be learned so far, however, seem to be that work for community sustainability cannot be postponed, not even in a municipal context such as the one Bräcke offers: societal experimentation that is quite extraordinary in the Nordic countries.

Fosieby (Sweden) – local ideals in the middle of metropolitan Malmö

Over the past ten years many regional chambers of commerce in Sweden have organized local groups of firms to explore different possibilities for cooperation. The prototype for these efforts, the Fosieby Industrial Group in Malmö, was established in 1973 in the industrial part of the same name. It is located about ten minutes by car from the centre of Malmö. The city has a population of approximately 250,000, which makes it the third largest metropolitan area in Sweden. It is also closest to the European continent. Today Fosieby park includes about 300 firms, of which well over 200 are members of the industrial group. Any local firm can apply for membership (the annual fee is only 700 Swedish crowns); most of the non-members are firms run by the self-employed. Out of the 6,000 or so people working in the park, 5,000 are employed by companies that are members of the group.

The joint activities organized by the industrial group originally had two aims. One was to operate as a pressure group on the municipal authorities and other institutions and organizations. This led to the opening of a post office in the area and prompt attention being paid to individual firms' requests for public services covering, for example, technical matters. Furthermore, other infrastructural activities such as restaurants, motels, banks and accounting services have been attracted to the industrial park. Second, cooperation has been promoted in order to seize opportunities for saving costs, such as by splitting overheads or getting quantity discounts through joint purchasing or contracting: oil for heating, security services, cars etc. A cooperative company health-care centre has been established. Currently, all members will be furnished with a file containing addresses, telephone numbers and additional business information concerning all members. Over the years some of the cooperative activities (such as the location of a petrol station in the industrial park and helicopter services to Copenhagen) have been abandoned for various reasons.

The group has successfully organized some extraordinary projects. One, a joint venture involving exports to Oman, now represents a significant share of Sweden's total exports to that country. Further joint export activities are planned, to USA via production units in Mexico, for example, and, perhaps more naturally, to the neighbouring Scandinavian countries.

Fosieby Industrial Group has over the years been organized by enthusiastic people, often with a businessman/administrator and a businessman/politician teaming together. Although cooperation with municipal authorities and others has been very close, no recourse has been made to public support for the development of individual firms within the industrial group. The municipal authorities do, however, supply premises for the group administration and co-finance a brochure about the group jointly with its board.

Originally, twenty out of thirty firms in the park were members, which made for very close cooperation with few formal meetings. Until 1986 such meetings were restricted to monthly board meetings and annual member meetings. The new board and its chairman then decided to survey the need for more frequent meetings. The result was that every third or fourth month lunch gatherings are organized at a restaurant in the area and equally frequent information meetings are held at different member companies within the area. Just before these activities started an independent survey confirmed that spontaneous interaction between members in the group was quite limited. Internal and external reviews have drawn the group's attention to the needs of the new and potential members.

The most vital resource of the group is the members' commitment. So far contributions have mainly come from members of the board but now the group systematically looks for ways to broaden its organizational base, for instance by setting up various project committees.

The Malmö area is a region which in the 1980s has been hard hit by structural crises, mainly within the shipbuilding industry. Although much of the expansion in Fosieby basically implies a relocation within the region, the symbolic value of an industrial park with successful businesses should not be underestimated. Originally, the group was set up to create an image to match the attractiveness of the city centre. Later several multinational companies moved their Malmö offices to Fosieby in order to trade on its image. Over the years only a couple of members have left the organization and only two or three have gone bankrupt.

3.4.3 A CONTEXTUAL REVIEW OF THE CASES

Before the cases are analysed in terms of their faithfulness to the territorial strategy, the 'egocentric' presentations will be summarized with respect to the model of the Nordic economies as 'organization economies' (cf. Figure 3.1). It has been argued that both the cooperation and the discretion offered by the societal context are vital conditions for the initiation and maintenance of local economic processes. The results of the analysis are presented in Figure 3.2.

The figure indicates first of all that the only case where local development efforts have so far (1987) failed, Pilgrimstad, contains the only change process in which no representatives of the market have participated (although several companies have studied the location conditions in the village because of the available public support). In Holstebro and Lannevesi, the *market parties* were until 1986 not very active, but strong local leaders have been at hand in these areas to condition participation by any non-local party. In all cases the *(local) political system* has been involved, either as sanctioner or as initiator/active participant. The *public administration* above the local level has played a subordinate role. The regional and national organizations specifically set up to implement the regional and industrial policies in each country have generally been inactive. The *interest groups* that have taken part in the mobilization processes are mainly connected with the trade union movement (Pilgrimstad, Måleås) and organizations within the agricultural sector (Selbu, Lannevesi).

Figure 3.2 also suggests various ways of *organizing local development*. In Målerås and Pilgrimstad, the two cases where mobilization reflects acute crisis, Målerås has both a charismatic social entrepreneur and an elaborate local organization, whereas in Pilgrimstad a strong, externally oriented leader autocratically administers the future of his community. In Selbu the municipal administration supplies both a social entrepreneur and an organization for economic development, while in Holstebro the municipal authorities have only recently institutionalized a local organization for business promotion. Both Fosieby and Lannevesi have efficient local organizations, directly (the board of the industrial group in Fosieby) or indirectly (the local co-operative bank) aiming at promoting territorial interests. In both organizations, individual persons, the board chairman and the bank manager respectively, play significant roles as administrators and information brokers.

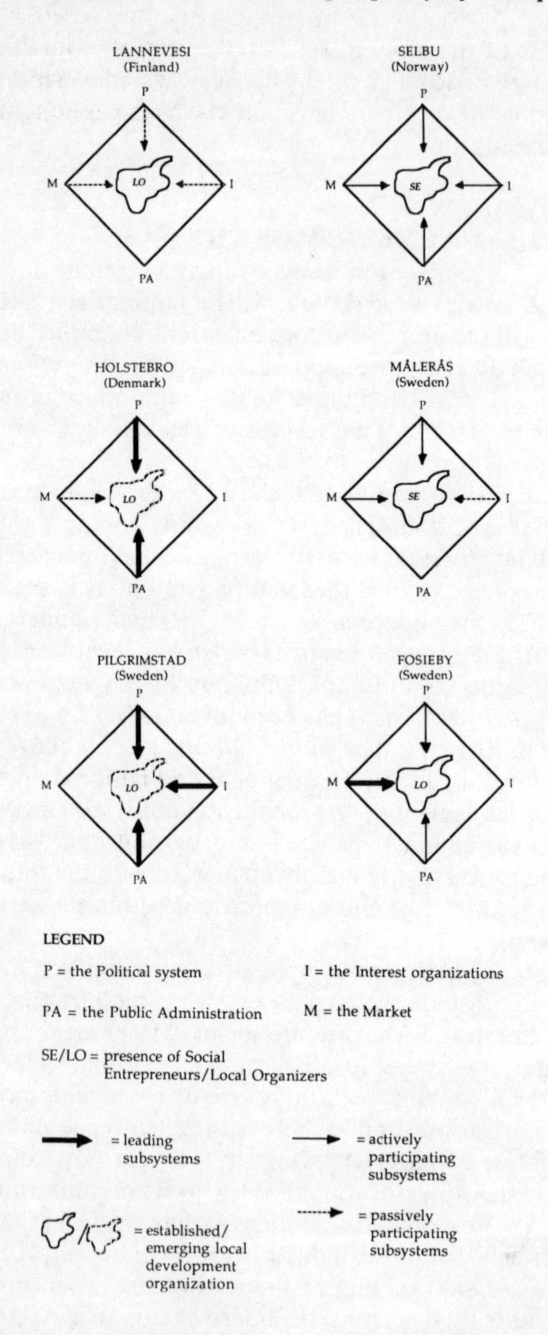

Figure 3.2 The case studies - a summary of local action structures.

3.5 TERRITORIAL STRATEGIES IN PERSPECTIVE – LESSONS FROM THE CASES

The comparative analysis introduced in section 3.4.3 will be continued here, before focus is shifted to loyalty to the territorial strategy in the individual cases. The analysis will follow the characterization of that strategy as illustrated in Table 3.1.

The territorial strategy designates the community as the *acting subject*. This means that local people and institutions should involve themselves in economic processes on behalf of the community, not only as representatives of a firm or some other sectoral/functional interest. In cases where the territory corresponds to that of the municipality (Selbu in Norway and Holstebro in Denmark) it has been natural for local politicians and officials actively to pursue a territorial strategy. In Fosieby (Sweden) on the other hand, the leaders have had to encourage a community feeling; local entrepreneurs participated in the group as businessmen primarily for the purpose of saving costs and gaining publicity for their own businesses. In Måleràs and Pilgrimstad (Sweden) as well as Lannevesi (Finland) economic activity was initiated in the interests of the community concern but only in the first case was it linked to indigenous entrepreneurship. In Lannevesi commuting seems to be accepted as a supplementary long-term answer to local business needs. In Pilgrimstad importing jobs is considered the only realistic way of ensuring the community's survival.

Sustainability as an *overall goal* of local economic development means the parallel development of the non-business community sectors and the creation of variety within the economic sector. The two (entire) municipalities studied approached this objective in different ways: Selbu, furnished with a private and public service structure that corresponded to local needs, focused on revitalizing the economic sector, deliberately nurturing both indigenous traditional industries and immigrant high-tech industries. The authorities in Holstebro, where the local structure was designed to reflect cosmopolitan values, invested heavily in the cultural and educational sector. Non-interference has until recently been the implicit strategy adopted in creating variety in the economic sector. In Lannevesi such variety was carefully and deliberately developed by introducing supplementary activities into the dominant agricultural sector. In Måleràs, although the glassworks was and still is dominant in practical and symbolic terms, the explicit aim of the mobilization project was to encourage broad-based entrepreneurship. A flexible approach in local venture-making was believed, in the long run, to generate a sustainable business community. Furthermore, the social and economic systems were developed in parallel, with the emphasis varying during the process. In Fosieby the joint efforts are now primarily devoted to consolidating and activating the present membership body. Concern for local colleagues has broadened to include commitment beyond the strictly instrumental. In Pilgrimstad the main aim so far has been to recreate jobs, whether local or non-local. Regardless of whether the jobs will be remote or supplied by prospective new externally controlled local firms, the approach adopted in Pilgrimstad will not create a sustainable community.

In many of the small communities surveyed *overview and scope* follow naturally from their history and their limited population and/or size. Examples are Målerås and Pilgrimstad, physically very closed communities. The potential for joining forces that this implies is, however, very differently exploited by each main actor. In Målerås the social entrepreneur, explicitly aiming at a self-organizing community, participates in different arenas in order to encourage people to interact, thus increasing individual overview of collective resources. In Pilgrimstad the focal actor operates in many ways as a gatekeeper with respect to the local people. In Fosieby, where geographical distances are small as well, the recent measures taken by the board of the industrial group indicate an increased awareness of the economies of scope. Local-firm cooperation is encouraged, for example. In geographically relatively isolated places like Selbu and Lannevesi, people have turned quite naturally first to personal and other resources available locally. Quite typically all the firms in Selbu, except the larger companies, were customers at the local cooperative bank; in the Finnish village the corresponding financing institution supplied all the local firms with the credit they required. In Holstebro, however, neither municipal actors nor representatives of the individual firms seem to recognize the need for even a survey of local resources, let alone actual cooperation with other local firms.

Obviously, the *initiation of the mobilization process* varies throughout the cases. In Målerås and Pilgrimstad an external shock – the closure of the major working site – generated crisis awareness which was in both cases expressed by individuals. In Lannevesi a similar alertness grew out of a quiet but assiduous effort for survival. A strong community spirit nourished and multiplied individual initiatives. The struggle for the restoration of the industrial base in Selbu, administered by the municipal organization, was initiated by an autonomous entrepreneur who realized the advantages of exploiting the municipality as an intermediary in raising funds for his own business expansion. In both Holstebro and Fosieby, the realization of available opportunities rather than that of acute problems, seems to have sparked the change towards a territorial outlook.

In Table 3.2 the main features of the *social entrepreneur* were presented. The cases include at least one person who meets all the requirements: the designer in Målerås. Another dominant local actor – the politician/union representative in Pilgrimstad – obviously did not in all respects act as a social entrepreneur. He did not spend much time enhancing the self-respect of other community members, he operated in an autocratic manner and did not devote much time to encouraging others to start their own businesses. In the other cases the functions of the social entrepreneur are combined in a 'community team'. In Holstebro the mayor and the chief administrator complemented each other when the revitalization was started, although neither of them focused on the economic sphere. They both considered the development of the community to be important personal goals, but while the mayor concentrated on building self-respect in other community members, the chief administrator took on the role of broker and established various local arenas, such as arts centres and social associations. In Selbu the mayor supplies a bridge between the municipality and external parties, while the liaison officer builds local arenas and links different personal networks. The latter role was assumed in Lannevesi by the bank manager, while the veterinarian inspired many others to start their own businesses. In Fosieby the main objective of the present chairman of the board is to encourage others to participate in joint projects and to create meeting places for local and global exchange. Since the position of chairman is a revolving one, there are some guarantees that the federative structure will be kept intact. In Pilgrimstad, as well as in Lannevesi and Holstebro, the local organization has instead adopted quite autocratic forms.

In most of the cases a shared history, perception of problems and opportunities and expectations of the future – that is, a local culture – seem to have been the *pivotal competence* in the mobilization process. In no case have experts been hired to lead the change process. Although researchers have been involved in the revitalization processes in both Målerås and Pilgrimstad, they have in neither case imposed their own solutions on the communities. In Målerås the community and its social entrepreneur were aware and self-sufficient enough to define the terms for external support. The main contribution of the researchers in Pilgrimstad was to help the local community to find ways of expressing how it envisaged its future. Inasmuch as the national and regional organizations implementing regional policies are swayed by the functional perspective, they were in neither case considered to be focal partners.

While loyalty to local values is desirable as an energizer in the mobilization process, this does not automatically mean that a strong local culture is always favourable to the main objective: self-reliance and sustainability. Local values in Pilgrimstad meant belief in help from above, which was certainly counterproductive from a territorial point of view. Some local values have proved dysfunctional: in Lannevesi difficulty was experienced in recognizing competences outside the agricultural sector, in Fosieby and Holstebro in acknowledging the importance of local exchange systems. But whether or not local values counteract or support local development, they must be taken seriously. They define what the community members themselves consider to be important and worth acting upon.

The *change process* itself is of course easiest to follow in the two cases where mobilization was initiated in response to an acute crisis – Målerås and Pilgrimstad (although the process in the latter never took off as a collective effort). The organizer of the job-creating efforts and all community members would have had to unlearn the established local values in order to be able to discover their own innate potential. In Målerås this unlearning phase was supported by the trade unions in the study-circle project and by the researchers who joined the mobilization effort. Once started, the mobilization process in Målerås continued through several distinct phases: exposure to external parties for the purpose of attracting resources, balancing of functional influence, and consolidation of territorial values. Once the first change cycle is completed the two latter phases alternate, the business community and the social community come into focus periodically. This repeated change of perspective encompasses the *ad hoc* and dialectic character of a genuine local economic development process.

In the other four cases the change process has been either lengthy and obscure or sectorally oriented. The latter certainly characterizes Fosieby, where a functional perspective was applied in the sense of the identification of appropriate goals and their corresponding implementation. While preserving sustainability in Lannevesi is a rather indistinct process based on an implicit community spirit, it none the less attracts immigrants and incorporates their functionally oriented competences within the territorial value system. In Selbu the need for sustainability has even deeper roots than in Lannevesi. However, as an independent administrative unit, the Norwegian community has involved local politicians and local administrators in the mobilization process. Basically, they have developed the municipality in an impressionistic way; it is only recently that the vitalization process has become 'regulated' by goals and plans. In Holstebro such plans explicitly guided the mobilization process, although they did not primarily concern the industrial sector. The spread into the latter sector from the focused cultural and commercial sectors has been a genuinely spontaneous one and does not lend itself to systematic enquiry.

Local identity as a focal resource, outstanding in several of the cases, is perhaps most obvious in Målerås and Selbu. Both are relatively isolated places. Local resources have always been important: in Målerås the glassworks was organized as a cooperative back in the 1920s and the local handicrafts in Selbu had a traditionally practical base. The designer in Målerås refuses to leave his community except for daily excursions and collects all his artistic inspiration within the neighbourhood. The local newspaper maintains local values for Selbu's emigrants and many returnees now have influential positions in the community. In contrast, much effort has been, and will have to be, invested in Holstebro and Fosieby in order to (re)create a local identity. In Lannevesi the local identity is subtle but has still succeeded in attracting qualified people who may in the future contribute significantly to the development of the community. In Pilgrimstad local identity is a more doubtful resource: it embodies a cultivated helplessness and many inhabitants seem to be prepared to leave home for a secure job in any neighbouring community.

The argument has been that there are limits to *formal procedures for organizing* local economic development. Although development in Holstebro has certainly been very efficiently managed, the elaborate strategy was aimed not at the economic sector directly, but primarily at the cultural sector. Social exchange is organized by the many local associations. In Fosieby industrial park, territorial cooperation was for a long time 'managed', but at the same time restricted mainly to projects with obvious cost advantages. The present management, however, intends to invest more effort into establishing arenas where social networks can be created. Such networks may subsequently, as in Målerås, function as platforms for industrial networks organizing routine economic exchange or even joint development projects. In Selbu much of the social exchange is divided between separate villages. However, the business and the municipal community have each established special arenas for exchange with the other party. Local associations and their social activities constitute an elaborate and diversified organizing structure in Lannevesi, while social life in Pilgrimstad is highly concentrated around the sports association and other groups reflecting the dominant working-class culture.

The territorial strategy suggests that, once a strong local identity is established, joint projects can be set up with external parties. The *context* can then be approached as a resource bank on terms that pay due respect to local values. In Målerås the application of that strategy increased as the environment's legitimation of the project grew. In Selbu the municipal authorities actively mediated between the firms and regional support structures in order to gain access to national power centres. The management of the Fosieby Industrial Group also actively cooperated with the municipal authorities once the group had become established. In Holstebro one of the ambitions of the recently installed liaison officer is to start cooperation with other municipalities in the region. In Lannevesi the external resource networks include, for example, the regional university, the cooperative bank movement and the municipal authorities in Saarijärvi, who consider Lannevesi to be a prosperous rural suburb. In Pilgrimstad the local organizer seems to have put all his efforts into cooperative strategies, albeit with modest success so far, possibly because the terms for cooperation have been defined by others.

It should be clear from this comparative analysis that there are many roads to choose in endeavours to promote local economic development. The lesson is that some roads do not create local self-reliance and that most roads include serious obstacles. The case histories taken from different national contexts also reveal that factors other than various national settings define the initiation, course and success of local economic development. Some lessons even go beyond societies which, like the Nordic ones, are extremely highly regulated 'organization economies'. These more general experiences will be briefly commented upon in the concluding section.

3.6 BEYOND THE SCANDINAVIAN SCENE – SOME GENERAL CONTRIBUTIONS TO THEORY

Three issues covered in the Nordic study seem to be of more general concern for local economic development in a European setting: *difficulties and inappropriateness in separating the economic and socio-cultural aspects of community development; the mobilization process; the need for bridging local and global perspectives.*

On all levels *economic and socio-cultural factors intermingle* – the individual combines them in his or her personal lifestyle, the community integrates individual styles into local life and on the national level networks of institutions bridge sectoral boundaries. Socio-cultural, environmental and economic phenomena are furthermore integrated over territorial/administrative levels. The mass media supply standardized images of society, communications tie regions together and, last but not least, financial dependences are considerable. A recent Swedish study revealed that about 90 per cent of state transfers to lower societal echelons with territorial implications were *not* included in the special measures encompassed by the industrial and regional policies. These indirect regional measures include, for example, unemployment support, general subsidies to municipalities and investment in higher education (cf. Ds I, 1987, no. 6).

However, the focus here is on the integration – at the local level – of different societal spheres. At this level the need for and realism of a holistic approach in community development is inherent in the notion of economies of scope (cf. above and Johannisson, 1987). As a matter of fact, the cooperative ideology, integrating individual and collective values, seems to provide a more suitable frame of reference in a region or community with a strong identity than in an isolated business venture. The Mondragon achievement in northern Spain illustrates how workers' cooperatives can thrive if the local/regional context is consonant with ideology (cf. Johannisson and Spilling, 1983, for example).

The implications of this integrative perspective of business and community development are several. First, any development should start with joint attempts to meet a *common need*. Secondly, a local economic development process may take off in *any sector of the community*, although preferably one encompassing unique local characteristics and hence both drawing upon and adding to the identity of the community. The strengthening of distinctive community attributes becomes both an end and a means in the development process. Only such a strategy will be able to combine individual motivation and potential for action with collective social support. Thirdly, local characteristics may be *at once supportive, insufficent and prohibitive* to the development process itself, embedded as they are in both the concrete and the value structure of the community. This means that a minimum concrete structure has to be built and attitudes such as reliance on help from above have to be unlearned. Premises for production activities may be lacking and/or local norms may conflict with local initiatives and entrepreneurial action. The restructuring of the material and institutional settings and unlearning of counterproductive attitudes and obsolete skills is preferably achieved by confronting the mobilizing community with other local development projects.

Once the *development process* has started it should be guided by the local 'theory' for solving joint problems within the area and not by general models proposed in general planning theories and policy implementation programmes. Such *ad hoc* approaches are preferably organized as loosely linked socio-economic networks which may include both 'social entrepreneurs' and task forces, formally organized groups specially set up to deal with acute economic problems. Thus, any formal structure will be secondary to informal ways of solving both individual and joint problems. Politicians or established entrepreneurs who take part in a local development project will therefore be expected to check, and counteract if necessary, their own functional responsibilities and interests.

The inclusion of established organizations, local or non-local, in the mobilization effort seems to be double-edged. On the one hand, co-opting such organizations may help to legitimize the development effort and thus speed up its pace and increase the likelihood of success. It may reduce or even prohibit the emotional recoil that generally seems to strike autonomous attempts to create local economic development, once the first enthusiasm has waned (cf. Almås, 1987). On the other hand, involving existing bodies and their professionals may affect both the direction, and the pace, of the development process in a way that works against the community members' ambitions. Neither 'local development' in an internal sense nor sustainability will then be created. A genuine local mobilization demands that the people concerned are given the right, and the responsibility, to define both the ends and the means of the mobilization.

The organization and operation of the development process are dependent on factors such as the depth and range of local awareness, the size of the community and the origin of mobilization (response to crises or aggressive seizing of opportunities, for example). This includes the internal, local organization as well as the external. The latter refers to the various network links with non-local individuals, organizations and government bodies. The basic principle is that the internal organization should be established before the external, thus guaranteeing that local interests are not eclipsed by global ones.

The external environment does not solely represent values and interests competing with those of the local community. There are also many individual and institutional actors who for various reasons *bridge local and global values*. In any local development strategy a focal ingredient should be to identify and take advantage of these potential brokers. Both their attachment to the community and their contribution to community development may be affected by factors such as time and space. Immigrants to and emigrants from the community usually have wider perspectives both in time and space than people who work and/or live locally. Returnees, having demonstrated a longstanding loyalty to local values, are especially efficient as boundary 'spanners'. Local exporting businesses or non-local firms with many local subcontractors also function as community brokers. According to the arguments presented here, such 'natural' brokerage mechanisms seem to be the most viable. In Sweden local development corporations and local venture-capital organizations have been set up for the explicit purpose of bridging local and global interests in the economic sector. Practical experiences show that such specialization is inappropriate. Only territorial strategies capable of integrating the economic and sociocultural dimensions of community life will make it possible to define, initiate and pursue strategies for local development. However, once a basic territorial strategy has been launched, lessons from functional strategies may be assimilated in the same way as the scope and need for incorporating global perspectives can be assimilated once local identity has been restored.

NOTE

[1] This section and the following are based on a preliminary version of Oscarsson and Öberg (1987) and internal reports provided by Arne Gaardmand at Planstyrelsen, Copenhagen. In addition, I am very indebted to Gösta Oscarsson and Arne Gaardmand as well as Walter Stöhr for comments on a preliminary draft of this Nordic report.

REFERENCES

Almås, R. (1987), 'New employment initiatives in marginal regions', in U. Wiberg and F. Snickars (eds), *Structural Change in Peripheral and Rural Areas*, Document D12:1987, Stockholm: Swedish Council for Building Research.

Brunsson, N. (1985), *The Irrational Organization*, Chichester: Wiley.

Ds I (1987, no. 6), *Geografin i politiken* (Geography in Politics), Stockholm: Department of Industry.

Friedmann, J., and Weaver, C. (1979), *Territory and Function: The Evolution of Regional Planning*, London: Edward Arnold.

Johannisson, B. (1986a), 'A territorial strategy for encouraging entrepreneurship', paper presented at the Babson College Workshop on Encouraging Entrepreneurship Internationally, Wellesley, Mass., 13–15 April.

Johannisson, B. (1986b), 'Local initiatives in a negotiated economy – lessons for sparsely populated areas', in E. Bylund and U. Wiberg (eds), *Regional Dynamics and Socio-Economic Change – The Experiences and Prospects in the Sparsely Populated Areas*, Working Paper from CERUM, 1986, no. 8, Umeå University, pp. 151–61.

Johannisson, B. (1987), *Entrepreneurship and Creativity – On Dynamic Environments for Small Business*, ser. 1, Economy and Politics 7, Växjö: Växjö University.

Johannisson, B., and Spilling, O. R. (1983), *Strategier för lokal och regional självutveckling* (Strategies for Local and Regional Self-Development), Oslo: NordREFO.

NOU (1982, no. 3), *Maktutredningen. Slutrapport* (The Power Investigation. Final Report), Oslo: Universitetsforlaget.

NU (1986, no. 4), *Yearbook of Nordic Statistics*, Stockholm: Nordic Council of Ministers and Nordic Statistical Secretariat.

Oscarsson, G., and Öberg, S. (1987), 'Northern Europe', in H. D. Clout (ed.), *Regional Development in Western Europe*, London: Fulton.

CHAPTER 4

The Celtic fringe: state-sponsored versus indigenous local development initiatives in Scotland and Ireland

John Bryden and Ian Scott

4.1 INTRODUCTION

4.1.1 HISTORY AND INSTITUTIONS. PROCESSES OF PERIPHERALIZATION AND DEPENDENCE

The areas from which our case studies are drawn – the Western Isles of Scotland and the west of Ireland – form part of the European 'periphery' sharing many of the general characteristics of peripherality common to other areas, remote from main centres of economic activity and populations, but also with distinctive features. Examples of common characteristics include those relating to economic structure (dominance of primary activities and services, especially tourist-related services, imbalanced labour-market characteristics, lower participation rates and per capita earnings), demographic patterns (migration, age-structure, population densities and settlement patterns), institutional structure (weak local structures, dependence on 'centres', fragmented political structures, cultural and linguistic minorities). In some cases marked changes have recently taken place in such areas – return migration, new rural industrialization, new service development, resurgence of local social 'movements', which reflect the external crisis, economic restructuring, changes in perceptions of 'work', 'urban life', 'the state' and so on.

Both the west of Ireland and the Western Isles of Scotland share a history of peripheralization. Both were important centres of a more ancient Celtic civilization. The institutions of that ancient civilization form an important part of the European heritage.

The processes of centralization of economic activity and government that accelerated after the industrial revolution had a serious debilitating impact on both areas. Population and labour force declined, the dependency ratios increased. These processes were aggravated by enclosures or 'clearances' undertaken by landowners in what had essentially become a feudal society in the eighteenth and nineteenth centuries – attempts to remove small landholders and create large agricultural holdings or 'sporting' estates for the pleasure of the 'new rich'. These processes were resisted first in Ireland (by the Land League) and then in the Western Isles. This resistance eventually led to the strengthening of tenure rights, land settlement, and action by the state aimed at stabilizing the population and developing new economic activities. Remarkably, this state activity was started in the nineteenth century; the Highland Congested Districts Board and the Irish Congested Districts Board were closely related phenomena. The first Crofters Commission was also established at this time to administer the Crofters Act of 1886.[1]

One outcome was the partial maintenance of small part-time agricultural holdings on land which was difficult to cultivate in an often hostile natural environment. Other sources of income – fishing, weaving, construction work, kelping – were always important, as they were in many other marginal areas.

Another outcome was a relatively heavy degree of state paternalism, which often replaced landlord paternalism. This state paternalism was, in retrospect, another feature of a declining 'self reliance': the main levers which could be operated to bring about change were controlled not in the west of Ireland or in the Western Isles but (at best) in Westminster, Edinburgh or Dublin.

When political philosophy and economic circumstances changed, and the state started to withdraw, a vacuum was left. Decision-makers and entrepreneurs had been sucked out. An elderly, dependent population, relatively poor in material terms, lacking natural resources which were locally controlled,[2] could not be regarded as promising ground for self-reliant development. Nor were such areas promising for the creation of new jobs by outside investment, even after fairly massive financial support from the state.

4.1.2 THE ROLE OF COOPERATION: IRELAND AS THE MODEL

As Commins (1983) points out, 'Indigenous forms of co-operation, based on reciprocal exchange, sharing and informal systems of mutual aid, have traditionally been important elements in the functioning and survival of local communities in the adverse conditions of life which typified the marginal regions.' This feature perhaps explained why *cooperation* was seen as one of the more appropriate institutional forms for local development in both Ireland and later in the Western Isles of Scotland. Community development was also being promoted in Ireland by the national voluntary organization Muintir na Tire, founded by a priest in the 1930s. In Ireland especially the rolc of local priests has been vital to the local development movement.

4.1.3 THE ROLE OF THE STATE: THE SCOTTISH CASE

In the Western Isles, on the other hand, the state played a much larger role. An important institutional change in the 1970s almost accidentally brought more opportunities to operate the levers of change to Western Isles people; an all-purpose islands authority – the Western Isles Islands Council – was established and its head office located in the islands (Stornoway). Previously, local government responsibilities for the islands had been split between Inverness-shire and Ross and Cromarty County Councils located in Inverness and Dingwall on the eastern mainland of Scotland. Secondly, the Highlands and Islands Development Board, established in 1965 by Act of Parliament, had been given wide powers to assist with development in the region and in 1975 took a decision to provide special support to encourage the development of 'community co-operatives' – along the lines of the Irish examples – initially in the Western Isles. This decision was taken firstly because the board was alert to the fact that its success would be judged, as its first chairman said, 'by its ability to hold population in the true crofting areas' (HIDB, Inverness 1967, *First Report*, p. 5) and secondly because the board had found more 'conventional' approaches to development in these areas wanting. The community co-operatives scheme was evolved not just to tackle the *economic* problems of jobs and income but also, significantly, to raise 'morale and self-confidence' (HIDB, Inverness 1978, *Twelfth Report*, 1977, p. 9).

4.1.4 PROCESSES v. PROGRAMMES. LOCAL INSTITUTIONAL DEVELOPMENT BETWEEN THE INDIVIDUAL AND THE STATE

Peripheral rural communities are not all the *same* even if at one level they have much in common. In particular, some, through economic and institutional dependence and a history of outward migration, lack the basic structures to permit local development initiatives. In such cases, the provision or creation of new formal or informal structures is a crucial precondition for local development. Without such structures, what pass for 'local development' projects, programmes or initiatives are more likely to end up as another layer of external domination and control! *Processes* become more important than *programmes* in such cases, and it is the analysis of these processes which seems likely to yield many interesting lessons for the development of so-called 'peripheral', 'backward' or 'underdeveloped' areas in general.

Until now such process questions were reserved for the most 'severe' cases of demoralized peripheral communities – the Gaelic-speaking areas of Ireland and Scotland being prime examples of communities subject to state initiatives which, as we have seen, were at least partly aimed at rebuilding local morale and self-confidence. Elsewhere in the north-west periphery, much reliance has been placed on oiling the wheels of the 'market' (grants and loans for private sector developments, infrastructure projects, etc.), emphasizing the 'project' approach to development and providing a theoretical and practical comparison with approaches placing greater stress on development processes.

The lack of a locally based institutional structure and the lack of local morale and self-confidence are important features, then, of the north-west periphery. Other features include:

- a history of state paternalism;

- incorporation in global market systems;
- external ownership and control of many key resources.

We are now at a stage where the response of the state to these features and characteristics can be assessed through the analysis of state agencies established to bring 'development' to such areas (the Highlands and Islands Development Board, the Scottish Development Agency, Udaras na Gaeltachta) and the analysis of local development initiatives undertaken either within or outside the activities of such agencies.

The idea of 'state-sponsored local development' is not, of course, without its contradictions. Some clues as to the practical consequences of these contradictions, which may never be entirely resolvable, are contained in the case studies examined below.

4.2 THE SCOTTISH CASE – STATE SPONSORSHIP OF LOCAL DEVELOPMENT

4.2.1 THE HIDB'S COMMUNITY CO-OPERATIVE SCHEME IN THE WESTERN ISLES – A PRELIMINARY ASSESSMENT OF LOCAL DEVELOPMENT PROCESSES

The Highlands and Islands Development Board was established by Act of Parliament in 1965. Its general functions and duties have been reviewed elsewhere (Bryden, 1981; Hughes, 1979; HIDB Annual Reports). We are here mainly concerned with the board's scheme for the encouragement of multi-functional community co-operatives, mainly in the Outer Hebrides, started in November 1977.

This scheme was heavily influenced by the experience of similar Irish co-operatives, and in its pilot phase offered local advice through two Gaelic-speaking field officers, with grants for establishing, managing and investing in projects to be run by the co-operatives. In the initial stages heavy emphasis was placed on 'multi-functionality', by which was meant a mixture of projects intended to earn revenue and provide employment and local services. Unlike many of the Irish cooperatives, however, these projects were not mainly based on infrastructure investment. Nor were they *predominantly* concerned with agricultural supplies and marketing, for which a cooperative structure already existed throughout the Hebrides, which featured heavily in the Irish case (Storey, 1982).

In the HIDB scheme, the function of the field officers in the early stages was crucial. They provided information and encouragement to local community groups, often groups involved in community associations or other pre-existing micro-institutions, and then to steering committees established to develop projects and initial fundraising, and thereafter to managers and management committees, although these field workers were not supposed to suggest 'initiatives'. The limits of their role were always problematic and unclear but in practice they played an important part in 'developing' many initiatives with the local committees, specially in the early stages.

Once a viable-looking set of ideas had been produced and local fundraising for the co-operative (or *co-chomunn*) had been started, the co-operative applied for an establishment grant from the HIDB. This grant matched local funds raised on a pound for pound basis. In addition, they applied for a management and administration grant to pay the salary of a manager and basic office costs at a level of 100 per cent for the first three years and 50 per cent in the fourth and fifth years. This tapering grant was to encourage co-operatives to plan ahead to meet all such costs by the sixth year.

These basic start-up funds were aimed at providing the framework for developing concrete economic projects which could be grant-aided by the board in the 'normal' way. The matching grant was aimed at 'securing working capital', which information on the Irish experience suggested was a problem (Storey, 1979). The aims of the board were to encourage indigenous development processes in remote marginal areas where more conventional approaches to economic development had failed. Income and employment creation were important, but so too were the provision of local services, and the creation of a 'learning process which improves the capabilities of the community to protect and advance its interests in the future' (Alexander, 1979).

To set what follows in perspective, by 1986 there were twenty-four community enterprises in the Highlands and Islands with 3,500 shareholders, 375 direct employees of which 52 were full-time staff, 100 trainees and an annual turnover of some £3 million (Pederson, 1987). If these figures seem modest, it must be remembered that most community enterprises are in extremely disadvantaged locations.

4.2.2 THE AIMS AND METHOD OF THE STUDY

In this study, we try to trace the development processes which occurred in two contrasting *co-chomunn* in the Western Isles. The first, Co-chomunn Nis, was the first to be supported by the board under its new scheme. The second, Co-chomunn Pairc, was established shortly afterwards. Ironically, Pairc, which is still trading, was for many years considered to be the most vulnerable of the *co-chomunn*, while Ness, which ceased trading in 1986, was in the early years considered the most stable.

During the interviews we were particularly concerned with the following issues:

1. The history, growth and changing activities of the co-operative.
2. The conditions which lead to the formation and development of new activities in the co-operative.
3. The key actors involved at local, regional, and national levels, in particular the role of any field staff employed by agencies.

4. What channels and networks these actors used directly and indirectly to achieve support, etc.
5. What collaboration took place with other local organizations.
6. What market relations the co-operative had as buyers and sellers of goods and services.
7. Effects of the co-operatives' activities on the following:
   (a) jobs;
   (b) equality of local employment;
   (c) mobilization of local political, social, cultural and economic activities;
   (d) the generation of new ideas; and
   (e) the removal or creation of 'resistance'.

Questions on these issues were intended to probe the broader impact of co-operatives and the processes behind their initiation and development. We are grateful to Carola Bell, Agnes Rennie and Ina MacKinnon for their help. Responsibility for the views and opinions expressed rests with the authors.

4.3 CO-CHOMUNN NIS (NESS COMMUNITY CO-OPERATIVE): EFFECTIVE LOCAL ACTION; FINANCIAL CRISIS; DEPENDENCE ON THE STATE

4.3.1 EARLY BEGINNINGS

The individual in the early community work in Ness during the mid-1970s (which provided a foundation for the community co-operative) was a field worker with the Community Education Project, funded by the Van Leer Foundation. This project was concerned with the mobilization of the adult population and had started a small vegetable project, which provided seedlings for crofters in the area, and also a small combine harvester which was used as a threshing mill. These projects had started before the board's scheme was announced. The community realized that it needed some local *structure* to put forward ideas for action although it had not at that time been thinking of a community co-operative.

Thc field worker was a local Gaelic speaker with good contacts in the new Western Isles Island Council. Provided with an office in Ness and paid by Van Leer, she had the time, energy and resources to push the Ness case. The Local History Society, set up in 1977, was also one of the products of the Community Education Project, as was a local playgroup, a community newspaper, and a football club. Most of these things happened after the reorganization of local government in 1975, which allowed the more effective operation of local networks. Three further key figures at a later stage were a local girl working with the Ness Historical Society in gathering material under a youth employment scheme and who later became an HIDB field worker and then *co-chomunn* committee member, an 'incomer' who was involved with the playgroup and ultimately became chairman of the *co-chomunn*, and a local schoolteacher, who was secretary of the *co-chomunn* in its early years. The origins of the local 'development movement' were thus genuinely *local*, rather than - as they are in many other British rural communities - largely associated with 'middle-class incomers'. The emphasis was very much on development from 'below', even if a widely shared philosophy of self-help and indigenous development was perhaps lacking. It is also worth noting that, although very much a minority throughout the *co-chomunn*'s existence, women played a vital role from the start.

In addition to having a very positive effect on the possibilities of influencing change at local levels, the creation of the Western Isles Islands Council in the local government reform of 1975 created new and responsible jobs for young Gaelic speakers. The high educational standards reached by island people, and the bilingual status of the council, brought back young local people with new ideas and new approaches. This was to be a vital influence in the decade that followed.

In these early developments the church played a very minor role - both positively or negatively. The divisions that did exist in Ness were less due to divisions within the church and their various congregations than the fact that Ness did not think of itself as a community but rather as a number of small communities. When, for example, the co-operative building was built in South Dell people from other townships did not regard it as 'their' co-operative and some even refused to use it purely on account of its location.

The first field officer appointed by the HIDB for its new community co-operative programme was another young Gaelic speaker who quickly realized that the Van Leer Project was a strong foundation for a community co-operative. The local community association set up a steering committee with one representative from each township. The population of Ness is around 2,400 and some 540 shareholders contributed £20 each. This was matched by the board. During the early stages of the steering committee's work a questionnaire was sent out to every croft with the steering committee's ideas seeking information about the potential use of the various facilities and services that the co-operative proposed to offer.

4.3.2 EXPLOITATION OF LOCAL MARKETS

These initial ideas included:

- the provision of agricultural supplies, mainly feed and fertilizer;
- contract hire of agricultural machinery;

- marketing vegetables and potatoes grown locally, including the supply to schools;
- the provision of a garage;
- making wooden implements;
- the provision of a fish and chip van using local fish and potatoes; and
- printing of material produced by the historical society.

Pressure from the board for a multiplicity of activities led the committee to consider a range of projects. In order to help sell shares in the community, promises were made by the steering committee regarding discounts for produce, etc. Those most enthusiastic found it necessary in the social climate of the time to raise expectations in the area, but this led to some of the difficulties which the co-operative faced subsequently.

The application to the board submitted by the steering committee in April 1978 included no fewer than thirteen possible projects, of which six were still at the discussion stage. These were vegetable selling, fish selling, a tearoom, hairdressing, a bakery and a chemist shop. The committee requested a matching grant from the Highland Board up to £15,000 and an administration grant of £7,500 for three to five years. An estimated employment of six full-time and four part-time was predicted by the third year. The original paper contained no cash-flow statement, although projected profit and loss accounts were provided for the first three years. The *co-chomunn* was registered in 1978, and the first management committee was elected in August, shortly after which the first manager was appointed, accommodation found in an old house and storage space in a garage, and trading commenced. Between December 1978 and June 1979 there was a net trading surplus of just over £4,000 *after* receipt of the HIDB administration grant of £7,500 and this was considered to be a very encouraging start. By the time the committee came to write its annual report the co-operative had been firmly engaged in the sale of feedstuffs and fertilizers, taken over the horticultural project, started the mobile snack bar, been awarded a Western Isles Islands Council contract for the erection of two houses in Ness, developed retail outlets for veterinary products and protective clothing, taken over the combine harvester from the Van Leer project and had become involved in the board's experimental lamb marketing scheme.

Agricultural supplies

The agricultural supplies enterprise ran into early difficulties with the established agricultural supplies co-operative on the Island, Lewis Crofters Ltd, which did not welcome competition in a limited market. In addition, although initial haulage to the community co-operative was arranged with local suppliers as back-haulage on fish lorries, these hauliers eventually went into business on their own behalf supplying feed on the island.

A further problem was created by inadequate storage facilities. Losses occurred because of the build up of unsaleable (damaged) stocks, although inadequate stock control meant that the full cost of this was not fully appreciated. The *co-chomunn* made regular deliveries of feed and fertilizer to crofts, partly because they had four different storage locations rented, and by delivering without charge they could avoid keeping a storeman in each of these locations. Sales of feed were also very irregular as, at that time, crofters would feed their animals if the weather was bad, making it difficult to predict demand. The full extent of losses through unsold and damaged stocks was not appreciated until towards the end of the co-operative's life. The committee tend to blame this on the lack of involvement by the management committee in day-to-day management decisions. The strong advice from the Highland and Islands Development Board was that the management committee should concern itself with policy decisions rather than day-to-day matters. Managers were therefore left very much to their own devices and did not always feel obliged to provide essential information on such things as stocks or unpaid bills to the management committee. In many co-operatives – and perhaps especially Ness – this problem was exacerbated by the lack of continuity of managers and the inadequacy of record-keeping and accounting, which is discussed below.

Construction

Encouraged by the first manager, a joiner, the co-operative started construction work when the opportunity to build two council houses arose. When the contract made losses, it transpired that costings had been prepared in present-day prices with no allowance for inflation, and the plumbing element was omitted entirely. This illustrates a lack of experience or training in costing matters at the critical point and also, again, the lack of effective control over management by the management committee.

Training problems

The problem of training managers and management committees was a persistent one. Although courses were provided, managers could not in fact take time off to attend them. Another HIDB field worker appointed or given responsibility for training in the early 1980s worked with a committee member to set up training sessions for management committees outside their home environment in an attempt to stimulate interchange. However, by this time the HIDB had reduced its field staff and there were problems in following up the issues raised in such meetings.

Cooperatives were generally very isolated from one another at that time and the HIDB's field staff felt that more collaboration between *co-chomunn* was necessary. An assembly of Western Isles *co-chomunn* was held in Ness in 1980 and this was followed by others. However, these assemblies were reluctant to set up an organization and it was not until 1984 that the Association of Community Enterprises in the Highlands and Islands (ACE-

HI) was established. One persistent problem with the assemblies (and it applies very often to other meetings and courses in remote rural areas) was that the people who would gain most from such events were unable to attend them; it is difficult to get away for two or three days from a croft or a young family.

Other projects

The mobile restaurant, or fish-and-chip van, did not make a profit and was dropped in 1981. The then chairman of the *co-chomunn* had the idea of purchasing a local grocery van from a local van owner who wished to sell the business as a going concern. The grocery van ran until 1985.

The machinery service, started with the combine harvester, received a boost from the Integrated Development Project (IDP) for the Western Isles in the early 1980s and spreading equipment was bought for land improvement. The machinery service was not a major factor in the losses made by the *co-chomunn*, but small-scale machinery was not found to be profitable. Bad weather meant that machinery could only work for about four months in the year.

4.3.3 THE IMPACT OF NESS CO-CHOMUNN

Employment at its peak was fifteen but dropped to seven when the *co-chomunn* was closed in 1985. Maximum wages and salaries in any one year were about £44,000 (in 1981 and 1984/5).

The overall effects of the *co-chomunn* are considered to be positive. Most people now regret that it is no longer there, even if the benefits were not realized at the time. The delivery service was particularly highly regarded, although experience in Pairc shows how difficult it is to make this profitable. People in general in the area now know more about the problems of business organization, and for those closely involved it has been a massive learning process. This particularly applies to the group of five on the management committee who wound it up. By that time the whole ethos had changed and the committee was forced to get involved in day-to-day matters. A very positive working relationship developed between the committee and the staff – for the first time. This experience led at least some committee members to conclude that the managers had prevented rather than encouraged good communication between themselves and co-operative staff. However, whether the level of personal commitment evident in the final committee at a time when the *co-chomunn* faced its greatest threat could be expected – on a voluntary and unpaid basis – as a 'normal' feature is highly questionable.

Roles of managers and committees

It is clear that the respective roles of managerial staff and committee members have to be very carefully appraised. In the early days of the scheme it was assumed that a paid manager was necessary and the model of other commercial co-operatives was used in assessing the respective roles of committees, managers and staff. There was a great deal of pressure to get 'commercially successful' managers, but bringing people in from 'outside' did not work, and experience of other businesses was rarely an adequate training for the business of running community co-operatives. One *co-chomunn* has recently decided to do without a manager - an interesting experiment which may be related to the Ness experience.

Increasing financial problems - analysis of financial accounts

The accounts prepared for the *co-chomunn* were totally inadequate as a basis for decision-making. The lack of enterprise costings and budgets made it impossible to be certain which enterprises were losing money and which were not. Although it is evident from the earliest years that trading margins were at best barely adequate to cover overheads, it is not now possible to say with any certainty which enterprises were the greatest loss-makers. This problem would not have been so great had the management committee been involved from the start in day-to-day matters or if there had been continuity of management. Even so, management records were never sufficiently transparent to allow the management committees to appreciate what was going on.

After a small profit in 1979 (after the HIDB grant), the *co-chomunn* made increasing losses in 1980, 1981 and 1982. The advent of the EEC-funded Integrated Development Programme (IDP) and the build up of IDP-related business led to an almost 300 per cent increase in gross sales in 1983, yet the *co-chomunn* only made a profit after an exceptionally large 'administration grant' from the HIDB. The apparent improvement in the accounts was, however, largely eaten up by a huge increase in debts (£60,000 or so) and a substantial decline in credits (£38,000), the combination of which led to a significant increase in bank lending.

In 1984, the first accounts showed a profit of some £3,000 but again a substantial increase in debts occurred (£30,000), along with a hefty increase in bank borrowing. It later transpired that the accounts were incorrect by almost £13,000, represented by credits. No wonder the committee began 1985 in a confident mood, unaware of the seriousness of the situation.

In reality (but not according to the information then available to the *co-chomunn*) the 1985 year started with debts in the order of £110,000, some of which were two years old, and stock in excess of £34,000, an unknown proportion of which were unsaleable. The total bank overdraft and other loans outstanding at that time were over £110,000, almost exactly matching the debtors' figure.

What was happening? In effect, the *co-chomunn* was financing part of the IDP programme because of the long delays between approval and undertaking of work, the receipt of grant funds by crofters and the settlement of debts with the *co-chomunn*. So far as can be seen, no interest was being levied on these outstanding bills, while the *co-chomunn* was paying high interest charges to the HIDB and the bank.

Not surprisingly, the *co-chomunn* paid for this in 1985 when interest costs were £13,460 and an allowance of £4,142 was made for bad debts. The main IDP expenditure had come to an end, business was shrinking, overheads increased and the HIDB administration grant ended.

Late in 1985, it was found that the accounts for 1984 were substantially in error. They had shown a break-even position but should have shown a sizeable loss. The 1985 accounts showed a loss of over £40,000 and at this point the HIDB argued that if the *co-chomunn* continued to trade they would be trading irresponsibly. The manager, who had been seriously ill for some time, left in the intervening period.

The administration grant – intended to help the *co-chomunn* to acquire 'adequate' management in the early years – could in retrospect be held to have encouraged high overheads for the level of business and its ending certainly left the *co-chomunn* vulnerable at a time when demands on management were increasing and business shrinking.

4.4 CO-CHOMUNN NA PAIRC (PAIRC COMMUNITY CO-OPERATIVE): SURVIVAL WITH INCREASING DEPENDENCE ON 'BIG CAPITAL'

4.4.1 ANTECEDENTS

The original steering committee for the Pairc *co-chomunn* arose out of the Pairc Community Association, which was set up in the early 1970s as a voluntary organization helping the elderly in the community, running a playgroup, organizing sales of work and so forth. A major triumph for the association was to persuade the new Western Isles Islands Council to provide sheltered housing for the elderly just after the council was established in 1975.

When the board established its Community Co-operative Programme, the board's first field officer came to speak to the community association. The steering committee was set up later in 1977 and initial ideas for activities included plant hire, knitting with island yarn, and sheepskin rugs. Of the 500-strong population of South Lochs, a high proportion joined the co-operative, which has 300 shareholders, including some from Glasgow and even New Zealand. £10,000 was raised from shareholders (each share cost £25) and the HIDB gave a matching grant. The co-operative was registered in 1978.

Initially, the plant hire service acquired a digger which was used on the new road to South Lochs being constructed by the island's council. Knitting had also been a traditional activity for some women in the area, but as out-workers for the mills. Those involved in this activity considered that by establishing knitwear within the co-operative any profit would then come back to the community. Although the impetus came from those who were out-workers, others subsequently joined in, and there are currently about sixty knitters working when the order book is full.

Field-worker support critical

The HIDB field workers seemed to have played a crucial role. The steering committee included nurses, schoolteachers and others, who did not at that time know much about marketing, cash flows or accounts. Field workers therefore helped with the costings for the various projects and with marketing ideas – particularly advice on which organizations to join (such as the Crafts Guild, an association of local craft producers). In addition they brought buyers and agents down to the *co-chomunn* if they heard they were in Stornoway. The normal procedure for knitwear is to knit against orders, providing only samples in advance. Marketing is now conducted in several markets. Since 1979 the *co-chomunn* has had an agent in France. They also sell in Japan and America: in the case of Japan the original sale was made as the result of a visiting buyer brought by one of the field workers. The American market is a recent one and arose through the recently established Hebridean Knitwear Association and a previous manageress. The Hebridean Knitwear Association was set up in 1985 and appears to be playing an important role with considerable potential in the marketing and design area. An influential figure in getting the association off the ground was the 'knitwear coordinator' with the Western Isles Council.

Knitwear for export

Knitwear is hand-knit to a traditional design and prices are not high given the amount of work involved. The average knitter would receive between 30 and 40 pence per hour for her work. Although there were fifty or sixty knitters involved only fifteen have the skills required for 'design' or high fashion knitwear. The co-operative has had no help at all on design and no HIDB grant on the knitwear side. However, the *co-chomunn* manager met a friend through the Hebridean Knitwear Association who is marketing high-fashion knitwear. This contact has offered to provide design assistance. The *co-chomunn* also now provides knitters for this designer when large orders come in and there appears to be the basis of an effective local network developing. The knitters are not all shareholders in the co-operative although shareholders get first choice of work if there are not many orders. Both groups receive the same payment. Quality control, of course, is a big problem for the co-operative and can cause ill-feeling in the community. The manager feels that training could help with this although most of the knitters are not young.

The *co-chomunn* was a founding member of ACE-HI, the Association of Community Enterprises in the Highlands and Islands, and has found this to be a useful source of network experience and a forum for discussing new ideas. The *co-chomunn* also looks to ACE-HI to provide intelligence on marketing in general, although the Hebridean Knitwear Association would be the best source of specific advice.

Local markets – the mechanical digger

The digger was also an idea generated by the steering committee for the *co-chomunn*. Originally the digger was used both to provide access to crofts, hill-grazings, peat-diggings and so on, and, importantly, on the new road which was being built by the council. However, in the early 1980s the council decided to contract work out to private contractors who had plenty of their own digging machines.

Shareholders v. consumers – service v. profit

In 1986 the *co-chomunn* management committee proposed to sell the diggers, but shareholders voted against this because the service was regarded as beneficial to the community. The enterprise makes losses, and shareholders have a special charging rate. This again illustrates the difficulties that arise in a small community where the membership is benefiting from the services provided and has an interest in their continuing. However, the management committee may have to force the issue and risk ill-feeling as a result.

The small agricultural supplies project has also been a loss-making enterprise conducted for the benefit of shareholders. In the early days five or six months' stock of feed was purchased at the beginning of winter but the *co-chomunn* found that up to eight different kinds of feed were required by crofters and the amount of each used in every year varied significantly. In about 1983, therefore, the *co-chomunn* started to obtain all its supplies through Lewis Crofters Ltd. Pairc then acted as an agent for Lewis Crofters, bulking orders for them, but delivering with their own vans. Small stocks of medicines and other everyday items are, however, kept at the local store. Clearly the *co-chomunn* is providing a valuable service for Lewis Crofters by bulking orders and carrying out the deliveries to individual crofts for them. The 5 per cent commission which Lewis Crofters allows to Pairc is probably not sufficient recompense for this service. On the other hand, this was always an activity conducted for the benefit of shareholders and it may be difficult to persuade them to pay enough to cover the real costs involved.

Fish farming: from local to external control

The *co-chomunn* got involved in fish farming – initially of rainbow trout – when at the board's suggestion they purchased the hatchery from the HIDB. For the first few years Pairc was the only rainbow-trout producer on Lewis and used to deliver and sell locally. The enterprise was profitable at that time. Later, the HIDB assisted other fish farmers to become established on the island, giving rise to strong competition for the local market. Prices fell and the fish had to be sold below cost price. At the peak production period, in the early 1980s, five people were employed on the trout farm.

In 1984 it was decided to stop trout production. However, most of the people who had been working or training there were able to find local employment in the new fish farms started nearby, particularly that established by Booker McConnell, a multinational enterprise.

In 1983 the *co-chomunn*, at the suggestion of the management committee and supported by a grant and loan from the HIDB, had put two small cages in the sea for salmon. When the trout enterprise collapsed they doubled the quantity of salmon smolts in the hope that salmon would fill the gap created by trout. However, the difficult financial situation of the *co-chomunn* meant that they could not expand to compete with large enterprises such as Booker McConnell and they then discussed the formation of another company in which Pairc would have an interest. These discussions were substantially helped by the HIDB Stornoway office and involved the injection of outside capital. Pairc Salmon was set up in 1985, with the *co-chomunn* owning 40 per cent of the shares, HIDB 20 per cent, Rolf Olsens (Norwegian Interests) 20 per cent and Dr Horobin (an entrepreneur involved with the production of 'natural' medicines) 20 per cent. The contribution made by the *co-chomunn* came from the sale of salmon which they had in their cages to the new company. Two directors are from the *co-chomunn* and the three other shareholders each have one.

The *co-chomunn* itself now has an incubation unit at Pairc Hatchery which produces young salmon smolts. In 1987 they planned to sell in the region of 105,000 smolts, half of which were to be sold to Pairc Salmon and the rest to other salmon farmers on the island. In 1988 they hoped to produce 130,000 smolts of which three-quarters will be sold to Pairc Salmon. The enterprise involves the stripping of eggs from hen salmon, their incubation for some four weeks in troughs until they hatch as fry, transference to an ongrowing unit and sorting into two groups to be sold in one or two years as smolts. The ongrowing stage is conducted in the fresh-water cages which were formally used for the trout farming. This enterprise is run by the wife of the manager of Pairc Salmon. The price which the *co-chomunn* receives for the smolts varies annually and there appears to be no contractual arrangement nor established methods for reaching agreement on prices.

In addition, the *co-chomunn* has laboratory facilities which provide a pathology service for the entire island's salmon and fish farmers. This does not make a profit, but does provide good contacts with market outlets and engenders considerable goodwill.

All the island fish farmers are getting together and setting up a company called Atlantic Sea Products. This company will have a factory which will buy bulk feed and also start processing salmon, farmed fish, vacuum packing, smoking, etc.

The involvement of the *co-chomunn* with capitalist large-scale enterprise has given rise to some concern in the community and may prove a divisive issue. The most complex substantive issues probably relate to transactions between the *co-chomunn* and such enterprises, the terms agreed for such transactions, and the prices for any trading between them. In addition, there is some uncertainty about what happens if the *co-chomunn* wishes to sell or purchase shares. The *co-chomunn* has had a measure of success in preventing fish-farming sites from being totally disposed of to interests outside the islands, but its direct involvement with external enterprises may provide backdoor access to such rights should they become very valuable. The issue would then be the terms under which disposals or transfers are made.

In addition, the *co-chomunn* has two profitable mussel rafts run with casual labour. The mussels are not fed artificially and stay on the rafts until the market is right. Demand is currently strong and marketing is carried out through the Shellfish Growers Association.

The *co-chomunn* has gambled much on the salmon-farming ventures. It is now unable to control the price of smolts, which will be largely determined by other enterprises, not generally of a co-operative nature. The involvement in Pairc Salmon is also a risk, but with a potentially large pay-off. Much depends on how the predicted over-supply of farmed salmon will work itself out in the marketplace and, in particular, how successful Atlantic Sea Products is in finding new markets and adding value. The *co-chomunn* has become firmly locked into what is now a major industry and is largely dependent on decisions made by others for success or failure.

4.4.2 THE IMPACT OF CO-CHOMUNN NA PAIRC

The *co-chomunn* had six full-time employees plus considerable casual labour and up to sixty knitters. The wage bill last year was £60,000, which compares with an HIDB loan and grant over the years of £150,000. Despite an overall profit in 1985 it continues to suffer from cashflow problems; a high bank overdraft gives rise to substantial interest payments. In addition, they need heavy insurance cover on the fish farming side.

The management committee for the co-operative needs training and induction to deal with the very major decisions which now have to be taken. It is hoped that ACE-HI will help with this. In the meantime there are monthly meetings of the committee to discuss budgets and cashflow projections. The local HIDB Officer and the staff officer in Inverness now responsible for co-operatives have been very helpful.

Some lessons from the Pairc experience

The manager feels that as a result of the *co-chomunn*'s activities, involvement in politics and decision-making has improved in Pairc. Specific examples are the fish farming which was started up following the initiatives taken by Pairc and the scale of decisions which have been made in the past few years. Local people have bought up fish-farming leases, individually as well as through the *co-chomunn*, and this has prevented sites from being dominated by outside interests.

There is local concern about the high stakes in which Pairc is now engaged and the lack of training of committee members to get to grips with the decision-making processes involved. The problem of training of both management committees and shareholders is obviously one which has not been adequately tackled. The consequences of any mistakes which *may* have been made in negotiating the terms with third-party enterprises will only show up in a few years time, but they could prove conclusive one way or another. Co-chomunn na Pairc was always considered the most vulnerable of the co-operatives supported by the board in the early years and indeed a meeting was called last year to see whether the liquidators should be called in. Of the original enterprises, two which were conceived of as providing local services – first, the digger and the plant hire and, second, the agricultural supplies – continue to make losses and are difficult to wind up in view of benefits to shareholders. A third – knitwear – seems to do reasonably well, although at the cost of very low wages, problems of quality control and a lack of new entrants. The economic future of Pairc currently depends almost entirely on the fish-farming side which, as we have seen, largely depends on factors beyond the *co-chomunn*'s control. It remains to be seen whether the terms negotiated with outside concerns will be such as to secure the interests of the *co-chomunn* in the long run given the potential volatility of the market for farmed fish.

Some hard lessons have been learned about the importance of cashflows and budgetary control, particularly during the era of high interest rates. The initial idea that the HIDB's contribution of matching funds would put the co-operatives in a position to handle cashflow problems proved to be naive. As in the Ness case both the committees and the board staff involved lacked appropriate commercial experience in the early years, although it would be a simplification to suggest that an input of 'experienced commercial people' was all that was required.

4.5 THE IRISH CASE: LOCALLY BASED COMMUNITY DEVELOPMENT IN THE WEST OF IRELAND

A number of rural communities in the west of Ireland have built up a considerable track record of achievement in the community development field over the past twenty years. Indeed, their experiences have had a good deal of influence elsewhere, as the Scottish cases illustrate, and have been well documented by observers throughout their existence (see Breathnach, 1984; Commins, 1983; Scott, 1985). The local development structures they have chosen have included development committees, community-owned companies and community co-operatives, generally aiming to address in varying degrees economic, social and cultural challenges, often, like the Scottish cases, in a multifunctional way. In this study we examine the long-term achievement of two parallel but in some ways contrasting approaches in this direction, one a community co-operative in west Kerry and the other a community-owned development company in west Connemara, and we come to some conclusions about their ability to continue to function successfully in the present difficult economic climate.

Two critical aspects of their progress at this stage are considered: first, the methods they have adopted to diversify and innovate once early targets were met, or as opportunities became limited; and, secondly, the relationship they have built up with state agencies and institutional frameworks. Of particular interest is the extent to which their past achievements afford them recognition and thus some stability in their negotiations with outside agencies and the extent to which this outside involvement tends to encourage them to move in certain directions and not in others.

4.5.1 TAPPING STATE RESOURCES AT LOCAL LEVEL

Unlike the Scottish case, where the community co-operatives were assisted into place by a single ready-constructed package of measures, the origins of the Irish examples largely predate the various institutional measures subsequently adopted to support them. In their early days they developed the skill to adapt forms of state aid not specifically intended to support community development (such as the means to fund managers of *Gaeltacht*[3] community co-operatives), but equally importantly they were able to tap a very strong tide of popular feeling in the west that local action and commitment were essential ingredients in revitalizing western communities and attracting resources to areas which they, and the rest of the nation, saw as being disadvantaged.

There was therefore a degree of political support in their favour for their early efforts to bring infrastructure (such as piped water and electricity supply) up to a basic standard enjoyed elsewhere. Overlapping this, the Irish economy was relatively buoyant in the 1970s, the *Gaeltacht* development agency – Udaras na Gaeltachta – being notably successful in attracting industry to western locations. Therefore, local development organizations felt they had some room to move in developing new ideas and projects and that although they were breaking new ground they were to some extent swimming with the tide.

4.5.2 CHANGING CIRCUMSTANCES IN THE CRISIS

These two factors have now changed. Basic infrastructure has now been installed in much, though not all, of the west and the deepening economic recession has brought into question state intervention that is seen to benefit unduly one part of the country at the expense of others. Also questioned is support of community-based activity that is seen to be competing with the private sector. These changes have affected some community development organizations more than others. Generally those in areas where private-sector provision is not an option or those which have built up a core of locally owned assets and activities not dependent upon recurrent injections of state aid have survived well. However, there have been some fairly spectacular falls.

4.6 COMHARCHUMANN FORBARTHA CHORCHA DHUIBHNE (WEST KERRY CO-OPERATIVE) – STATE DEPENDENCE AND DIVERSIFICATION EFFORTS

Of all Irish rural community development activity perhaps the best known, and most highly regarded outside Ireland, would be the *Gaeltacht* community co-operatives. The scale of their achievements is impressive. In 1981, when around twenty of the co-operatives were in existence, they had a combined membership of 7,000, were involved in a very wide range of activities, and were turning over approximately £3.5 million as well as providing 225 full-time jobs plus many more part-time and seasonal ones. By 1983 their turnover had reached approximately £5 million, although membership and employment provision were beginning to decline by then, a decline which has continued since. The reasons for their present difficulties can perhaps best be considered in this paper in relation to our case study *Comharchumann Forbartha Chorcha Dhuibhne* (CFCD – the West Kerry Co-operative).

CFCD was one of the first community co-operatives to be founded (1967) and has been one of the most consistently successful performers. Its achievements so far can be demonstrated fairly dramatically. One measure of its performance would be an examination of its land reclamation work, an area on which it has particularly concentrated. To date CFCD has assisted more than 600 of the 837 farmers in its area, added at least £4 million extra income for them through increased milk yields and, through the related reclamation/plant-hire workforce, itself generated employment over the years. Equally important, the psychological impact has been great.

'The impression of seeing hillsides which had previously produced nothing other than furze and rushes suddenly blossom in fields of good grass' was the first reflection of hope in the surrounding Dingle *Gaeltacht* according to Bishop Casey of Galway in 1982. However, the future of the co-operative is now uncertain. Several of its core activities at which it has performed well, including land reclamation, have had to be discontinued and the co-operative, being undercapitalized, has been forced to retrench. Its current labour force is reduced to around eight employees (from a peak of thirty or forty jobs during the most buoyant period) and a recent financial restructuring has led to assets being sold to the *Gaeltacht* development agency, Udaras na Gaeltachta, and then being leased back, together with a modest bank loan being written off. Despite these setbacks, the co-operative is still involved in innovatory work, which it hopes will provide it with a continuing role.

4.6.1 EARLY BEGINNINGS

The origins of CFCD can be traced back to a brief intervention in the area by the local (Kerry) county development team in 1967. After analysing the challenges the community faced, the suggestion was made that the community consider establishing a co-operative. Much to the bewilderment of the co-operative's newly constituted committee (then consisting principally of farmers and teachers), the development team withdrew shortly afterwards. None the less clear targets of employment creation, land reclamation, the building of community halls and the promotion of the Irish language were soon established. Having initially raised £10,000 in share capital locally, the co-operative then built two community halls, one of which has since housed its offices, and began to tackle the crucial question of the area's poor agricultural output. The key to this proved to be land reclamation, together with the founding of a demonstration farm unit.

4.6.2 LAND IMPROVEMENT

Chorcha Dhuibhne has a land area of some 77,000 acres supporting a predominantly agricultural community. The land is not as poor in quality as in most *Gaeltacht* regions (such as Connemara), but was not yielding anything like its potential before CFCD was started. The average farm size according to the County Kerry Agricultural Survey in 1972 was 35 acres with over half the farms fragmented. Farm buildings were often in poor condition and equipment inefficient. Two and a half acres of land were being used to support each cow, well above national targets. Alternative employment was difficult to find except through emigration. The fishing industry in the area collapsed in the 1930s and the present harbour at Dingle is inadequate for modern boats.

As a result, by the 1960s the population was ageing, with a high proportion unmarried. Since 1891 the population had halved from 12,783 to a present-day stable figure of around 6,200, and it has been estimated that more than 20,000 people lived on the peninsula before the famine of 1847 (Mac Giobuin, 1982, p. 15). During the 1960s nearly half the inhabitants were dependents, being either under 15 or over 65 years old, and there were considerably more men than women (Mac Giobuin, 1982, pp. 15–16). This was the by now familiar picture of *Gaeltacht* decline.

The success of the co-operative's *land reclamation* programme depended upon:

1. the discovery that an impervious layer in the soil approximately 60 centimetres under the surface restricted drainage. The co-operative discovered the means to correct this, by deep ploughing to a depth of a metre, and other more conventional treatments. More than £250,000 worth of excavators, bulldozers, tractors and ploughs were purchased and maintained whilst CFCD were engaged in land reclamation;

2. financial support from government sources of up to 70 per cent of the cost of reclamation available to individual farmers. By passing this on through work undertaken by the co-operative a considerable source of income for CFCD was generated.

Until recently this equipment has been constantly used in improving the peninsula's land resource. However, the point came where the programme reached the end of its viability, principally for four reasons:

1. levels of grant-aid have declined significantly;
2. present economic conditions in agriculture are not favourable and the outlook is not encouraging;
3. the amount of land remaining to be improved is small; the most viable reclamation tended to be carried out first;
4. three contractors have set up in competition to the co-operative.

Members of the co-operative were disappointed that each of these private operators was grant-aided by the state via Udaras na Gaeltachta despite the fact that the co-operative had a good record of achievement in this area and that the amount of further work that could be done was clearly limited.

Demonstration dairy farm

The demonstration dairy farm was set up following the recommendations of a survey undertaken by the Irish Agricultural Institute in 1975. CFCD decided to take on a typical 52-acre dairy farm, which included 30 acres of poor land, for demonstration purposes. It opened in 1980 and through good management successfully raised milk yields from an average 694 gallons per cow per year to over 1,000 gallons from each of its thirty-seven cows. The average milk yield on the peninsula at the time was 600 gallons per cow. Regular farm open days have been a part of the unit's programme to publicize its management techniques, but despite this the co-operative is not eligible for any financial support from the national farm advisory service, ACOT. Partly because the co-operative cannot afford to carry losses on a venture such as this without some such assistance, and after a poor year in 1986 due to a combination of very wet weather and milk quotas, the co-operative has decided in principle to pull out from the farm and lease it to the present farm manager.

Tomato growing

For several years the co-operative was successfully involved (through an independent company) in growing tomatoes in a 2-acre glasshouse. This was always bound to be a marginal activity because of competition from other growers (notably Dutch), but it did provide up to seven jobs locally during this period. Unfortunately, despite the excellent crop in 1986, the company has recently been liquidated. The two main reasons for this were:

1. To reduce fuel costs the co-operative entered into what it regarded as favourable negotiations with Udaras na Gaeltachta to convert its oil-fuelled heating plant to solid fuel. The cost of the conversion was around £100,000, of which the Udaras contributed £51,000, the

remainder being paid off at around £7,000 per annum on apparently generous terms. However, during the year an unforeseen drop in oil prices reversed the cost benefit with oil actually becoming cheaper than coal.

2. On top of this, the wholesale price of tomatoes dropped to an unusually low price in 1986.

4.6.2 GROWING STATE DEPENDENCY AND NEW DIRECTIONS

As a result, some £20,000 were lost on the venture, and negotiations are in hand for a takeover of the plant by the private sector. The problem CFCD faced once again was a tendency shared by most of the community co-operatives, a lack of liquidity to survive a crisis like this in one of its enterprises, combined perhaps with too great a dependence on one form of state aid for the project. Many of the community co-operatives thrived whilst they were involved in the development of infrastructure or the promotion of natural resources in partnership with the state. As has been said already, directly commercial ventures have proved more difficult. By 1984 the co-operative anticipated difficulties resulting from the winding down of several of their core activities and therefore negotiated a restructuring with the two main state agencies with which they (and all other *Gaeltacht* community co-operatives) were working. To some extent the co-operatives entered negotiations at this time together. For example, a joint document was produced by a confederation of the co-operatives suggesting, among other things, the following changes:

- a clearer definition of their status and objectives; they are still in fact legally registered under the antiquated 1893 Industrial and Provident Acts;
- a recognition that the social, cultural and economic development role of a co-operative is very likely to be non-profitmaking;
- a long-term commitment by the state to pay 75 per cent of the cost of management (100 per cent for the island co-operatives);
- that the state should use the co-operatives' offices as development offices for the area concerned; an extension of this principle is that the co-operatives should be adequately funded as local community development agencies;
- further training;
- long-term planning both for the co-operatives and for all *Gaeltacht* regions.

Nonetheless, negotiations were in effect conducted on a one-to-one basis. The deal that was agreed with CFCD was centred upon the state contributing £2 for every £1 the co-operative could raise itself from local shareholders.

The co-operative, in fact, raised £30,000, largely drawing upon the loyalties of those involved in its remaining activities (such as the Irish colleges, see below). In addition, the Udaras paid £75,000 for the co-operative's buildings, the community halls and offices, which are now leased back to CFCD for £3,000 a year. Although this seemed to be a good deal at the time, the present manager feels that these recurring payments will inhibit the co-operative's ability or inclination to become involved in new activities. In addition, negotiations with the co-operative's local bank led to a loan of £45,000 being written off.

The Irish colleges – a continuing enterprise

The chief areas of activity that the co-operative remains involved in are one of its early and continuing successes, the provision of Irish colleges, together with two innovatory projects, training in computer skills and an ambitious promotion of the very rich heritage enjoyed by the peninsula.

Irish colleges are by no means unique to the Dingle *Gaeltacht*. Indeed there are examples of them up and down the west coast of Ireland, organized in a variety of ways. Provided that the courses meet a required standard set by the Department of Education, the cost to students of attending is subsidised by the state. CFCD first became involved in the summer of 1970 when 650 students attended the colleges at Ballyferriter. The annual attendance figure is now much higher, with 1,750 students attending the 1985 summer three-week courses and a further 1,400 coming to the shorter weekend winter courses. CFCD's reputation in this field is considerable and can be gauged by the increasing attendance figures at a time when there is a great deal of competition between such colleges for a limited number of students.

This is a particularly important activity for CFCD for it both affirms its commitment to the Irish language and at the same time maintains local participation in its work. Some seventy households take in students during the tuition period, thereby distributing widely the further income brought into the area (estimated by Mac Giobuin at around £400,000 p.a.). Teachers are assigned twenty-five students each and are mostly local people, some of whom would be working outside the region during the rest of the year.

Heritage as a resource

The co-operative has also organized concerts, art exhibitions and other cultural events in the area and has published a number of books and a local newspaper. In 1978 it decided to establish a separate subsidiary company, Oidreacht Chorcha Dhuibhne (OCD), to undertake all the various cultural activities of the co-operative. *Oidreacht* is the Irish word for heritage, and by this OCD means 'the sum of all our resources for living' (quote from interview with Doncha O'Conchuir of OCD). At present three aspects of this heritage have been specifically identified:

- the natural landscape as it has evolved, its geology and marine biology;

- the flora and fauna; and
- human creativity, in providing both material welfare and cultural expression. This is perhaps the most complex aspect and includes the rich archaeology of the area, its history of settlement patterns and agricultural systems, together with less tangible expressions of creativity through language, music, dance, the visual arts, and so on.

The primary aim of OCD is to enhance appreciation of this heritage within the west Kerry *Gaeltacht* by means of research and the establishment of centres throughout the area, each centre concentrating on one aspect of the heritage. Members of OCD are determined that this should not be a sterile historical survey, a freezing of the past. Rather, it is intended to be dynamic and creative and, in the tradition of the co-operative, encouraging the participation of the local community. Throughout the co-operative's existence they have learnt that this participation depends upon the community being well informed and motivated.

Specific achievements of the heritage programme so far include, first, the publication of papers by various experts as an introduction to the further exploration of aspects of local heritage. Second, a comprehensive archaeological survey which underlines the international importance of the peninsula's antiquities has been published, identifying and documenting over 2,000 sites within the area. Incidentally, fieldwork for the survey may well have re-established contact with some local farmers, who became less involved with the co-operative's work after the early land reclamation work was over.

Third, the former village school at Ballyferriter has been renovated at a cost of around £100,000 to provide an impressive archaeological interpretation centre/café situated near to a concentration of the recorded sites. Plans to create further interpretative centres are well advanced. Various settlement patterns will be demonstrated through the restoration of a *clachan* (nucleated farmstead), a *rath* (single-homestead ringfort) and an Anglo-Norman manorial village. The history of the borough settlement of Dingle is already being presented at a new centre in the town. There are also plans for a centre concerned with marine fauna to be situated on the Blasket Islands.

Fourth, a research programme into attitudes to the Irish language has been established, and will draw conclusions about the best means to ensure its survival in the area.

Overall, there has been successful negotiation with outside agencies/funding bodies of financial and technical support for various parts of the programme. To a large extent the programme is dependent upon such outside support to fulfil its aims. OCD is in a strong position to repay such support by the unique contribution it can make to the study of antiquity and pre-industrial society through to the contemporary practice of social and political organization. The area has barely been touched by industrialization; attempts to introduce factories have generally failed and such economic development as has occurred through the co-operative's work has mostly concentrated on natural resources. Very substantial support for its work has been obtained by an impressive fundraising drive, with support obtained nationally (from the Employment Guarantee Fund, ANCO, the Electricity Generating Board, Kerry County Council, for example) and especially from an American foundation (responsible for the funding of the Ballyferriter interpretation centre). This investment should perhaps be seen as the building up of a local resource rather than the further increasing of dependency upon outside aid, as there is very strong direction coming from the local community via OCD. In terms of its developing relationship with the outside world the programme should, at the very least, help avert some of the worst excesses of mass tourism to which the area might well be especially vulnerable. However, throughout its cultural work CFCD primarily looks to the continuing experience and expression of heritage within the local community. This concern is perhaps best put in Father Noel O Donoghue's definition of heritage as quoted by O'Conchuir (the present chairman of OCD): 'Memory is the past in the present. . . . Heritage is the community's memory. This is why it is so important that a community be aware and informed of its heritage. The greater its awareness and appreciation the more vital the community is.'

New technology

The co-operative first purchased a computer in 1978, and all its activities are now computerized. In 1981 the national training agency, ANCO, agreed to fund a training course in computer programming based at CFCD's offices at Ballyferriter. This scheme, which has carried on successfully since then, attracts students from all of the *Gaeltacht* communities; once again this brings extra cash into the local economy. Each of the twenty trainees is allocated a £1,300 allowance by ANCO and stay residentially in the area for the twenty-week duration of the course.

4.6.3 CONTROVERSY OVER FISH FARMING

For the future CFCD looks to the Irish colleges to remain a mainstay through the three summer months, as at present. In addition it is looking for a second major new activity that can re-establish its roots in the community, as the land-reclamation work succeeded in doing in its early days. One possibility is in the area of fish farming. There are currently two separate controversial proposals regarding the establishment of a salmon-farming enterprise at Ballyferriter. CFCD may become associated with one of these, although the degree of local opposition to the proposals may put the co-operative in a difficult position if it does. There are worries about the environmental impact of such a scheme and the effect it could have on existing fishermen in the area. The future of the salmon market seems especially unpredictable because of the number of start-ups in this field in the west of Ireland at the moment. Given the sheer scale of capital required for its launch, the co-operative would most likely become a vulnerable junior partner in such an association, providing local knowledge and goodwill (if this can be obtained). On the other hand, if the co-operative does not accept the opportunity to become involved it may well face local criticism for *not* doing so, in effect becoming a 'whipping boy' either way. This vulnerability to criticism is almost inevitable now that its earlier activity in land reclamation is over. One way that this vulnerability might be reauced that would benefit other Western communities similarly involved in negotiations over their local natural resources would be if they had more to bargain with. For example, most of these schemes are dependent upon being licensed and upon being eligible for state aid. In return for this a measure of linked direct benefit to the local community could be required, as happens in some cases with mineral rights. Another suggestion for a future activity, possibly with more appeal, is to provide chilling facilities for local fishermen.

4.6.4 JOINT PLANNING WITH THE STATE

Despite its outstanding achievements and especially its ability to demonstrate the cost-effectiveness of its use of state aid so far, CFCD now needs to re-establish its direction. In common with other community organizations in the West it needs the benefit of joint planning with state agencies to agree a realistic development programme that both are interested in pursuing.

4.7 CONNEMARA WEST LTD – SUCCESS WITH SELF-RELIANCE

The second case study, by contrast, concerns a non-*Gaeltacht* community living around Letterfrack in west Connemara, a remote area of western Ireland of some 300 square miles with a population of 1,800 people. Partly because of its non-*Gaeltacht* situation, but more especially because of its established policy, the community's development arm, Connemara West Ltd, has avoided some of the traps that have snared community co-operatives recently. Together with several other local development organizations Connemara West has built up locally owned assets and participation in decision-making, thus boosting local confidence. At the same time opportunities have been taken to tap external funding for projects where this has been possible without compromising local development objectives. There has, however, been considerable frustration when attempts have been made to renegotiate successful projects; this was especially evident in the recent repetition of its outstanding craft training programme.

4.7.1 ORIGINS

The area became English-speaking before the turn of the century, largely through the anglicizing effect of an English garrison established at the nearby town of Clifden. None the less, its people are confronted with similar challenges to those facing their neighbours within the Connemara *Gaeltacht*.

Much of their land is unproductive with little potential for improvement; more than half their farm holdings are under 15 acres. As in the *Gaeltacht*, traditional occupations of agriculture, fishing and forestry have failed to hold the population. Over the past sixty years emigration has taken place on a large scale, particularly to the UK.

A good starting point for considering their present achievements is the formation of a Muintir na Tire (see above, p. 91) guild and parish council, founded in the area by the local curate in 1961. This first attempt at operating a local development group shared the weaknesses other communities experienced at the time using this model: it was based upon 'vocational' groupings in the community, it was weakened by party political divisions and the parish council found itself acting in a frustratingly unfulfilling pressure group role with a lack of clarity about specific objectives and about the practical action required within the community to achieve them. It is therefore significant that when a new approach to community development emerged in 1970 all three of the local development groups that followed had specific well defined targets.

4.7.2 THE CREDIT UNION

The first of these three groups was the Tullycross Credit Union Ltd, founded in 1970 with the encouragement of a new curate. A credit union is a co-operative association whose members can obtain low interest loans out of their combined savings and participate fully in management. The credit union has flourished and from its modest start has now built up a substantial annual turnover, taking a full part in the development process in the area. Recently it has been in a position to make a crucial contribution to the funding of a new farmers' co-operative.

4.7.3 HOLIDAY COTTAGES AS A SOURCE OF SURPLUS

It was the building and subsequent management of nine self-catering holiday cottages at Tullycross that led to the formation in 1971 of the second of the three development organizations, Connemara West Ltd. This is a community-based and controlled development company of which 80 per cent of households in the area are shareholders. The cottages were some of the first of their kind to be built in the west and have since provided a solid foundation for the company's further development, with income generated reinvested elsewhere. The high occupancy rate of the cottages has meant that substantial reinvestment has been possible in other Connemara West projects, although the stage has now been reached in their life where they require fairly substantial maintenance work (such as rethatching).

Over 5,000 people from the area around Letterfrack contributed a total of £13,000 in share capital, which, at one-sixth of the total cost of the cottages scheme, was the proportion required in cash at the time. They were each limited to a maximum of £100 investment to ensure a broad base for the company. The remainder was borrowed and the debt was fully paid off in 1981.

4.7.4 LOCAL INSTITUTIONAL DEVELOPMENT

This core group of people remained together, becoming a development committee and then, in December 1972, with the encouragement of a local Muintir na Tire part-time organizer, a fully fledged community council again using the Muintir na Tire model. Thus, by 1973 there were three separately constituted local development organizations in the Letterfrack area: the Credit Union, Connemara West Ltd, and what became known as Ballinakill Community Council. Ballinakill Community Council has since acted successfully on behalf of the community in several directions, for example by negotiating the designation in 1976 of 1,500 acres of local land as a national park, and especially, as far as Connemara West is concerned, in looking after the community's takeover of an important local building that came onto the market in 1974. This former industrial school, a huge Victorian building of some 40,000 square feet, was eventually purchased by Connemara West as its centre in 1978.

Connemara West Ltd today is a vigorous community-controlled company. It has a managing board of fourteen directors who are elected by shareholders from amongst their number, with a policy ensuring a degree of turnover in board membership. Apart from major decisions, such as the setting of budgets, day-to-day responsibility for projects is delegated to a series of committees, each of which has at least one director from among the members.

4.7.5 CULTURE AS A RESOURCE

Following the successful construction and letting of the cottages, the company was soon able to use its income to further effect. In 1977 it built a cultural centre at Tully - The Teach Ceoil - for £17,000. This provides classes in traditional music, song and dance throughout the year which it supports through income generated by summer events such as a traditional Irish cabaret, which attracts visiting tourists. Then, in 1978, the company purchased the industrial school at Letterfrack for £21,000; this has since become the centre of its development activities and is now known as the Connemara West Centre.

4.7.6 TRAINING

Throughout its existence Connemara West has endeavoured to tackle the related issues of employment provision and training. After repairing and converting the centre building between 1978 and 1981 the company let and serviced craft workshops within the centre complex as a first step towards its goals of creating employment and providing training.

This policy was not completely successful. Because of the isolation of the area from markets and the subsidized competition provided by state agencies elsewhere the first craft industries that came did not have the stamina to stay. Furthermore, they failed to provide the training in craft skills that Connemara West had hoped would result from their arrival. They tended to be independent of local labour requirements, with little prospect of growth in this direction. When this became clear, the company decided to change its policy. It encouraged replacement enterprises as the first ones started to leave, and began helping them by providing comprehensive back-up services and publicity. In addition, the national Industrial Development Authority was invited to take over responsibility for five workshop units in the centre complex. A new component was then added – the design and implementation of Connemara West's own training scheme, tailor-made to suit the needs of local young people. This scheme, known as the Craft Training Programme, took advantage of a government intiative introduced through the then newly established Youth Employment Agency (YEA). Through the agency community organizations became eligible to act as sponsors of government-funded youth training schemes. Connemara West's proposals to launch its new programme fulfilled the agency's requirements very well. There were a large number of young unemployed people in the area, many of whom had poor academic qualifications but unfulfilled potential in other directions. Local opportunities for employment were extremely limited. The Craft Training Programme at Letterfrack in fact offered a high standard of training in woodworking skills and design as well as business management. From Connemara West's point of view what was especially attractive about the Youth Employment Agency's backing was that it could take full responsibility for implementing the programme, using its own staff and premises and directly employing a skilled craftsman of its own choice as supervisor. It could also develop training techniques it thought suited to the locality. Fourteen trainees, all local and previously out of work, completed the course in June 1985 to a high standard, assessed by the City and Guilds Institute of London and the Regional Technical College, Galway (see *Report on the Training Programme in Wood* published by Connemara West and including external assessments). Evidence of the course's achievement remains at the Connemara West Centre, both in the form of well-designed, well-made products and a new woodturning business set up locally by one of the trainees in a centre workshop.

It was therefore particularly frustrating that negotiations begun at least a year before the course's close for the purpose of repeating this approach should initially have proved fruitless. Ideally a follow-on scheme should have started in the autumn of 1986 with a period beforehand for the recruitment of trainees. This would have meant that the same two skilled trainers could have been retained by the scheme and continuity ensured. As it was, because of a long series of unsuccessful discussions with various agencies which might have been able to fund at least an element of future programmes (the YEA having early on indicated that it had only funded the first scheme as a *pilot* programme and could not continue it), Connemara West found itself falling between their individual stated responsibilities; for some agencies the programme was too educational, for others too vocational. They mostly agreed that the programme was of high quality but could not agree on a funding package. The result was that despite Connemara West's intense lobbying and detailed submissions the programme had to be postponed and the trainers laid off.

Connemara West was able to maintain considerable pressure for the scheme to be revived a year later by allocating some staff time from the new Community Resource and Education Project which it had launched in November 1985. Lobbying and negotiations were kept up through the year with the result that at the very last minute, late August 1986 , agreement came through for a revised two-year Fine Woodworking and Design programme, starting in January 1987. The funding for the programme is from the education sector, channelled through the Regional Technical College at Galway and drawing upon European Social Fund contributions. The lateness of the decision made recruitment of trainers and trainees particularly difficult, but the programme is now well established with fifteen trainees (aged 17–23) recruited from the two counties of Galway and Mayo. Despite the enormous difficulties of negotiating the new scheme, which Connemara West's staff say made them as community workers feel 'marginalized', having to repeat basic negotiations over and over again, there is a degree of satisfaction that the new basis is educational which they feel better reflects the nature of the programme and should allow for simpler repetition if the present scheme again proves successful.

The educational base from which it is now operated and the shortened time scale have led to a reshaping of the programme with more emphasis on design skills. As Connemara West's press release marking the scheme's launch puts it,

> The central aim of the course is to train young people to become highly skilled and creative craftspeople in wood. Specifically, the course will aim at equipping students with the highest possible proficiency in a range of core woodworking skills (cabinetmaking, woodturning and allied skills) and training them in the theory and practice of basic design skills ... this course gives young people from the area a unique access to third level education and training within their own community which they would not otherwise get ... nationally it fills a major gap in this type of training. There has been very little emphasis on intensive structured training in fine craftsmanship in wood combined with training in experimentation and innovation in relation to techniques, processes and products opposed to conventional training either in machine operation or in straightforward carpentry/joinery for the construction trade.

Connemara West now faces two tests of this current scheme: firstly, a measure once again of the quality (an especially challenging test given the precipitate start); and secondly, whether, if successful, a continuation can be smoothly renegotiated without the fairly desperate attempts that have had to be made to stitch together a package of financial support from different agencies. That the essential integrity of the training scheme has not been compromised during these protracted negotiations is very much to the credit of Connemara West. Asked whether they would favour the introduction of a national community development agency to ensure continuity at times like this, the staff's view is that it would be more helpful if existing agencies would recognize community development and take notice of the track record and situation of such local development groups as their own; they could then contribute to mainstream training activity rather than be pushed to the side-lines until it suits one or another agency to work with them. They point out that a small local development group requires continuity and a sense, when it is successful, of having established some ground for it to be able to play its part in the development of what is clearly a disadvantaged area.

4.7.7 COMMUNITY EDUCATION

By contrast with the uncomfortable launch of the woodworking and design programme, Connemara West's other major new initiative, the Community Resource and Education Project, was launched successfully in November 1985 and is able to draw entirely upon Connemara West's experience in the field.

Funded under the current European Combat Poverty Programme it will initially have a life of four years with three main stated aims:

1. to encourage and assist in the development of the resource base of the community with community participation in planning and management;
2. to investigate and promote educational, cultural, information and training opportunities and services, some of which may lead to economic spin-offs;
3. to increase awareness within the community of issues and activities concerned with development.

In a sense then it is tailor-made to continue Connemara West's general development role in the area, being broad enough in scope to accommodate new work beyond the present core activities. As part of its first year's operation it has run a wood-sculpture symposium, successfully contributed to the launch of the wood-skills programme (see above), completed and published a very substantial community-information directory, hosted the first meeting of the ten integrated rural projects within the Combat Poverty programme Europe-wide and organized local evening classes.

The interesting thing about the programme, which illustrates Connemara West's ability to attract external funding for local development work, is its place within the historical development of Connemara West's activities. It can be traced back directly to a previous local community action project funded under the first European Combat Poverty Programme (1975–80). Unlike the first scheme, the current project is locally managed, modestly funded and more directly connected with the other rural projects within the programme. Following the completion of the first scheme, Connemara West hosted an international conference in 1984 to discuss the shape of a successor. Thus, their part in the present programme is the product of steady, coherent contact with the European programme over a number of years, a programme which to some extent has been shaped by local development organizations. This is in sharp contrast to the various national schemes with which Connemara West has dealt.

4.7.8 CENTRAL SERVICES FOR SMALL ENTERPRISE

At the core of Connemara West's experience of local development lie two elements. First, a considerable voluntary commitment on the part of its board of directors, who both look after projects individually (through the delegated committee structure already mentioned) and develop, through their collective experience, a coherent philosophy of development against which they can measure proposals as they arise. This first element is complemented by the second, an increasingly well-equipped and confident office and paid staff base for current and future activities. Services provided by the office for enterprises at the Connemara West Centre include typing, telephone, telex, photocopying, duplicating, binding and an impressive display area. Combined with these facilities, it is a feature of Connemara West's development that its staff have shown great loyalty and commitment to the project, helping to maintain continuity through the more difficult times.

4.7.9 CONFIDENCE AND DEVELOPMENT OF SELF-RELIANCE

A test of the extent to which their work over the years has succeeded in countering dependency came in 1983 when the local agricultural supplies store closed. Local farmers realized how vulnerable they might become if private enterprise were to fill the vacuum created by the closure. Historically, they had seen the damage that could be done in such a remote area as theirs by uncontrolled market forces; the higher prices, poorer service and dependency they could expect as a result. Crucially, they now felt sufficiently confident to explore alternatives. They therefore acted very quickly, calling a public meeting at which there was a very large measure of agreement that the community should set up its own farmers' co-operative and a committee of twelve was formed, publishing a leaflet explaining their decision. Within three weeks of the public meeting, collecting points in local primary schools were set up. In a single day around £8,000 share capital was raised, which, together with the additional help of local development groups, provided a sufficient sum to enable the store to be established and stocked and a fulltime manager appointed. One hundred and twenty local farmers became founder members of the co-operative and the new store opened within six weeks of the closure of its predecessor.

4.8 CONCLUSIONS: SOME LESSONS FROM THE FOUR CASE STUDIES

The co-operative and community development organizations highlighted by the four case studies were established with a range of objectives and hopes. All regarded the generation of new sources of income and employment as important. In most cases the provision of new or improved local services or, in the Irish case particularly, infrastructure, was also important. But so too were the broader objectives of furthering self-reliance and local involvement in development processes.

One of the founder members of the West Kerry Co-operative, its first chairman, Doncha O'Conchuir, recently concluded a summary of his view of the progress CFCD has made over the years with the following quotation from George Kent: 'People become weak by acting weak. In much the same way people can gain power by acting as if they had power'. In general the case studies tend to support this view; in the Scottish cases the level of interest and involvement in local development issues has increased as a result of the *co-chomunn*. That is not to say that all such initiatives have been successful in financial terms – clearly they have not. Mistakes *have* been made, as much by supporting state agencies as by local communities, although blame tends to focus on the local personalities involved or to be subsumed in a general critique of local development initiatives. In fact, community organizations often have to mould their ideas for action into an inappropriate and often 'unfriendly' system of state support.

In appraising the experience of the four case studies the following main issues seem to be important:

- local action and the role of the state;
- the vulnerability of local markets;
- the importance of education and training;
- management style and structure;
- local, second-tier and external networks;
- dependency relations.

4.8.1 LOCAL ACTION AND THE ROLE OF THE STATE

The role of the state in seeking to encourage the development of community co-operatives in the Scottish Highlands and Islands directly through the HIDB's programme of advice and financial support contrasts markedly with the Irish case, where direct community initiatives have sought to adapt their objectives and activities to a variety of support schemes. Beyond that basic difference, there are many similarities.

In their attempt to become a network of local development offices, a first port of call through which state intervention is directly channelled into their areas, it would appear that the Irish community co-operatives have not succeeded. The evidence strongly suggests that state agencies establish lines of communication quite separate from the co-operatives, and indeed can undermine a core activity of a co-operative by directing support elsewhere. For example, although predictably finite, the end of CFCD's land reclamation work was hastened by grants paid to competing private-sector operators. The absence of legislation to improve the legal definition and working practices of the community co-operatives is also striking. Efficient practices such as the creation of subsidiary companies have had to be learnt through experience. However, as the perception grows that the current recession is national in scale, that Ireland as a nation feels the pressure again of being peripheral to world markets and economic indicators show serious problems on the east coast as well as in the west, recognition of the special needs of remoter western communities is becoming increasing scarce. This in turn reduces the potential for community development activities. What apparently keeps community development activity alive is the continuing involvement and commitment of those at the local level who have interests and expectations in this direction, not some national policy built upon earlier beneficial experiences. In fact, long-term community development aspirations appear to be expressed increasingly through educational programmes and local associations rather than directly through economic policy.

Although not set up through state initiative, the Irish co-operatives have been highly dependent on indirect state support. In the case of CFCD the future interaction between their own efforts and state aid seems less clear, although they will continue to channel resources into the area through the Irish-language colleges and computer training schemes. In seeking another major activity they will continue to emphasize the inherent qualities of self-determination within their experience. The manager of Connemara West sees this in practice as meaning that the State should respond to local initiatives.

> Most people feel the state could and should do a lot more, that would generally be the view from a local perspective. In our case, we have felt from the beginning that the onus was on people here to take the initiative, and having taken the initiative to ask the state to respond by complementing what we were doing, rather than asking the state to start something and we would assist them in the process. I think it is an important distinction. (Kieran O Donohue, personal communication)

Recent experience demonstrates that community development organizations do not operate in a vacuum and indicates the pronounced effect of national and international economic trends upon their performance. National pressures, such as increasing levels of unemployment or emigration, bear heavily. Both CFCD and Connemara West have been able to adapt their policy to take account of these trends and have alleviated some of the local effects of adverse conditions. Connemara West's use of the Youth Employment Agency's financial support for their craft training programme has both directed resources towards a part of the national unemployment challenge and done so in a way which enhances community development. In the absence of a community development perspective in the working of the main national agencies and government departments, this *ad hoc* or opportunistic shaping of their schemes by local development groups is probably the most that can be achieved.

In Scotland, too, the role of the co-operatives has been more limited than some local activists had earlier hoped. No direct role was given to them in relation to the Integrated Development Programme for the Western Isles, which could have greatly strengthened their position as local level development institutions. State support has also been ambiguous. Despite the larger social role envisaged for co-operatives, the tendency has been to stress financial returns and the need to compete in the local or external market on equal terms with private industry.

For most of the period under review, the HIDB were learning just as much as the staff and committee members of the *co-chomunn*, yet this symmetry was never officially recognized. The tendency was to blame mistakes on 'managers', 'lack of training', 'committees' and even nefarious practices of one kind of another. This is illustrated by the fact that the HIDB inadvertently led Ness into a problem of high overheads in relation to the sale of the business by offering administrative grants – part of the package of support measures originally offered by the board. The HIDB is now very reluctant to offer this grant to a new *co-chomunn*.

Co-operatives are but one of many 'meso-level' institutions which lie between the individual and the state and 'macro-level' institutions. As Bryden (1987) points out, this level of social action is one on which little useful academic enquiry has been conducted, and the tools to analyse the interrelation with the state are underdeveloped. Nevertheless, it is clear that too great a dependence on state funding both leads to vulnerability and constrains room for manoeuvre.

Although local government has been relegated to an extremely minor role in Ireland, in the Scottish cases the creation of a much more 'local' and cohesive all-purpose Islands Council in 1975 clearly gave much impetus to the community associations. Further, by steering contracts to the *co-chomunn* the council was able to help provide early markets: road works in Pairc and housing in Ness. Finally, by its active role in developing the Craft Guild and Knitwear Association, the council helped to build important local networks which were and are instrumental in marketing and possibly also in ensuring that design problems are raised and tackled.

Local markets are therefore a natural focus for community enterprise – the interests of members as investors, employees and consumers can be made in some degree coincident and customer loyalty might be expected. In both Scottish and Irish cases, however, there has been some understandable pressure to avoid competition with existing local private sector activities. The co-operatives (a good example is Aran Mor Co-operative in Ireland) have tried to fill the gaps in demand and supply which exist. They have also sought to provide for social needs not met by private enterprise; this has usually brought immediate tangible benefits to local communities. However, on the whole, such activities tend to be the least profitable and the interests of the consumers/members is to keep margins as tight as possible.

4.8.2 VULNERABILITY TO LOCAL MARKETS

Local development seeks to 'capture' as many of the 'levers' of change as possible. Local markets and control over local resources tend to be regarded as important. Of our two Irish examples, one (CFCD) has benefited from relatively stable professional management over the years and the other has chosen to share the direction of work more obviously between voluntary and paid members. Both have resisted any attempt to impose a management structure upon them and have thought long and hard about the meaning of self-management. None the less, economic decisions relating to the running down of a job-creation service (e.g. CFCD's land reclamation) or the future of an asset that requires reinvestment (such as the Tully Cross cottages) cannot be decided purely on economic grounds.

In the Scottish cases, the agricultural supplies enterprises at Ness and Pairc and the local contracting services provide similar experience. The markets for farmed fish and knitwear are less personal and less likely to turn up and vote at meetings!

4.8.3 EDUCATION AND TRAINING

The cases demonstrate that at the local level as well as in the wider arena it is in the area of education and training that one clue as to future directions emerges. There have been notably successful examples of 'self-education', including exchange visits between communities involved in developmental approaches, and broadly educational programmes within communities, and these have given those involved with community development activities a realistic view of the challenges they are likely to face. Although not nearly enough of such training opportunities are thought to be available and funding for them remains scarce, their potential has been demonstrated and there is a determination that they should continue. Relationships between national or European programmes and local action (in this respect a more favourable treatment appears to be available at present from a European programme) may well continue to be difficult. Responsibility for maintaining continuity and a coherent approach, especially links between communities, remains in the meantime at the local level and assistance should be made available at the very least to those established local development organizations with a demonstrable track record of success.

An outstanding example of the impact of this educational approach has been the work of University College, Galway. It has built up a strong involvement with communities in the west, including Connemara West and other rural communities involved with the current Combat Poverty programme.

For several years the college has been offering local community development courses, the contents of which include the 'nuts and bolts' of successful local action and the translation of case studies from one context to another. This is itself an example of effective local organization. More recently it has launched a Master of Rural Development course, which aims over a two-year period to enhance the skills of people directly involved in development activities. Unlike the earlier courses whose entry is unrestricted, the masters course is looking for people with both proven academic abilities and at least three years' practical experience. In fact it is a measure of the authority of the college in this field that its first intake for this course (1985–7) included people with considerably more experience than this, who were interested in the opportunities the course offered them to extend their involvement. The Gaelic college in Skye, Sabhal Mor Ostaig, has also sought to emphasize the community dimension of local development in its courses, although it has not yet been able to develop an outreach programme comparable with that at UCG.

Nevertheless, the Scottish experience of training courses for managers and management committees suggests that it is probably much more important to learn from the experience of other similar businesses than from 'experts' brought in from outside. The training schemes and support services put into the co-op by the HIDB (including intensive 'after care' by their own Management Services Unit) failed to have any noticeable impact on the key problems. On the other hand, it is evident that other co-ops did share some of these problems, especially in the Western Isles.

These failures partly reflect weaknesses in targeting the problems and individuals involved, partly the fact that training courses were sometimes designed for other purposes and partly the general problems of delivering training into remotely located groups and organizations. Both Scottish cases show that much learning was required, just as much by staff of the board as by the managers and committees of the *co-chomunn*. This key fact was not fully appreciated by either side, and insufficient resources were applied to this essential learning process.

4.8.4 MANAGEMENT STYLE AND STRUCTURE

The structure of Co-chomunn Nis demonstrated that management reporting and accounting needs to be more rigorous than is necessary in other small businesses – even where these are multi-enterprise businesses – because no single person is continuously responsible in the sense of having 'mud on his/her boots'. The general 'feeling' of how the business is running, which most small entrepreneurs rely upon heavily, is absent. Community businesses have important strengths, but this is a key weakness. Although interpreted as a weakness of training or experience, in reality it was the accounting and reporting *systems* which were fundamentally inadequate. Managers in such businesses need to develop sales, customers, good suppliers; they are not, and probably should not be, clerks or accountants. The established means of record keeping, management reporting, budgetary control and so on were totally inadequate for the specific needs of small multifunctional enterprises, and this situation seems still to prevail a decade after the HIDB's scheme was started.

Relations between manager and management committees have also been a serious problem in Ness and Pairc. Connemara West, however, faced the issue squarely. It has developed sufficient confidence and independence within the local community to be able to negotiate possible activities or enterprises with which it may become associated from a position of strength. This is partly due to its policy of resisting the potential watering down of its directors' voluntary involvement as its activities grew. The manager of Connemara West sums up this key element of their philosophy in this way.

> As enterprises got off the ground and became more complex management-wise, we saw that we had two main options in terms of efficient management. One was to employ more and more professional, full-time staff and therefore have a larger central office; that's the way most groups respond. We saw that way fraught with danger, in that what tends to happen is that professional managers run away with the whole policy-making function of the enterprise. It can then become more and more divorced from the community and the people who effectively have set it up, managed it and given that emotional or moral commitment to it and whose support is essential in the future. We wanted to avoid that. We also wanted to avoid a central hierarchical system of management where you have supervisors supervising workers, and managers supervising supervisors and a general manager supervising managers. It becomes a crazy system in this kind of development which is really about people. So we felt as an alternative that what we would do was to develop a stronger committee system which demanded more voluntary work by Board members; it does involve a fairly heavy commitment of time for some Directors, depending upon the projects they are in, but they clearly recognise that this is an option we have chosen. (See Scott, 1985, p. 38)

Added to the evident commitment of the directors, the Connemara West office is a centre of activity and creativity. In this way a 'peripheral' community is redefining its relationship with the outside world.

4.8.5 LOCAL, EXTERNAL AND SECOND-TIER NETWORKS

Local networks in a dependent society, accustomed to exporting primary products and external decision-making at political and administrative levels, tend to be weak. Local authorities and development agencies have an important role in helping to build such networks. Locally based organizations, on the other hand, may have weak links to external markets. Such wider networks are also crucial and must be developed by associations or similar more formal institutional developments.

The development of associations or confederations of community co-operatives has been important in both Ireland and Scotland. The Association of Community Co-operatives in the Highlands and Islands was some time gestating. In Ireland it took the 1983 crisis to create a Confederation of Community Co-operatives which has played a representational or lobbying role *vis-à-vis* the state.

It is difficult in the 'periphery' to separate 'economic' from cultural and social activities and changes. The impetus behind community action has had a powerful economic motivation. However, the action taken has also been related to the creation of local institutions, the strengthening of local network structures both formal and informal and the development of the media. In both the Western Isles and the West of Ireland, the advent of local radio (in Ireland it was originally a 'pirate' radio station broadcasting in Gaelic) has been crucial, enabling a certain focus on local issues and concerns, a certain very sophisticated handling – usually in Gaelic – of issues to be mediated with the 'external' world. Educational efforts, formal and informal, are beginning to become a central focus of local development efforts in Ireland and Scotland.

The effectiveness of local networks is related to the nature of the local 'community', a difficult and often ambiguous term. In Scotland Ness was not a single 'community'. The 'community solidarity' which was said to exist at the start of the project was in reality less than solid. The state, through the HIDB, supported those in the community who *wished* to bring about change and saw in the *co-chomunn* the opportunity to provide a vehicle for this change. Many were younger people seeking ways of staying within a culture and environment which was and remains at least as important to them as financial gain. They were dedicated, enthusiastic and intelligent. But they did not and probably do not represent the majority. Without such a group with such vision, a community is more likely to die than to live. However, in dealing with a *business* which relies primarily on *local* markets any consequent divisions – be they based on attitudes, social class, geography, religion or whatever – will limit the extent of those (already small) markets, while local rivalries and cliques will limit the effectiveness of local networks.

Since the development of effective meso-level institutional structures is evidently crucial for the success of local development efforts, greater efforts to understand them seem an urgent priority.

4.8.6 DEPENDENCY RELATIONS

The importance which local communities place on controlling local resources is highlighted by some of the attitudes to fish farm development in both the Western Isles and the west of Ireland. Development agencies – Udaras na Gaeltachta and the HIDB – are planning such development on fish farming and although both have tried to involve local co-operatives and individuals the big investment pressure comes from international and multinational private enterprise. The incorporation of Co-chomunn Pairc into multinational capital has not been without rancour and debate. In an Irish case, 2,000 signatories were obtained for a petition *against* an application for fish-farming development by private enterprise. The fear is ostensibly to do with secondary effects on traditional fisheries etc., but underlying this is a general concern about loss of control over what looks – at least in the short term – to be a profitable resource.

Dependency relations of different kinds developed in the two Scottish cases. In Pairc they are with 'big capital' and the agreements reached can only be assessed in years to come. In Ness the dependency relations were with the HIDB, especially its management grant. Other evidence suggests that only very strong co-operativist philosophy can counter or control such 'dependency' trends or tendencies – and perhaps that philosophy requires nurture and support.

4.8.7 FINAL REMARKS

As Bryden and Hart (forthcoming) point out, the 'meso-level' of institutions between the state and the individual has been largely ignored by the social sciences since classical political economy. This level is clearly vital – as our case studies demonstrate – to the understanding of local development processes.

NOTES

1. See Hunter 1976. Crofting is a special type of land tenure with small areas of arable ground being held individually and larger areas being held in common. Pluriactivity is a common feature of crofting householders.
2. Rich fisheries resources were, typically, regarded as a national or EC resource. Local control through a regional fisheries strategy was rejected by Dublin and Westminster.
3. *Gaeltacht* regions are those officially designated as Irish-speaking. Situated mainly on the north-west and south-west coasts of the Republic, their communities are eligible for State aid specifically related to the maintenance of the Irish language. Such assistance is mostly channelled through the Gaeltacht Department (Roinn na Gaeltachta) and the development authority Udaras na Gaeltachta (formerly Gaeltarra Eireann).

REFERENCES

Alexander, K. (1979), *The Work of the Highlands and Islands Development Board with Particular Reference to the Role of Education and Training*, The Arkleton Lecture, 1978, Langholm: Arkleton Trust.

Bassand, M., Brugger, E. A., Bryden, J. M., Friedmann, J., and Stuckey, B. (1986), *Self Reliant Development in Europe*, Aldershot: Gower.

Bell, C. (1987), 'Co-Chomunn Nis – its failure and success', *Community Enterprise*, no. 4.

Breathnach, P. (1983), *Rural Development in the West of Ireland: Observations from the* Gaeltacht *Experience*, Occasional Paper No. 3, Maynooth: Department of Geography, St Patrick's College.

Breathnach, P. (1984), 'Popular perspective on community development co-operatives, findings from surveys in Galway and Kerry', paper to Society for Co-operatives Studies in Ireland conference, Department of Geography, St Patrick's College, Maynooth.

Breathnach, P,, and Regan, C. (1981) *State and Community: Rural Development Strategies in the Slieve League Peninsula, Co. Donegal*, Occasional Paper No. 2, Maynooth: Department of Geography, St Patrick's College.

Bryden, J. M. (1979a), 'Core periphery problems: the Scottish case', in Seers, Schaffer and Kiljunen, op. cit.

Bryden, J. M. (1979b), *A Speculative Look at Prospective Developments in Co-operation in the Scottish Highlands and Islands*, Oxford: Plunkett Foundation.

Bryden, J. (1981), 'Agriculture and regional development in Europe: the case study of the Scottish Highlands and Islands', invited paper for the Third Congress of the European Association of Agricultural Economists, Belgrade, September 1981. Published in *Manchester Papers and Development*, 1982, and in the *European Review of Agricultural Economics*, vol. 9, 1982.

Bryden, J. M. (1987), 'Crofting in the European context', *Scottish Geographical Magazine*, vol. 103, no. 2.

Bryden, J. M., Commins, D., and Saraceno, E. (1985), *Education, Training and Rural Development. Summary Report on a Collaborative Programme between Rural Areas in Italy, Ireland and Scotland, 1982–83*, Langholm: Arkleton Trust.

Bryden, J. M., and Hart, J. K. (forthcoming) *Land Use and Economic Development*, Balbury-Longman.

Bryden, J. M., and Fuller, A. M. (1987), 'Pluriactivity as a rural development option: the emerging policy and research agenda', paper for the European Association of Development and Training Institutes General Assembly, September.

Clarke, R. (1982), *Our Own Resources – Co-operatives and Community Economic Development in Rural Canada*, Langholm: Arkleton Trust.

Commins, P. (1981), *State, Co-operatives and Language in the Gaeltacht*, Dublin: Economics and Rural Welfare Research Centre, An Foras Taluntais.

Commins, P. (1983), Development in less favoured rural areas: problems, policies and emerging issues in Ireland', paper given to Arkleton Trust Stornoway Conference, May/June.

Committee on Language Attitudes Research (1975), *Report: Oifig Dhiolta Failseachain Rialtais*, Dublin: CLAR.

Crotty, R. (1979), 'Capitalist colonialism and peripheralisation (the Irish case)', in Seers, Schaffer and Kiljunen, op. cit.

Curtin, C., and Varley, A. (1984), 'Marginal men? Bachelor farmers in a west of Ireland community', paper presented to the 1984 annual conference of the Sociological Association of Ireland, available from the authors at the Social Science Research Centre, University College, Galway.

Curtin, C., O'Donohue, K., and Varley, A. (n.d.), *Leadership and Brokerage: Community Councils and Community Development in Two West of Ireland Rural Communities*, Galway: Department of Political Science and Sociology, University College.

Gaeltarra/SFADCO Working Group (1971), *An Action Programme for the* Gaeltacht. A government-commissioned review, with recommendations for a comprehensive development programme for the *Gaeltacht*, Galway: Gaeltarra Eirean.

Henry, M. (1983), *Comharchumann Forbartha Chorcha Dhuibhne* (Community co-op with flair and purposes), 84 Merrion Square, Dublin: Plunkett House, pp. 35–9.

Herlihy, J. (1984), 'Community development courses: the Donegal experience', paper presented to the 1984 European Seminar on 'Rural Poverty – The Problem and the Challenge', Connemara West Centre, Lettefrack, Galway, 13–15 September.

Hughes, J. T. (1979), 'Evaluating the work of a Regional Development Agency', a paper for a Regional Studies Association (Scottish Branch) Conference on Aspects of Development in a Peripheral Region', 28 September.

Hunter, J. (1976), *The Making of the Crofting Community*, Edinburgh: John Donald.

Johnson, M. 'The co-operative movement in the *Gaeltacht*', *Irish Geography*, vol. 12, pp. 68–81.

McCleery, A. (1982), 'The persistance of co-operation as a theme in marginal development', in Sewel and O'Cearbhaill, op. cit.

Mac Giobuin, M. (1982), 'Rural Development in Chorcha Dhuibhne', in record of Prizewinners' Addresses: Justus-Liebig Preise 1982 Agrarwissenschaftliche Fakultät der Christian-Albrechts-Universität, Kiel. Available from Comharchumann Forbartha Chorcha Dhuibhne, 1982.

O'Cearbhaill, D. (1982), *Development through Self-help – The Achievements of Killala Community: Co-operatives and Community Development: A Collection of Essays*, Galway: Social Science Research Centre, University College.

O'Conchuir, D. (n.d.), *Community Endeavour*, available from the author via Comharchumann Forbartha Chorcha Dhuibhne, Ballyferriter, West Kerry.

O'Conchuir, D. (n.d.), *The Irish Oral Tradition*, available from Oidhreacht Chorcha Dhuibhne, Ballyferriter, West Kerry.

O'Donohue, K., O'Neill, M., and Roddy, M. (1984), 'Connemara West Ltd – community development project', paper presented to European Seminar on 'Rural Poverty – the Problem and the Challenge', Connemara West Centre, Letterfrack, Galway, 13–15 September.

Pederson, R. N. (1987), 'Community enterprise in the Highlands and Islands: a decade of evolution', paper presented to the 9th International Seminar on Marginal Regions, July.

Schaffer, B. (1979), 'Regional development and institutions of favour (the Irish case)', in Seers, Schaffer and Kiljunen, op. cit.

Scott, I. (1985) *The Periphery Is the Centre. A Study of Community Development Practice in the West of Ireland 1983/84*, Langholm: Arkleton Trust.

Seers, D., Schaffer, B., and Kiljunen, M. L. (eds) (1979), *Under-developed Europe. Studies in Core–Periphery Relations*, Brighton: Harvester.

Sewel, J., and O'Cearbhaill, D. (1982), *Co-operation and Community Development: A Collection of Essays*, Galway: Social Sciences Research Centre, University College.

Storey, R. J. (1982), 'Community co-operatives – a Highlands and Islands experiment', in Sewel and O'Cearbhaill, op. cit.

Toner, J. (1955), *Rural Ireland: Some of Its Problems*, Dublin: Clonmore & Reynolds. This is a history of Muintir na Tire.

Trevelyan, M. (1980), *A Study of Community Co-operatives. North Wales Employment*, Llanrwst: Resource and Advice Centre.

United Nations (1956), *Concepts and Principles of Community Development*, New York: UN.

PART II

RESTRUCTURING IN OLD INDUSTRIAL CORE AREAS

CHAPTER 5

Local economic development in England and Wales: successful adaptation of old industrial areas in Sedgefield, Nottingham and Swansea

Peter Roberts, Clive Collis and David Noon

5.1 INTRODUCTION

Local economic development has in recent years become a major area of public-, private- and voluntary-sector activity in many of the communities of England and Wales. This rapid growth of local economic development initiatives results partly from the failure of previous policy measures, including traditional regional industrial policy, to arrest or alleviate the economic decline of many local communities; and partly from an increased awareness amongst local authorities, community groups, trade unions, chambers of commerce and other organizations of their potential role in and contribution to the design and implementation of local policy. The onset and spread of recession into previously prosperous regions has outpaced the ability of traditional regional policy to create new activities and employment. The extensive incidence of industrial decline, especially in areas dominated by primary and manufacturing activities, has also accentuated the social, environmental and political divisions which exist in British society. Although local economic problems are virtually endemic in the traditional industrial areas of the north of England and south and north Wales, the current recession has witnessed the widespread occurrence of economic decline in the West and East Midlands regions and the rapid development of significant pockets of decline in the more prosperous southern regions of England. Local economic development is now a priority area of activity in all regions and receives significant support from private business, local government, the church, trade unions, community and neighbourhood groups and financial institutions.

This new localism in economic development policy is a reflection not only of the failure of traditional regional and industrial policy, but also of a growing awareness that community development implies more than simply promoting economic growth. Benington (1985) has outlined the response of local authorities and other agencies to the challenges of local restructuring as relating to:

- the failure of inward investment to provide a sufficient or effective response;
- the need to support and develop indigenous industries;
- a shift away from a property-led approach towards direct investment in the production process;
- the growth of concern with the quality as well as the quantity of employment;
- an increasing direct involvement of local authorities with local industry and trade unions; and
- the desire, of many local authorities, to increase democratic control over work and over the economy.

Although not all forms and styles of local economic activity display all of the above characteristics, it is possible to identify certain general directions of policy and the development of measures for the implementation of policy. There has been a shift away from approaches which stress the role of inward investment and a movement towards the stimulation of indigenous activities; this has been accompanied by an increasing emphasis upon longer-term social restructuring. In parallel there has been an increased emphasis upon initiatives and upon the creation of opportunities for proactive intervention. These advances in the theory and practice of local economic development have resulted in the design and implementation of policy packages which are better targeted than their predecessors and which frequently integrate the actions of a number of authorities, organizations and agencies. In England and Wales in a typical district or county it is likely that a programme of local initiatives will be supported by:

- the EC: through, for example, the Social Fund, the Regional Development Fund, the Agricultural Guidance and Guarantee Fund, the European Investment Bank, European Coal and Steel Community and New Community Instrument loans;
- national government: through, for example, the Urban Programme, the Manpower Services Commission, nationally funded agencies such as the Welsh Development Agency, the Development Commission and various Department of Trade and Industry programmes;
- local government: including, for example, the activities of economic development, education, estates, treasurer's, chief executive's, planning and recreation/tourism departments;

- the private sector: both in terms of financial support through, for example, investments by pension funds, banks and insurance companies, and through the direct involvement of local companies and chambers of commerce in, for example, local enterprise trusts and agencies;
- other hybid organizations: often closely linked to local government, such as enterprise boards, or specially created companies formed to tackle specific problems resulting from industrial closures, for example, British Steel Corporation (Industry) or British Coal (Enterprise);
- community and trade-union organizations at local, national and regional levels, including, for example, church and voluntary sector groups.

The aim of this chapter is to provide a range of examples of successful local economic policy; these examples have not been selected as typical; rather they are indicative of what can be achieved. Much of the material contained in these case studies reflects the position outside of the metropolitan centres of England and Wales; the position in metropolitan areas is reported by Marshall elsewhere in this book. Local economic development initiatives have generally been seen to originate in metropolitan areas and spread gradually to the more rural shires (Collis *et al.*, 1987).

The first case study is of Sedgefield's National Programme of Community Interest, an example which brings together a series of actors engaged in a variety of tasks with the intention of providing a programmatic basis for the long-term development of a disadvantaged region. A second case study is derived from the experience of the Nottingham textile and clothing scheme, a sector initiative undertaken by a local authority in an intermediate-sized urban area. The final case is of the Swansea Maritime Quarter redevelopment, a linked economic and environmental initiative in an old industrial region. All of the cases reveal the importance and influence of local circumstances and actors, but they also yield certain common lessons.

5.2 SEDGEFIELD'S (NORTH-EAST ENGLAND) PROGRAMME INITIATIVE: RESTRUCTURING A PUBLIC-SECTOR CRISIS AREA

5.2.1 BACKGROUND

Sedgefield District, a constituent local authority of Durham County, is located in one of the most economically disadvantaged regions of the UK. The north-east of England has experienced both a long-term decline in its traditional basic industries (coal, iron and steel, shipbuilding and engineering) and in recent years a considerable loss of jobs in more modern industries which have responded to falling demand by seeking productivity gains and demanning. The current recession has further accelerated this process of economic decline with the resulting increase in the incidence of plant closure and redundancy and a rise in unemployment. Evidence of the disadvantaged position of Durham is provided by the European Commission's synthetic index which in 1981 placed the county at a level of 61 in relation to an EC average (nine member states) of 100 (EEC, 1984); by 1985 the Durham position remained below the average for the Community (twelve member states) at 88 (EEC, 1987). Within the county a number of districts have experienced a severe collapse in traditional employment on account of the rapid decline or closure of dominant industries, including, for example, Easington District (coal), Derwentside District (steel) and Sedgefield District (railway engineering). This continual and cumulative process of decline in all traditional industries has placed a great strain upon the financial and organizational ability of the county council. The response of the county council to any particular closure will inevitably be conditioned and constrained by the ever-increasing number of calls upon its limited resources. This places a greater degree of responsibility for the formulation and implementation of policy upon district councils and other agencies and organizations. At both county and district level the same political party (Labour) has long held control, making the process of seeking and maintaining political agreements relatively easy. It is against this general background that the Sedgefield initiative should be judged. The district experienced a significant loss of jobs in coalmining in the 1950s and 1960s, but the growth of manufacturing employment, especially at Newton Aycliffe New Town (mainly in firms attracted to the area under traditional regional policies) at least partially compensated for the loss of traditional activities. The industrial structure, in 1981, of the area which was subsequently defined under the Shildon–Newton Aycliffe–Bishop Auckland Programme reflects the imbalanced nature of local employment and, more importantly, the vital contribution at that time of the Shildon Wagon Works. Table 5.1 indicates the high proportion of employment in manufacturing and the extremely underdeveloped nature of service employment. Unemployment in the area increased steadily throughout the 1970s at a level above both national and regional averages, and by 1985 was, at 26.8 per cent, some 10 per cent higher than the GB rate and 3.7 per cent higher than the average for the Northern Region. The closure of the Shildon Wagon Works of British Rail Engineering Limited (BREL) precipitated a jobs crisis in the local economy: in 1978 there were 4,900 jobs in Shildon; the loss of 2,600 jobs at the BREL works and a reduction of employment in other activities caused total employment to fall to 1,800 by 1984 and unemployment to rise to over 35 per cent. The closely interconnected labour markets of the adjacent small towns of Bishop Auckland and Newton Aycliffe, where other job losses had also been experienced, were also directly affected by the closure of the Shildon Wagon Works. Taking into account the job losses at BREL, together with other smaller plant closures and job losses in Shildon, Bishop Auckland and

Table 5.1 Industrial structure: employed residents by industrial sector

	Programme area		*Northern region*	*Great Britain*
	Numbers employed	*% employment*	*% employment*	*% employment*
Agriculture	360	1.2	2.1	2.2
Energy and water	360	1.2	5.4	3.1
Manufacturing	13,490	44.6	28.7	27.0
Construction	2,380	7.9	7.4	7.0
Distribution and catering	4,690	15.5	18.4	19.2
Transport	1,460	4.8	5.7	6.5
Other services	7,510	24.8	31.3	34.0
Total	30,250	100.0	100.0	100.0

Note: Totals may not sum owing to rounding.
Source: Census of Population.

Newton Aycliffe, it was estimated in 1985 that the area required over 4,300 jobs to be created immediately if the rate of unemployment was to be reduced to the national average (Sedgefield District Council, 1985). The tasks of economic restructuring and job creation were seen as vital if the prospects for the unemployed were to be improved from the position in 1985 when there were over forty unemployed persons competing for each vacant job.

5.2.2 RESPONSE AND ORGANIZATION

North-east England generally, and especially the county of Durham, has experienced the effects of the full range of central-government regional policy measures: the earliest assisted areas were designated in the Durham coalfield in the 1930s, the first major modern industrial estate was established at Team Valley near Gateshead and three New Towns (Washington, Newton Aycliffe and Peterlee) have been designated within the county. Other traditional regional policy responses have also been implemented, including the improvement of transport infrastructure, the diversification of the region's industrial base through the attraction of mobile industry and attempts at service sector stimulation through the decentralization of central-government departments, and the support of a major programme of environmental improvement. Even this range of measures proved insufficient to meet the major challenges of restructuring that faced Sedgefield District in April 1982 when BREL announced its intention to close the Shildon Wagon Works. This threat of closure prompted a range of responses from local authorities, trade unions, central-government departments and agencies, BREL and a variety of other organizations. The problems of the area were debated in the House of Commons where the government promised that 'all government agencies which can help will be working with BREL and others locally to stimulate new jobs' (Local Authority Associations, 1986). Although the local authorities, the trade unions and other groups continued to fight the proposed closure (Shildon Joint Shop Stewards Committee, 1983), Sedgefield District Council also started to plan a number of major local economic development initiatives in the event of the closure taking place. In April 1983 the Shildon Action Group was formed with representatives from BREL, government departments of Trade and Industry and of the Environment, the Manpower Services Commission, Durham University Business School, the Confederation of British Industry, trade unions, Durham County Council, Shildon Town Council and Sedgefield District Council. The district council prepared, for the Action Group, the Shildon Interim Action Plan in 1984 which laid a basis for the coordination of policy initiatives and provided a strategic framework for other measures. A number of specific actions emanated from the work of the Action Group. Three detailed studies were prepared: a business survey, an action plan for the redevelopment of the former Wagon Works site and a study of the potential for economic development initiatives. These studies proved to be of considerable value both in the preparation of the subsequent programme bid to the EEC and in the development and implementation of other economic and social measures. Chief among these other measures was the creation in October 1984 of the Shildon and Sedgefield Development Agency (SASDA), a limited company jointly established by BREL and Sedgefield District Council. SASDA operates a series of economic development schemes in tandem with the district council and has also developed a close working relationship with the local business community as was suggested in the study of economic development initiatives (JURUE, 1985). Other complementary measures include the provision of new industrial floorspace by English Estates (a central-government-sponsored agency which is responsible for the provision of advance industrial premises), the development of a number of local training initiatives, the provision of a centre to help respond to the social and community problems associated with unemployment and the creation of an information technology centre to provide high-technology industrial training (sponsored jointly by the district council and the private sector and financed on a reducing basis by central government).

The above initiatives and actions, while providing a comprehensive and clearly targeted package of responses to the restructuring crisis that faced and still faces Sedgefield, are individually less significant than the broader programme that was prepared in 1984 and is currently in operation. The programme, based upon the Interim Action Plan of 1984 and the studies of the redevelopment potential of the Wagon Works site and of economic development initiatives (which were both funded by the EC under Article 24 of the European Regional Development Fund), was prepared under EC Regulation 1787/84 (EEC, 1984b), which encourages the development of National Programmes of Community Interest (NPCI). The regulation states that a NPCI should 'be defined at national level and shall consist of a set of consistent multiannual measures corresponding to national objectives and serving Community objectives and policies'; in addition it indicates that 'such measures may concern, jointly or separately, infrastructure investment, and schemes for industry, craft industries and services and operations to exploit the potential for internally generated development'.

The major features of a NPCI are:

- it is multiannual and provides a basis for a medium-term programmatic commitment to a series of linked projects;
- it involves a number of agencies and organizations who agree to provide support for the programme on a continuing basis;
- it provides for the integration of a series of coordinated economic, social and physical initiatives.

These characteristics ensure that a NPCI is exceptionally well defined at the outset with clear and quantifiable objectives being specified and a financial programme being agreed between the partners.

5.2.3 THE SHILDON–NEWTON AYCLIFFE–BISHOP AUCKLAND PROGRAMME

The NPCI represented the culmination of a series of initiatives undertaken in the area around Shildon following the closure of the Wagon Works in 1984 (Litherland, 1987). These initiatives, which have been referred to above, were acknowledged by and incorporated within the NPCI as complementary or specific measures. The area that was defined for the NPCI covered some 100 square kilometres and contained a resident population of 70,000, while the majority of the defined area was within Sedgefield District, the western section of the NPCI, was within Wear Valley District. The whole of the NPCI was within the Bishop Auckland Travel to Work Area (the local labour market area as defined by the Department of Employment). Although the formal submission of the NPCI was undertaken by Sedgefield and Wear Valley District Councils, Durham County Council and Newton Aycliffe Development Corporation, it also included a programme of projects developed by BREL, the Opencast Executive of the National Coal Board and the regional water, gas and electricity boards. Sedgefield District took responsibility for the coordination and monitoring of the NPCI. Support for the NPCI was provided at national-government level by the Departments of Trade and Industry and of the Environment. The Commission of the European Communities agreed to provide financial support for the NPCI in November 1985 (EEC, 1985). As required under Regulation 1787/84 the submitted programme document specified a number of objectives for the NPCI. These objectives were:

- to coordinate the activity of local authorities, other public-sector and central-government bodies and the private sector;
- to increase job opportunities within the defined area;

- to provide an advance supply of industrial floorspace and serviced industrial land;
- to improve services to existing industrial sites;
- to encourage private-sector investment, especially in indigenous activities, housing and commerce;
- to improve transport infrastructure within and outside the NPCI area;
- to reclaim derelict land and improve the quality of the local environment; and
- to attract private-sector investment through environmental improvements on or adjacent to industrial areas.

These objectives were also translated into a schedule of activities for the period 1984 to 1989 (see Table 5.2), the total cost of the NPCI being estimated at £36.1 m (Sedgefield District Council, 1985). Seven major elements were identified in the programme:

- transport infrastructure, mainly road improvements but also some public-transport schemes considered important because of the low level of car ownership;
- industrial infrastructure, industrial access, sewers, water and electricity supply;

Table 5.2 Financial summary of programme

	Estimated cost in £000s						
	Total	*1984*	*1985*	*1986*	*1987*	*1988*	*1989*
Transport infrastructure	6,827	-	95	232	1,600	2,100	2,800
Industrial infrastructure	7,030	-	269	1,816	4,340	415	190
Reclamation	1,344	42	469	583	250	-	-
Waste disposal	2,184	49	80	1,223	832	-	-
Industrial development	13,903	553	1,720	3,375	2,950	2,905	2,400
Article 15 measures	4,727	276	1,203	1,120	994	567	567
Administration	130	-	30	25	25	25	25
Total	36,145	920	3,866	8,374	10,991	6,012	5,982

Source: Sedgefield District Council (1985).

- reclamation and small-scale environmental improvements aimed at enhancing the attractiveness of the area and encouraging tourist development;
- waste disposal, including a waste-treatment facility;
- industrial development, the provision of industrial floorspace within the NPCI and through the complementary activities of English Estates, and the redevelopment of the Wagon Works site;
- development of enterprise and indigenous potential, the stimulation and diversification of existing firms and the encouragement of new businesses, and the promotion of inward investment; and
- programme administration, the responsibility of Sedgefield District on behalf of the NCPI Coordinating Committee.

In total the NCPI anticipated that some 65 hectares of serviced industrial land would be provided during the programme period together with 88,000 square metres of new or refurbished industrial floorspace. An overall job target of 4,000 to 5,200 jobs was specified. However, some of these jobs were not expected to materialize until after 1989 (Sedgefield District Council, 1985). The central theme of the programme was to achieve an overall uplifting of the physical, social and economic condition of the area. One recent commentary (Johnston, 1986) has summarized the initiatives as having three distinguishing features:

- the 'coordination by the district council of the efforts of many other agencies';
- the 'pioneering use' of ERDF funds; and
- the 'development of enterprise support measures, with the aim of creating an enterprise economy in an area formerly dominated by major public sector employers'.

5.2.4 IMPLEMENTATION AND PERFORMANCE REVIEW

Although the NPCI was not approved until November 1985, the overall programme period extends from June 1984 to December 1989. In 1987 the first monitoring statement was published (Sedgefield District Council, 1987). Of the approved total expenditure of £36.1 million, some £17.1 million had been identified in applications submitted to the Coordinating Committee (see Table 5.3); actual expenditure to the end of 1987 was expected to total £12.5 million. The shortfall in programme expenditure was attributable mainly to an underspend within the factory-building sector but also to a shortage of capital available to Durham County due to a 24 per cent reduction in the county's capital allocation from central government. Progress in each individual element of the programme from June 1984 to June 1987 was variable:

Table 5.3 Shildon NPCI - expenditure summary

	Approved expenditure (1984/1989) (£m)	*Expenditure identified in applications submitted (to June 1987) (£m)*	*% of approval*
Transport infrastructure	6.83	3.06	44.8
Industrial infrastructure	7.00	5.9	84.3
Reclamation and environmental improvements	1.34	1.13	84.3
Waste disposal	2.18	2.25	103.2
Factory building	13.90	1.74	12.5
Articles 15/19	4.73	3.05	64.4
Administration	0.13	0.05	38.0
Total	36.15	17.15	47.4

Source: Sedgefield District Council (1987).

- 44.8 per cent of the approved allocation for transport infrastructure had been spent on road and public-transport improvements;
- 84.3 per cent of the industrial infrastructure total had been used in order to service new and existing industrial sites;
- 84.3 per cent of the available finance for reclamation had been committed to a variety of schemes;
- all of the expenditure on waste disposal had been allocated;
- only 12.5 per cent of the total available for industrial development had been spent because of the disruption caused by the cessation of the Newton Aycliffe Development Corporation, the reduction in the central government's capital allocations to Sedgefield District, a reduction in factory size eligible for ERDF assistance and a higher than expected private-sector participation;
- 64.4 per cent of the available funding for the support of enterprise had been utilized in the funding of small- and medium-sized firms.

In relation to the original objectives of the NPCI a high level of coordinated expenditure had occurred with certain specific achievements:

- 55 hectares of serviced industrial land had been provided or was in progress;
- 125,000 square metres of new or refurbished industrial floorspace had been created or was planned;

- an estimated 1,700 jobs had been created within the programme area with a medium-term job potential estimated at 3,000;
- unemployment had fallen and was converging with the regional and national rates;
- 90 per cent of firms assisted were locally based and private-sector investment had risen;
- the reclamation of 60 hectares of derelict land was complete or under way;
- considerable improvements had been made to the environmental quality of the area.

5.2.5 ASSESSMENT

The Shildon–Newton Aycliffe–Bishop Auckland NPCI initiative is at the leading edge of the practice of local economic development. It clearly demonstrates the potential benefits that can be gained from the coordination and integration of the actions of a variety of agencies within the context of a medium- to long-term economic development programme. The adoption of a programmatic approach implies that the contributing agencies are all agreed upon the definition of an explicit strategic direction for future action. The Sedgefield initiative is clearly not an *ad hoc* collection of assorted projects; it is an integrated programme for growth. Other benefits can accrue from the adoption of a programmatic approach: the contributing partners are *de facto* committed to a specified level of financial contribution over a given number of years; it is possible to link associated and complementary measures and actions to the main programme; and it is possible to exert considerable leverage on activities outside the programme. A number of problems can arise, including the inability of one of the partners to make a full financial contribution or genuine disagreements about the implementation of the programme. In all but the most exceptional circumstances, such problems can be anticipated and minimized. The Sedgefield initiative has been hindered by the decision of national government to accelerate the cessation of the activities of new town development corporations. In this particular case Newton Aycliffe has ben unable to contribute fully to the programme. A final lesson from the Sedgefield experience is that it is important to create a simple and accountable administrative system at the outset and to use such organizational arrangements for a variety of purposes and tasks.

This case study demonstrates the very considerable potential that exists for small local authorities to design and implement substantial programmes of economic development. Indeed, it is all the more important as an example because Sedgefield District is a small authority located in one of the most disadvantaged regions of the UK. The upper-tier authority, Durham County, was and still is beset with a multitude of economic, social and environmental problems and it fell to Sedgefield District to take the initiative to develop and implement the NPCI. This programme, with its emphasis upon indigenous development, demonstrates the vital role of the public sector in creating the climate for private-sector investment and in enhancing the social welfare of disadvantaged localities.

5.3 NOTTINGHAM (EAST MIDLANDS): RESTRUCTURING A TEXTILES AND CLOTHING AREA

5.3.1 BACKGROUND

Nottingham, a city with a population of over half a million, including its suburbs, lies on the southern edge of one of the UK's most productive coalfields and is also at the centre of the East Midlands textile and clothing industry. The city is also well known for its tobacco (Players), pharmaceutical (Boots), cycle (Raleigh) and telecommunications (Plessey) industries. Changing market conditions and associated restructuring have highlighted the dependence of Nottingham upon a limited range of industrial activities and in recent years substantial job losses have occurred in manufacturing industry. These reduced employment opportunities have only been partially offset by the expansion of the city's service functions.

The textile and clothing industry is a good example of how rapidly fortunes change, although the fundamental problems of the industry are deeper rooted. Between 1971 and 1981 it is estimated that in the Greater Nottingham area a total of 13,500 jobs were lost in the industry, leaving a total employed workforce in 1981 of 17,200. This sector nevertheless still accounts for some 24 per cent of all manufacturing employment in the city and 11 per cent of total female employment. The main threat to the local industry has come from overseas competition, particularly the Far East and increasingly other EC countries. In 1971 import penetration into the UK clothing market was some 11 per cent; by 1984 this figure had risen to 32 per cent.

The problems faced by the Nottingham textile and clothing industry have been the subject of detailed analysis and debate at the local level. This has led to the introduction of a range of initiatives designed to assist in restructuring industry in the city. Locally based measures to address the problems faced by the textile and clothing industry in Nottingham reflect a strong commitment to intervention by the city council and a growing willingness of Nottinghamshire County Council to pursue initiatives within the city.

5.3.2 RESPONSE AND ORGANIZATION

Political changes in 1979, with the election of a Labour-controlled city council, marked a significant turning point in the form and nature of economic development policy within the city. Previously intervention in the local economy by the Nottingham City Council had occurred primarily through the City Estates and Treasurer's Department. The development of industrial land and premises and some modest financial-assistance schemes were the main areas of activity. The establishment of a new committee and organization structure for economic development, an increase in the funds made available and a greater political commitment marked a significant shift in policy aimed at assisting and regenerating the city's economy. New opportunities were also provided through changes at national level: additional powers and finance became available following Nottingham's designation in 1979 as an Urban Programme Area under the 1978 Inner Urban Areas Act.

In developing new initiatives the city council was concerned to ensure that their scarce-resource base, which is largely reliant upon the product of a 2p rate (local tax), was used effectively to deal with the problems of the local economy through properly targeted initiatives in key sectors such as clothing and textiles. Therefore, in order to inform the development of a local economic strategy, the city council in 1982 commissioned a study of the local economy (Trent Polytechnic, 1983). The terms of reference were:

- to provide immediate guidance to the city council in their task of formulating, coordinating and implementing policies to relieve existing levels of unemployment in the city; and
- to ensure mutual understanding of the characteristics of changes in the city's economy leading to a framework within which the Economic Development Sub-Committee can evaluate policies to promote employment growth.

This study formed the basis of the local Labour Party's manifesto for the city elections in 1983, which contained the following proposals:

- the creation of a major new unit to coordinate economic intervention within the city council accountable to a new Employment and Economic Development Committee;
- a sectoral approach to intervention based on detailed analysis of key industries and the development of clear objectives;
- community-based employment initiatives targeted at the most vulnerable groups within the city's workforce; and
- the creation of an arena through which the city council, trade unions and the community could confront key issues.

A strong research base was therefore seen as an essential element in underpinning local policy initiatives. The study identified a number of key sectors upon which the city council's activities should be focused. Textiles and clothing was a sector highlighted for specific attention and where an industry-wide approach was considered appropriate. The main problems facing the textiles and clothing sector were identified by the report:

- the central significance of small- and medium-sized undertakings;
- cumulative underinvestment in many undertakings;
- poor working conditions and poor wages;
- the failure to introduce new technology;
- the need for a greater design input and greater specialization in production;
- the need for effective marketing strategies and cooperation between producers;
- the over-dependence upon a small number of major retailers;
- the need for collaboration in production;
- the lack of training for operators and managers; and
- poor management and production practices.

A number of specific policy recommendations were drawn from the analysis including the provision of industry-wide services, through the establishment of a fashion centre and the pursuance of a more targeted approach towards financial assistance. Following the return of a Labour majority in the 1983 city-council elections a sector-based strategy for textiles and clothing was developed in accordance with the recommendations of the Trent Polytechnic study of 1983.

During the early 1980s political changes also occurred within Nottinghamshire County Council. The Labour Party's success in the 1979 city elections was followed in 1981 by the election of a Labour-controlled county council which was committed to giving high priority to action in the local economy. Again organizational changes followed to reflect the increased importance attached to economic development policy. For example, in 1982 a new Economic Development Unit servicing a new Economic Development Sub-Committee was established. Whilst the county council continued to be involved in the land development process, albeit mainly on sites outside the city, two major new areas of activity developed: financial assistance schemes, providing a range of soft loans, grants and equity finance, and increased promotional activity to attract mobile investment, particularly from abroad.

The county council's activities in the immediate post-1981 period were not informed by a strong research base to the same extent as the city council's initiatives. However, with increasing concern being expressed by some local politicians as to the effectiveness of certain initiatives, particularly industrial promotion and inward investment, there was increasing pressure from elected members for the development of a clear county council strategy towards economic development and the targeting of initiatives. The city council's approach was being cited as an example of best practice and a model for the county council to follow. Following the county elections in 1985 there was a much stronger sectoral focus and a greater willingness to work jointly with the city council on both research and joint initiatives. This increased activity by the county council since 1985 has been supported by a higher level of spending on economic development and by raising the status of economic development within the county council by giving the activity full committee status.

Commitment from the county council to a sector-based approach is reflected in the direct funding of a separate research unit within its Economic Development Unit. In conjunction with the Community Projects Foundation, a central-government-sponsored body, the county council committed a total of £60,000 over a period of two years to support the project. This new unit, which started work during 1986, is currently undertaking further work on the textile and clothing industry as well as examining other sectors, particularly engineering.

The new work on textiles and clothing is primarily concerned with the county's other main textile area, Mansfield and Ashfield, where 13,200 people were employed in the industry in 1981. The industry in this part of Nottinghamshire has somewhat different characteristics from that of the City of Nottingham with larger plant sizes and a strong representation by the hosiery sector. Detailed sectoral analysis is currently being undertaken and it is already clear that a different package of initiatives is likely to be required to address the problems of restructuring in the textile and clothing industry of Mansfield and Ashfield. This underlines the point previously made by the authors that initiatives in one area cannot simply be transplanted to other areas in the absence of detailed assessment (Collis *et al.*, 1987). The project is also concerned with community development initiatives and the promotion of new forms of trading organization, particularly in areas of high unemployment. In carrying out this work the unit is working closely with a range of key agencies including local businesses and community groups.

It is apparent from an examination of the development of policy in Nottingham and the shift towards a sector-based approach that the election of a sympathetic local council represents a precondition for concerted action. This pattern of local political action is also reflected in other recent studies (Mawson, 1983). Second, the development of a sound research base is essential to inform and direct policy initiatives. However, it was also apparent that the process of learning and adaptation has, on occasion, been constrained by the reluctance of key officers to become involved in new areas of activity. Experience suggests that organizational inertia in this form can only be overcome through the persuasion and advocacy of key politicians and the appointment of new personnel to implement policy.

5.3.3 LOCAL INITIATIVES – IMPLEMENTATION AND ASSESSMENT

The remainder of this case study examines the form and nature of specific initiatives, their effectiveness and the potential for the further development of these schemes. To date, the city council have been the main driving force behind the bulk of initiatives within the city, but in doing so they have drawn upon a wide resource base. Central-government policy has been important through the Urban Programme, the Manpower Services Commission and the Department of Trade and Industry, while county, business and trade-union support has been important in the development and implementation of initiatives.

Policy developments in Nottingham to assist the textile and clothing industry have been informed and influenced by the experience and lessons of initiatives elsewhere in the UK and overseas. The main problems which have been addressed and which reflect the weaknesses identified by the Trent Polytechnic local economy study (1983) are: the fragmented structure of the industry; the need to develop more effective marketing; the importance of responding rapidly to changing demand and tastes; and the desire to improve working conditions in the industry.

Elsewhere similar problems have led to a number of alternative approaches being advocated. The approach pursued by the Greater London Enterprise Board (GLEB) has been to restructure the textile industry in favour of labour with the quality of employment assuming a high priority. It is argued by GLEB that this can only be achieved through increased local ownership and control of the industry. Direct investment linked to 'enterprise' and 'planning agreements' are regarded as the most effective means of achieving this, through making the improvement of working conditions a major objective of financial assistance.

The resource base available to Nottingham City Council and Nottinghamshire County Council, together with concern by local politicians about the accountability of local enterprise boards, has generally precluded such an approach. Local authorities in the Nottingham area have taken the view that the provision of collective infrastructure is the most appropriate form of intervention supported by the careful targeting of other economic development initiatives.

Based upon the model of the Hackney Fashion Centre, the opening of the Nottingham Fashion Centre in 1984 represented a major step forward in the provision of collective services for Nottingham's textile and clothing industry. The Fashion Centre, funded largely by the Urban Programme, houses permanent showroom facilities, acts as a resource centre for local industry and promotes links between firms in order to provide opportunities for collaborative marketing and production ventures. The Fashion Centre was formed as a separate company with business, the city council, trade unions and those involved in training being represented on the board. More recently the city council has taken more direct control, a change which reflects the increasing dependence of the Fashion Centre on the city council for funding.

The concept of collective infrastructure has been further developed through the construction of twelve fashion enterprise workshops, located close to the Fashion Centre, where sewing and pressing machinery is available and advice is on hand from the resident manager. The primary aim of this initiative is to assist young designers to start up and become established in the city. Funding was again primarily through the Inner Area Programme.

Davenport and Totterdill (1986) have pointed to the lessons to be gained from the Fashion Centre initiative. They argue that the main benefits from the Fashion Centre arise through the provision of technical advice. The provision of showroom facilities has been less successful because of the reluctance of London-based buyers to travel out of the capital. It is also argued that there is a need to ensure that the benefits to those firms who use the Fashion Centre should start to filter down to the workforce. This demands setting up monitoring procedures to ensure that firms making use of the facilities are operating as responsible employers.

These criticisms have been recognized by the city council and have influenced the development of policy. It is now proposed to bring together the Fashion Centre and the fashion enterprise workshops under one roof on the site of the present enterprise workshops. Greater emphasis will be placed upon the provision of small workshop units, with an additional twenty-four units being constructed, and support services will be extended. Lower priority will be given to the provision of showroom facilities.

The county and city councils provide a range of financial assistance schemes, but until recently the decision to invest in a local firm has generally been based upon strictly financial criteria with little regard to industry-wide issues, for example, the importance of the firm to the local economy. A sector-based approach has provided the opportunity to target assistance more clearly within the context of a more proactive strategy of financial assistance. Over one-third of all rent and interest relief schemes have been targeted at the textile and clothing sector with those involved in design being particularly favoured. Within the industry the dyeing and finishing sector has been identified as a sub-sector where the major contractions of recent years have caused alarm. Cooperation between employers, trade unions and the city council has assisted in the development of a number of financial-assistance packages and one company is now benefiting from an urban-development grant and soft-loan facilities. The Department of Trade and Industry has also been involved through the provision of consultancy grants to assist in the application of new technology. The city council estimates that as a consequence of these initiatives in the dyeing and finishing sub-sector fifty new jobs have been created and 320 jobs have been saved.

A sub-sector focus within the industry is also an important feature of training initiatives. In collaboration with the former West Midlands County

Council, Nottingham City Council has developed a clothing-operatives training scheme at the Fashion Enterprise Workshop. This is aimed at providing high quality training for machinists and is a response to the industry's need for greater versatility and the desire to improve the career prospects and bargaining power of those employed in the industry. This initiative is funded jointly by the European Social Fund, the Department of the Environment through the Urban Programme and the city council; training allowances are paid by the Manpower Services Commission. A dyeing and finishing training scheme is also funded along similar lines and both schemes have proved highly effective with those completing the courses readily finding permanent jobs.

An important role for both the city and county council is advocacy, designed to inform and influence national and EEC policy. The county council was involved in major representations on the importance of the Multi-Fibre Agreement, a trade quota agreement, before it was renewed during 1986. The city council has been involved in a range of activities to influence government policy towards the sector. It has played a key role in establishing a link between local authorities with a common interest in the textile and clothing industry and this has now been formalized as an organization called Local Action on Textiles and Clothing. In addition to being a collective voice for those local authorities with a textile and clothing industry, it also provides an information exchange service for promoting new initiatives and provides a basis for collaborative schemes, e.g. training packages.

5.3.4 FUTURE POLICY

The future development of a sector-based approach in Nottingham will depend upon a continuing commitment to intervention in the local economy backed by resources. At present there is considerable uncertainty as to whether this support will be forthcoming, following the election of a Conservative city council in May 1987. In the absence of such support from the city council greater demands will be placed upon the limited resource base of the county council.

On the assumption that future funding will be forthcoming it is becoming increasingly evident that a broad sectoral approach is likely to prove inadequate. More detailed sub-sector research and associated policy responses will ensure that resources are being targeted on the most vulnerable groups and workers and on those sectors of key importance to the local economy. It may also be necessary to establish a stronger link between access to local authority support, such as commercial facilities, and the quality of employment.

In summary, the Nottingham textile and clothing initiatives demonstrate the importance of establishing a sound knowledge and understanding of the characteristics of the local economy upon which to base a programme of action. It also highlights the wide range of organizations and agencies which can be drawn together as a collective resource in order to address the problems of economic decline and restructuring providing there is a strong commitment from key organizations and actors.

5.4 SWANSEA (SOUTH WALES): AN ENVIRONMENTAL AND ECONOMIC INITIATIVE

5.4.1 INTRODUCTION

Swansea, situated in West Glamorgan in South Wales, was described by its poet, Dylan Thomas, in his 'Reminiscences of childhood', as 'a large Welsh industrial town . . . an ugly, lovely town . . . crawling, sprawling, slummed, unplanned . . . and smug-suburbed by the side of a long and splendid curving shore. . . . This sea was my world' (Thomas, 1954).

A statue of Dylan Thomas now looks across Swansea's new marina within the Maritime Quarter by the seashore. The quarter is a major element in the transformation of the former metallurgical centre of the world located in the lower Swansea Valley. The old 'industrial town' is now a city with a diverse economic base. The city acts as the commercial and shopping centre for the sub-region, functions which have been enhanced through the redevelopment of the city centre. The development of the Maritime Quarter, between the new central shopping area and the seashore of Swansea Bay, has added to the natural tourism resource provided by the Gower Coast to the west of the city.

West Glamorgan is a county which is located at the western end of the South Wales coalfield. Physically, the county consists of an inland upland area, subdivided by the Swansea, Neath and Afan Valleys; a coastal belt in which the main urban centres are Swansea, Neath and Port Talbot; and the Gower coast, an area of outstanding natural beauty located at the western end of the county and falling within the boundaries of Swansea City Council.

In 1981 the population of West Glamorgan was 363,619 and that of Swansea, the largest urban centre, was 183,484. In the 1960s the population of West Glamorgan was growing, mainly as a result of an increase of 11,000 people in Swansea, which offset the population decline in Neath, Port Talbot and the valleys. During the 1970s the population of both the county and Swansea declined and the forecast to the end of the 1980s is one of continuing decline in the county and population stability in the city.

The main features of economic change between 1945 and 1975 were a decline of coalmining in the valleys; industrial growth on the coastal plain associated with tinplate, steel and non-ferrous metal manufacturing and petro-chemicals; and an increase in the sub-regional importance of Swansea city as a commercial and shopping centre. Since 1975 not only has coalmining continued to decline, so that now only two mines operate, but there has also been a rapid decline in manufacturing. Between 1976 and 1982 the jobs lost in manufacturing totalled 20,000, of which 14,000 were in metal manufacturing. The decline in mining and manufacturing employment combined with virtual stagnation in the formerly growing service sector led to a fall in total employment in the county from 159,100 in 1976 to 139,300 in 1982. The share of services in the reduced level of total employment rose from 60 per cent to 70 per cent between 1976 and 1982 with extractive and manufacturing employment declining from 40 to 30 per cent. By 1986 total employment had fallen further to 130,000, of which manufacturing's share was 20 per cent. Swansea has shared in the decline of manufacturing employment but it has benefited from its sub-regional role as a centre for shopping, commercial, cultural and administrative services. As a consequence female unemployment in Swansea at 10 per cent is below the county and national figure. However, male unemployment at 20 per cent, while comparable to that in the county, is above the national average of 16 per cent.

5.4.2 POLITICAL AND ORGANIZATIONAL STRUCTURE

In both West Glamorgan and Swansea City the councils have for many years been controlled by Labour members. In West Glamorgan seven committees report to the county council. For economic development functions Policy and Resources is the key committee. Industrial Promotion is one of six sub-committees reporting to the Policy and Resources Committee. In Swansea there is also a Policy and Resources Committee which advises the city council on all major policy matters. For specific policies and activities Policy and Resources operates through sub-committees such as that for the city centre and Maritime Quarter.

In West Glamorgan the senior local government officer is the county clerk and chief officer. His department includes the industrial development officer, a post created in 1975. Although there is no freestanding Economic Development Unit, the industrial development officer calls on the services of two officers in the Central Research Unit of the county clerk's department and can call on officers from the Planning Department. In Swansea a separate planning department was created in 1974 after local government reorganization. In 1985 the Development Department was created through amalgamation of the Planning and Environment Departments. Swansea appointed an Industrial Development Officer in 1979 and from this beginning the Swansea Centre for Trade and Industry developed. The centre is a section of the Chief Executive and Town Clerk's Department and is responsible for a wide range of local economic development initiatives.

5.4.3 THE NATURE AND GROWTH OF INITIATIVES

For West Glamorgan the only detailed published statement on local economic development is contained in the 1984 *Alteration to the County Council's Structure Plan of 1980*. The latter saw the development of the physical environment, particularly through the provision of land, as making a contribution to economic development. The 1984 *Alteration*, however, regards land-use policies as just part of the county's total economic development package. The aim is to develop and further diversify the local economy through the attraction and effective use of both private and public resources.

In order to stimulate investment, land and premises are made available, the county's attributes are promoted and business information and advice are provided. While direct financial assistance to industry is not provided by the county council, financial support is given to support the activities of the West Glamorgan Enterprise Trust and the West Glamorgan Common Ownership Development Agency. The county council also pursue an advocacy policy. This reflects the need felt by the council to obtain a larger share of public assistance from central government and its agencies and from the European Communities. A recent example of the success of the advocacy approach is the decision of British Coal to open a deep mine at Margam in the east of the county.

Swansea City Council published their Economic Development Policy (with a Tourism Development Supplement) in 1980 followed by a Tourism Development Policy Statement in 1984. The central features of the Economic Development Policy are the provision of an environment which encourages private- and public-sector employers to remain and expand, the encouragement of new-firm formation and the attraction of inward investment. The Tourism Development Policy of 1984 confirmed and extended the 1980 objective of increasing the flow of visitors to the city through publicity and promotion and by developing new attractions, events and accommodation.

The specific local economic development initiatives currently operated by the Swansea Centre for Trade and Industry and financed under section 137 of the 1972 Local Government Act from a 2p rate (local tax) derive from those outlined in the 1980 Economic Development Policy. These include business development services in the form of a small-capital grant, a soft-loan and an overdraft-guarantee scheme, a business-information service and various consultancy and training activities.

The Economic Development Policy also provides for the continuation of the city council's development land supply policy. After local government reorganization in 1974 and faced with an outdated land-use plan the city reviewed its development strategy. The creation of a separate planning department within the new city council reflected the greater emphasis placed on the planning role. Interim planning statements were prepared for the city, covering major land-use issues and economic and social issues such as industry and employment and shopping, housing and recreation. This procedure enabled major proposals to begin immediately and the council's city centre Quadrant shopping precinct and the leisure centre were started in 1975. A major problem which the city faced was how to regenerate the derelict Lower Swansea Valley. The 1975 Interim Planning Statement for the Lower Swansea Valley defined the strategy of regeneration for an area of 2,000 acres running from the present M4 motorway to the river mouth and docks.

5.4.4 THE LOWER SWANSEA VALLEY

The industrial history of the valley goes back to 1717 with the opening of a copper smelting works. By 1830 the valley was the location of over three-quarters of British copper smelting works. A process of diversification into other metals meant that by 1891 there were 150 metal works of thirty-six varieties employing 30,000 people in the Lower Swansea Valley. The dominant industries in the late nineteenth and early twentieth centuries had become steel and tinplate, but then these activities moved to the coastal plain. The last metal works closed in the valley in 1980 thus completing its transformation from metallurgical centre of the world to one of 2,000 acres of industrial dereliction.

In the 1960s a group of representatives from the University College of Swansea, the Welsh Office, Swansea Council and local industry began a study which resulted in the 1967 Hilton Report on the Lower Swansea Valley project. The council then embarked on a major programme of land assembly and reclamation in the Lower Swansea Valley and after local government reorganization in 1974 the council published an Interim Planning Statement on the Lower Swansea Valley (Swansea City Council, 1974). This statement defined two parks: an industrial park and an urban park. This was developed into the 'Five Parks' concept. The enterprise park was designated as the UK's first enterprise zone in 1981, being an enlarged version of the original industrial park. To the south of the enterprise park is the leisure park which accommodates a major sports complex within a woodland setting. The tree planting of the formerly derelict area was one of the first positive responses to the Lower Swansea Valley project, and playing fields were laid out on the site of a tip from a former copper works. The city park is being joined to the leisure park via a riverside park leading to the seashore of Swansea Bay where the maritime park is located. The riverside park, to which a barrage proposal now relates, is a linear park being created along the banks of the River Tawe linking the other four parks of the Lower Swansea Valley.

5.4.5 THE MARITIME QUARTER WITHIN THE LOWER SWANSEA VALLEY

The Maritime Quarter is centred around the former South Dock. The dock was opened in 1859 to enable the port to cope with larger ships of the time which brought in copper ore and took out copper and coal. With the demise of the Lower Swansea Valley's metal industries the South Dock was closed in 1969 and purchased by the city council. The original intention was to build a city-centre relief road through the area. However, the 1975 Interim Planning Statement for the South Dock (Swansea City Council, 1975) recommended the development of the dock and half-tide basin as a marina with associated activities and for the land surrounding to be developed mainly for residential purposes with the area interrelating with the city centre developments of the leisure centre and Quadrant shopping precinct.

The Maritime Quarter consists of a water area of 7 hectares, a development area of 26 hectares and a 6-hectare conservation area (Swansea City Council, 1983a). Prior to the building of the South Dock in the mid-nineteenth century the area had been a fashionable resort. When plans to redevelop the area were put into effect in 1975 the area still contained numerous Georgian, Victorian and Edwardian buildings. These were conserved and rehabilitated by the city council and are now used for a variety of purposes including dwellings, artist studios and craft workshops and a yacht chandlery to serve the marina. An old shipping company warehouse was converted into a maritime and industrial museum, a pumphouse became a restaurant and a garage was converted into the Dylan Thomas Memorial Theatre. To the south and west of the conservation area is the marina which has been developed out of the derelict South Dock and the Tawe half-tide basin. The marina is the focal point of the Maritime Quarter, and currently provides 380 berths. Surrounding the marina is the development area. Here the predominant land use is housing, primarily for permanent residence. There is also holiday accommodation, and a 120-bedroom hotel with conference facilities is to be built between the western end of the marina and Swansea Bay. In addition to accommodation, the development area includes a major recreational facility in the form of the leisure centre, opened by HM The Queen in 1977, and numerous small shops, bars and restaurants. A large supermarket has been built at the north-eastern end of the development area, providing a major shopping facility both for city residents and long-stay visitors.

The policy of the city council towards the Maritime Quarter has been one of land-assembly and infrastructure provision aimed at providing the impetus for private-sector investment. City-council expenditure between 1974 and 1987 of £13 million has been topped up by £3.3 million of other public expenditure, from the central government regional and urban policy and from European Regional Development Fund grants. Private expenditure to date totals £36 million and is projected to rise to over £63 million by 1990. Total employment in the Maritime Quarter in mid-1987 stands at 1,245. Thus the development of the quarter represents a good example of the stimulation of private investment and employment creation resultant upon city council investment supported by central government and EC funds.

In order to build upon the success achieved so far the intention is to build a barrage across the mouth of the River Tawe adjacent to the existing Maritime Quarter (Swansea City Council, 1987). A barrage would increase the tourist potential of the Quarter by allowing an extension of the marina into the river mouth and facilitating pleasure trips along the river. It would also enhance the development potential for residential, commercial and industrial purposes of sites adjoining both banks of the River Tawe. It is estimated that extension of the Maritime Quarter consequent upon building the barrage would attract £37.6 million in private investment in property development and, in addition to construction employment, the barrage project would lead to the creation of 605 permanent jobs. The environmental impact of the barrage in raising the level of the river would be to remove from view the mudbanks and residues of industrial dereliction. Additionally, the riverside park would be enhanced by the retention of water in the river and the introduction of river-based activities. Thus, not only would the barrage enhance the recreational facilities of residents, it would also provide a major addition to the city's tourism resources through the extension of the marina into the river mouth area and allow for further residential, commercial and industrial development of both banks of the lower river which runs into Swansea Bay.

5.4.6 ASSESSMENT

The development of the Maritime Quarter reflects the approach adopted by Swansea City for the redevelopment of the whole of the Lower Swansea Valley, this approach being to transform the disadvantage of an industrially derelict environment into a positive asset. This is exemplified clearly by the redevelopment of the derelict South Dock into a marina which is the focal point of the Maritime Quarter. Not only has the Maritime Quarter added to the city's tourism attractions, the leisure centre and large supermarket contained within the quarter also provide leisure and shopping resources for the city's permanent residents. The development of residential property surrounding the marina provides housing for the local population and short-stay tourist accommodation in a high-quality environment. The quarter's shops, bars and restaurants and the marina-related commercial activities serve local residents as well as tourists. The city council has been the prime mover in the development of the Maritime Quarter, with additional public funds coming from central government and the EC. A modest investment of public money has acted as a catalyst for considerable private investment and the creation of 1,245 permanent jobs in the Maritime Quarter itself.

The Lower Swansea Valley was a demonstration site for the European Campaign for Urban Renaissance, launched in 1980. An International Campaign Seminar was held in Swansea in 1981 sponsored by the city council, the Council of Europe and the UK Department of the Environment. An international exchange of experience was furthered by a Planning for Enterprise Seminar in Swansea 1982 as a contribution to the Council of Europe's Urban Policy Programme. It was suggested that 'visitors studying urban renaissance should return to the Valley to assess progress, preferably in less than two years if the pace of reconstruction is to be properly appreciated' (Swansea City Council, 1983b). The development of the Maritime Quarter at the mouth of the valley exemplifies the pace of reconstruction which makes full use of the sea so beloved of Dylan Thomas.

5.5 CONCLUSION: LOCAL MODELS FOR FUTURE REPLICATION

Although it would be unwise to attempt to draw detailed conclusions from these three cases, which could not be claimed to be typical of all situations in England and Wales, it is possible to reach certain general conclusions which can be used to provide guidance to theorists and policy-makers alike. In all three cases there has been a recent and rapid collapse in the traditional industrial base of the area, and in all three cases new initiatives have proved essential in providing a basis for economic regeneration rather than simply enhancing the prospects for the further extension of existing opportunities for economic growth. A further characteristic of all three cases is that successful economic regeneration has been viewed as just one, albeit vital, element in a broader programme for area renewal and community development. Local economic policies are increasingly viewed as being directed towards the achievement of political, social and environmental goals and, by implication, to the encouragement of local solutions to local problems. Indigenous development, which is aimed at the longer-term objective of restructuring the local economy from within as well as the shorter-term aims of job and wealth creation, is slowly but surely replacing traditional policies which rely upon encouraging an ever-shrinking pool of mobile industry to relocate and generate inward investment (Roberts and Noon, 1987). Two of the three cases (Sedgefield and Swansea) are located in regions which have been the recipients of regional policy since the 1930s. As can be seen from these cases the previously accepted solution of relocating mobile manufacturing has simply proved to be a short-term and partial respite in the inexorable decline of the traditional extractive and manufacturing industries of these areas. It is only with the inception of policies which seek to rectify the fundamental weaknesses of an area's economy that a foundation for lasting success is likely to be created. Chief amongst the weaknesses experienced in traditional industrial areas is the suffocating economic dominance exerted by a small number of major firms or sectors. This dominance is reflected in a restricted range of labour skills, low levels of local entrepreneurship, limited diversity in the local economy and in other social and political manifestations which are seen at their extreme in the company town or village (Checkland, 1976).

Local economic development, in the three cases reported, is aimed at encouraging the restructuring and revitalization of a local community through the implementation of an integrated package of measures and initiatives which have been designed and developed as part of an overall strategy for the area; this approach clearly differs from the *ad hoc* application of general national or other policies irrespective of the characteristics, needs or potentials of a particular locality. The distinguishing features of a successful local initiative would appear to be:

- the presence or creation of adequate but not unnecessarily complicated organizational arrangements which can be utilized in the design, development and implementation of individual initiatives;
- the existence of a research base upon which to develop a local economic strategy and specific sector or area programmes;
- the generation of a local economic strategy aimed at the achievement of clear objectives and targets and for which explicit political and financial commitments can be sought and given;

- the development of clear links between economic policy and other aspects of local action which are aimed at the social and physical regeneration of a locality;
- the design and operation of an implementation programme, often involving a wide range of authorities and agencies, which seeks to integrate and coordinate actions and activities; and
- the existence of a mechanism for monitoring initiatives and evaluating their performance.

The above characteristics might suggest that success in local economic development is simply a function of clarity of policy and good management. In reality, other elements more random or difficult to quantify are often of great importance. First, the role played by key actors which is reflected in all three case studies; the importance and catalytic effect of perceptive and supportive local actors cannot be underestimated. Local political leaders often figure as key actors, as do trade unionists, local businessmen, local clergy, community activists and professional officers employed by local authorities. Second, a further factor of importance is the ability at local level to recognize potential opportunities for development and the vision to translate what are often viewed as negative elements into positive assets. The Swansea Maritime Quarter is an example of an environmental and economic liability that has been transformed into a local recreational resource, an important tourist attraction and a focus for investment. Symbols of success become important elements in local economic development, they raise confidence in an area and inspire further investment in difficult ventures. Third, it is important to stress that economic initiatives should be locally derived and should be relevant to the particular needs of the local community; initiatives should reflect local potentials, aspirations and resources. This is all the more important in the older industrial areas where traditional regional policy often relied upon importing jobs, which were poorly suited to the skills and aspirations of residents, in branch plants which had few loyalties to the area and a limited interaction with the rest of the local economy. A local economic strategy based upon indigenous development is also more likely to achieve the broader social and environmental objectives of a local community.

Although it is often difficult to initiate an indigenous development strategy, since this often implies the creation of completely new activities, the benefits from successful local economic development are lasting in nature and accelerate rapidly following an initial success. The growth or retention of employment in the three case studies has been matched by social, educational, environmental and infrastructure improvements. In all three cases the local authority proved to be of great significance in the initial formulation of an initiative. However, successful implementation was only achieved through complex and often difficult negotiations with a wide range of public and private actors. The key roles performed by the local authority were in providing pump-priming finance and support and in exerting leverage upon the other participants. The case studies demonstrate that the fundamental requirement for successful local economic development is the existence of an integrated and coordinated local mechanism for the design and implementation of policy which is supported and maintained by explicit political and organizational commitments. Given such a mechanism and the necessary commitments, then, other localities can also achieve success. The severe economic difficulties that faced the three case-study areas should not be underestimated; neither should the task of obtaining strong political and institutional support for novel and often risky initiatives. Replication of success elsewhere will depend upon the willingness of local actors to work together in the best interests of the community and to perceive opportunities which can be translated into action through an agreed and properly resourced programme of initiatives.

REFERENCES

Benington, J. (1985), 'Local economic initiatives', *Local Government Policy Making*, vol. 12, no. 2.

Checkland, S. G., (1976), *The Upas Tree*, Glasgow: University of Glasgow Press.

Collis, C. M., Noon, D. M., Roberts, P. W., and Barton, P. (1987), *Economic Development in Shire Counties*, Manchester: Centre for Local Economic Strategies.

Davenport, E., and Totterdill, P. (1986), 'Fashion centres', *Local Economy*, no. 1.

European Economic Communities (1984a), *The Regions of Europe*, Brussels: Commission of the European Communities.

European Economic Communities (1984b), *Council Regulation 1787/84 on the European Regional Development Fund*, Brussels: Commission of the European Communities.

European Economic Communities (1985), *Commission Decision on the Shildon–Newton Aycliffe–Bishop Auckland Programme*, Brussels: Commission of the European Communities.

European Economic Communities (1987), *Third Periodic Report from the Commission on the Social and Economic Situation and Redevelopment of the Regions of the Community*, Brussels: Commission of the European Communities.

Hilton, K. J. (1967), *The Lower Swansea Valley Project*, London: Longman.

Johnston, B. (1986), 'Sedgefield strikes silver on Shildon success', *Planning*, no. 698.

Litherland, J. (1987), *Economic Development: Policy and Implementation*, Preston: Lancashire Enterprises.

Local Authority Associations (1986), *Crisis and Opportunity: Local Authorities' Response to Decline of Major Traditional Industries*, London: Association of County Councils.

Mawson, J. (1983), 'Organising for economic development', in C. Mason (ed.), *Urban Economic Development*, London: Macmillan.

Roberts, P., and Noon, D. M. (1987), 'The role of industrial promotion and inward investment in the process of regional development', *Regional Studies*, vol. 21, no. 2.

Sedgefield District Council (1985), *Shildon–Newton Aycliffe–Bishop Auckland ERDF Programme Area Submission 1984–1989*, Sedgefield: District Council.

Sedgefield District Council (1987), *Shildon–Newton Aycliffe–Bishop Auckland ERDF Programme: First Monitoring Statement*, Sedgefield: District Council.

Shildon Shop Stewards Committee (1983), *Shildon – A Case of Unfair Competition*, Shildon: Shildon Works Joint Shop Stewards Committee.

Swansea City Council (1974), *Lower Swansea Valley: Interim Planning Statement*, Swansea: City Council.

Swansea City Council (1975), *South Dock: Interim Planning Statement*, Swansea: City Council.

Swansea City Council (1983a), *Maritime Quarter: First Monitoring Report*, Swansea: City Council.

Swansea City Council (1983b), *Planning for Enterprise: Council of Europe*, Swansea: City Council.

Swansea City Council (1987), *Tawe Barrage Project Appraisal*, Swansea: City Council.

Thomas, Dylan (1954), 'Reminiscences of childhood' in Thomas, *Quite Early One Morning*, London: Dent.

Trent Polytechnic (1983), *Nottingham Local Economy Study: A Report*, Nottingham: Trent Polytechnic.

CHAPTER 6

Regional alternatives to economic decline in Britain's industrial heartland: industrial restructuring and local economic intervention in the West Midlands conurbation

Michael Marshall

6.1 INTRODUCTION: BACKGROUND TO THE DEVELOPMENT OF LOCAL ECONOMIC INTERVENTION IN THE 1980s

The UK is not alone among West European economies in having experienced a pronounced economic slowdown during the past two decades. Industrial decline in the UK has been particularly severe, however, owing to the country's acute crisis of competitiveness as a manufacturing nation. The decline of the UK manufacturing sector has accelerated with each successive downturn in the economic cycle since 1966, the sharpest contraction occurring during the last recession of 1979–81. By the end of 1986, UK manufacturing output remained 5 per cent, investment 19 per cent and employment 27 per cent below their peak levels in 1978.

The effect of the last recession has been to accentuate the broad North–South divide which has been an enduring feature of British economic, social and political development throughout its industrial history. Recently published government figures revealed that between 1978 and 1984 some 72 per cent of the job losses in UK manufacturing occurred outside the relatively affluent southern regions while the South benefited from 62 per cent of the net UK job gains in service sectors (Department of Employment, 1987).

Since 1979 no UK region has suffered a more rapid and pronounced industrial collapse than the West Midlands. The region centres upon the metropolitan conurbation of Birmingham, Coventry and the Black Country – so called because of its dense industrial landscape – surrounded by the rural counties of Hereford and Worcester, Shropshire, Staffordshire and Warwickshire. At the last census in 1981 some 52 per cent of the region's 5.1 million population and 57 per cent of the region's 1.9 million employed workforce were in the metropolitan area.

The conurbation is the largest spatial concentration of manufacturing industry and employment in the UK. In 1981 some 43 per cent of the conurbation's total workforce was employed in manufacturing compared with 28 per cent for Great Britain as a whole. Within the conurbation's manufacturing sector the four key industries of motor vehicles, mechanical engineering, metal goods and metal manufacturing accounted for 67 per cent of the total manufacturing workforce as illustrated in Figure 6.1.

Traditionally regarded as Britain's prosperous industrial heartland, until the mid-1970s the West Midlands enjoyed employment levels and living

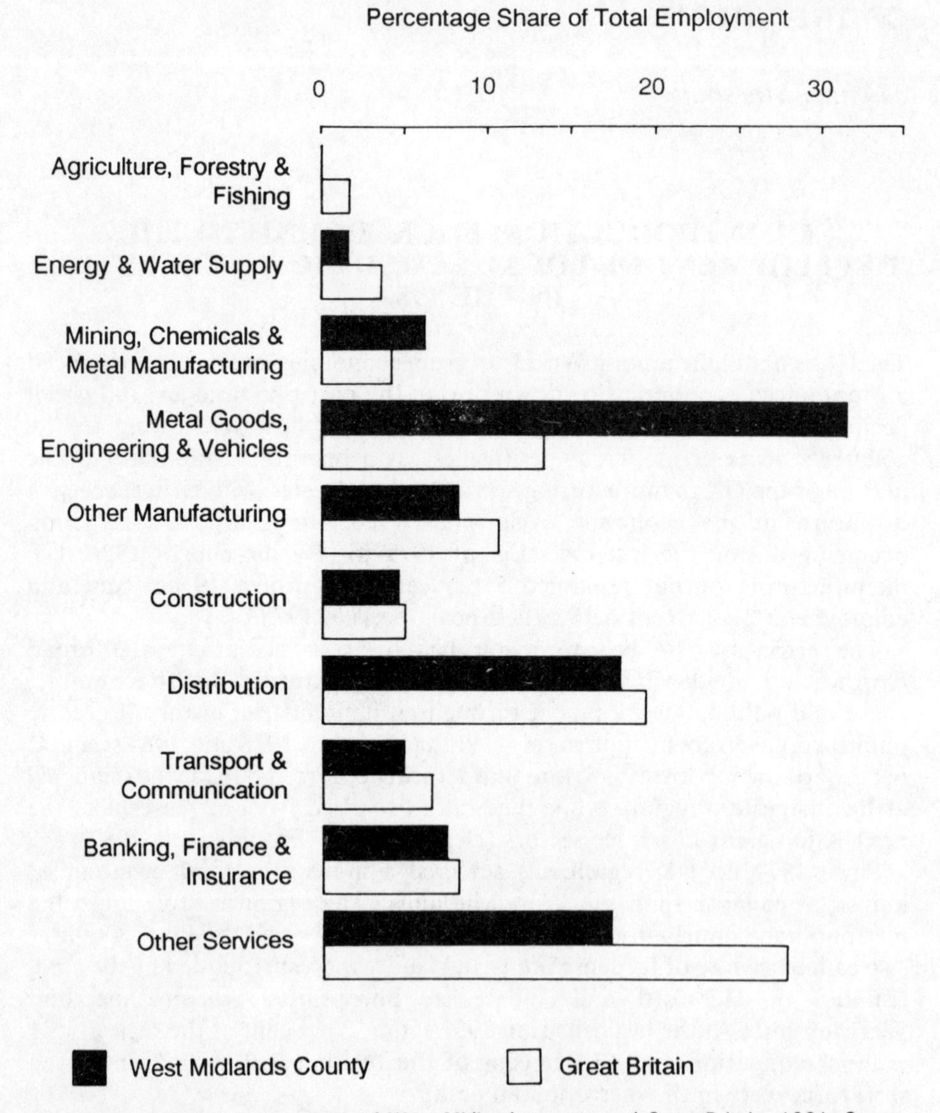

Figure 6.1 Employment structures of West Midlands county and Great Britain, 1981. Source: Census of Employment.

standards second only to the South-East. Since the onset of the recession in 1979, however, the West Midlands' former strengths in interlinked and diverse metal-based manufacturing have been dramatically undermined. Between 1979 and 1984 the region lost 283,000 manufacturing jobs amounting to a 29 per cent decline in employment in the manufacturing sector. The *Financial Times* (1981) was not exaggerating when it described the West Midlands at the height of the recession as a 'region in a state of shock'.

The year 1979 was a turning point in British development, not only because it marked the beginning of the most severe recession since the interwar years, but also because it witnessed a major watershed in British political life with the election of the present Conservative government. The ensuing period of 'Thatcherism' broke with the previous postwar tradition of Keynesian economic demand-management policies and social consensus politics (Martin, 1986). While the official UK unemployment total rose from 1.5 million in 1979 to 3.1 million by 1986, the Conservatives continued to prioritize anti-inflationary measures. The Conservatives' preoccupation with restraining the growth of state expenditure continued the new direction in economic policy begun by their Labour predecessors in the monetary crisis of 1976, contrasting with the preceding period of public-expenditure-led economic management. The contraction of manufacturing employment and capacity was accepted as a necessary step towards achieving a 'leaner and fitter' industrial base through market selection. The privatization of nationalized industries and state assets created new outlets for profitable financial investment, contributed to the reduction of public spending and accelerated the withdrawal from state intervention in industry.

It was against this background that, following the 1981 local elections, increasing numbers of Labour-controlled local authorities embarked upon wide-ranging programmes of economic intervention in their areas (Boddy, 1984). The ensuing wave of local economic initiatives emerged at this time due to a confluence of interrelated economic, social and political factors. They included:

- the deepening economic decline, mounting unemployment and social distress suffered by the inner-urban areas and industrial conurbations;
- the major civil disturbances in several inner-city areas during the summer of 1981, the worst in mainland Britain in modern times;
- the apparent central-government indifference to the socio-economic crises of the conurbations;
- the radicalization of the local government arena since the mid-1970s in response to central-government erosion of local services;
- the changing political complexion of Labour local authorities accompanying a shift to the Left within the Labour Party as a whole in

response to dissatisfaction with the past performance of Labour governments;

- the development of a new tradition of local-community-based politics, partly resulting from central-government-initiated inner-city programmes such as the Community Development Projects of the mid-1970s;
- the assimilation of radical urban and regional political and social theory among a new generation of local councillors and officers with academic backgrounds.

Styles and forms of economic intervention by local authorities vary widely, but among the so-called 'New Left' authorities three main areas of activity can be distinguished (Benington, 1986):

1. resourcing initiatives by the trade union and community movements;
2. using the economic leverage of the local authority as a major employer and economic agent in its own right; and
3. direct intervention in private-sector job creation and business development.

This chapter is principally concerned with the third area. Many of the most ambitious policies and initiatives in this field were pursued by the Labour administrations controlling the Greater London Council and the six metropolitan county councils, covering the major conurbations, subsequently abolished by the government in March 1986. A number of authorities established enterprise boards as the focus of their employment and economic development initiatives. The five major enterprise boards operating on a regional or county-wide basis cover the West Midlands, Lancashire, West Yorkshire, Merseyside and Greater London. The organizational structures, political perceptions and functions of these agencies vary although their core activity in each case is investment in local jobs and industry. Some enterprise boards were developed purely as investment agencies, complementing other economic development activities carried out by their parent authorities. Others were established with a broader remit extending to skill-training, technology initiatives and economic analysis. The common principles upon which the five major enterprise boards are based can be summarized as follows.

- Traditional local authority arm's-length support to private industry, most notably through property development, is rejected as an insufficient response to local industrial decline. Where enterprise boards *do* engage in property development this is usually part of a wider economic package.
- Support is directed at the needs of *indigenous* industry, rejecting traditional forms of redistributive regional policy as well as 'beggar-my-

neighbour' local-authority promotion of their areas' attractions for inward investment.

- Priority is given to *manufacturing* industry in recognition of the importance of this activity in the authorities' local economies, rejecting the notion that service-sector development can compensate for the contraction of the industrial base.
- Support for local industry is focused on direct capital investment, generally in the form of shareholdings and long-term loans. Historically low levels of investment are regarded as the key reason for the decline of UK manufacturing relative to its chief international competitors. This is seen as partly the result of unfavourable investment policies pursued by the major financial institutions, hence the need for regionally based investment agencies to provide development capital for local manufacturing industry.
- It is argued that public investment in industry should be publicly accountable. Enterprise boards are themselves accountable to elected local authorities who provide the majority of each board's directors. Investments are made in the form of equity or loans, rather than grants, which are expected to yield a return and are governed by planning and investment agreements covering 'social' aspects of firms' business plans.
- Past central-government intervention in industry is criticized as piecemeal and reactive, divorced from any analysis of the implications of intervention in a particular enterprise for the local economy as a whole. Enterprise board investments are set within the context of an analysis of, and strategy for, whole industrial sectors rather than the individual enterprise. The approach to industrial intervention based on 'sector strategy' has been particularly strongly developed by the West Midlands Enterprise Board and is examined in detail later in this chapter.

Some of the most pioneering local economic initiatives since 1981 have been developed in the West Midlands. In the five years to March 1986 the West Midlands County Council (WMCC) developed and implemented a wide-ranging strategy for regenerating the conurbation's ailing manufacturing base. Since 1986 this strategy has been preserved and extended by the West Midlands Enterprise Board (WMEB), which was originally established by the County Council but is now controlled by the seven district councils within the conurbation. During recent years the enterprise board and former county council have gained increasing national recognition for their approach to economic development and the lessons this might yield for future national interventionist economic strategies.

6.2 THE MAKING OF A NEW DEPRESSED REGION: ECONOMIC CHANGE AND DEVELOPMENT IN THE WEST MIDLANDS

The West Midlands' recent economic problems are the latest phase in a long-term process of structural change in the regional economy which began to emerge in the mid-1960s but which has accelerated sharply over the past decade. Although the West Midlands was particularly badly affected by the international recessions of 1974–5 and 1979–81, the region's difficulties are of more than temporary or cyclical nature. The West Midlands' recent economic decline, and local responses to the region's crisis, can only be understood in the context of these long-term changes.

6.2.1 INDUSTRIAL STRUCTURE AND DEVELOPMENT

The present-day industrial structure of the West Midlands is the product of successive historical long waves of economic change and adaptation illustrated in Figures 6.2 and 6.3 (Marshall, 1987). From the late eighteenth to the mid-nineteenth century the initial development of the regional economy

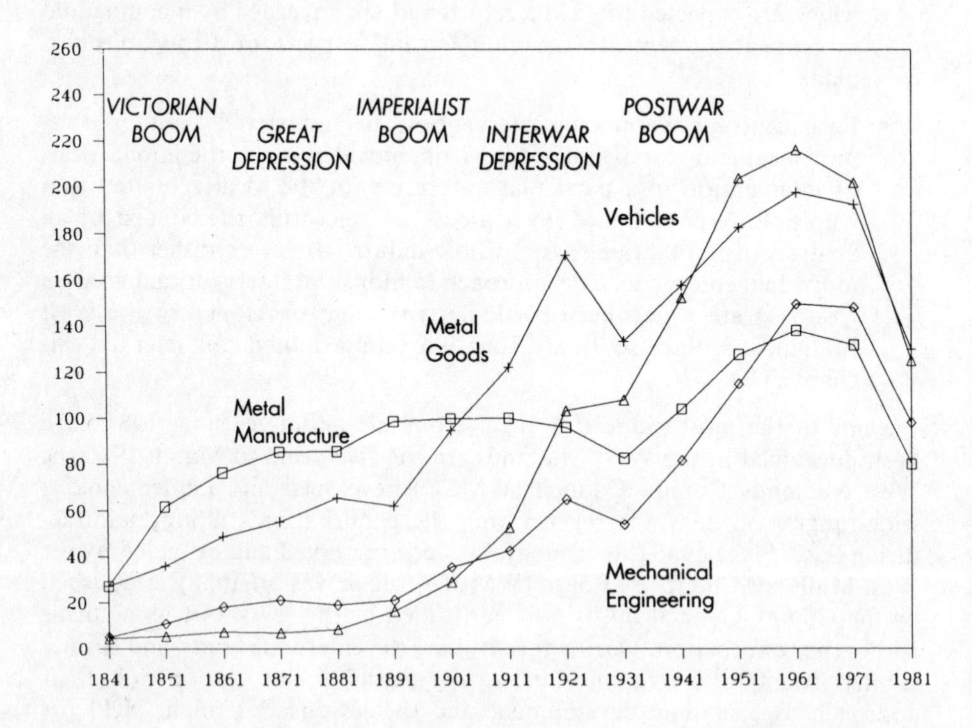

Figure 6.2 West Midlands region: employment in metal-based manufacturing industries, 1841-1981 (thousands). Source: data from Lee (1979) and Census of Employment.

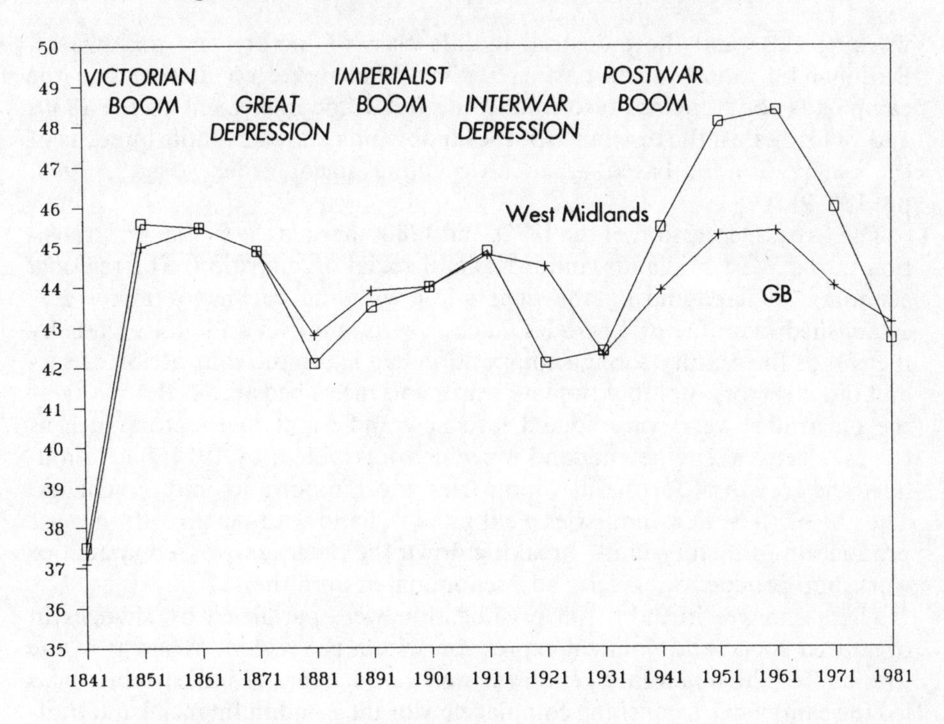

Figure 6.3 West Midlands region and Great Britain: employment activity rates (per cent), 1841-1981. Source: Marshall (1987, p. 126) and Census of Population.

was founded upon its basic industries of coal, iron and steel together with a complex of metalworking trades. During the first half of the nineteenth century, Birmingham developed into the industrial, financial and commercial focus of the regional economy. The growth of Birmingham was accompanied by the decline of many of the cruder metalworking trades in the Black Country where production became directed towards the demands of Birmingham manufacturers for raw materials and semi-finished products. The boom in railway investment and construction after 1850 led to the formation of the Birmingham Stock Exchange and consolidated the city's function as a distribution centre through its geographical position at the heart of the rail network.

A distinctive characteristic of industrialization in the West Midlands during this period was the predominance of small-scale, workshop production largely based on craft skills through which the workforce exercised considerable day-to-day control over the labour process. The West Midlands' social cohesion and productive organization contrasted with that of other industrial regions, most notably the cotton textiles region of Manchester and the north-west, where there was a glaring social contrast between the urban working class and the industrial middle class of factory and millowners. Birmingham industrialists partnered their Manchester counterparts in the campaigns for franchise reform and the free-trade movement of the 1830s and 1840s against the resistance of the landowning and mercantile interests of the older, London-based aristocratic ruling social order (Briggs, 1968, pp. 186–9).

The Great Depression of the 1870s and 1880s marked the first major transition in the West Midlands' industrial and social organization. The regional economy suffered during the depression with the decline of many old-established manufacturing trades, such as wrought iron and nails, while by the end of the century foreign competition had led to the contraction of several other sectors, notably tinplate wares and metal bedsteads. But many of the old trades were consolidated into new and expanding sectors such as cycles, electrical engineering and machine tools which, by 1914, had stimulated the growth of further new industries, most importantly motor vehicles. The rise of these new industries went hand in hand with the growth of mass production in factory units, breaking down the older craft-based system of workshop production which had predominated until the 1870s.

These changes in industrial organization were paralleled by changes in organized social and political expression within the region. Whereas in the first half of the nineteenth century it was the Manchester industrialists who led the campaigns against the complacency of the London financial and trading interests, by the late nineteenth century that role had passed to Birmingham. During the Great Depression of the 1870s and 1880s for the first time British industrial capitalism began to lose ground to overseas competitors, notably Germany and the United States. Unlike Germany, where there was an unparalleled fusion of industrial and banking capitals, one of the most prominent features of British capitalism remained the division between London's financial and commercial oligarchy and provincial industrial capital (Ingham, 1984). This division widened further during the imperialist expansion of the 1890s when Britain experienced an unprecedented flow of capital overseas at the expense of domestic industry (Hobsbawm, 1969, p. 192) while the banking system became increasingly concentrated in the City of London (Crouzet, 1982, pp. 331–2). It was against this background that the Birmingham 'social imperialists' campaigned for a defensive policy of tariff barriers to protect the British domestic and colonial markets from foreign competition (Gamble, 1981, pp. 166–74).

During the late nineteenth and early twentieth centuries the West Midlands consolidated its position as the leading industrial centre of the leading international industrial and imperialist power. The flexibility and adaptability of the region's industry and skill base that had been demonstrated during the Great Depression continued to insulate the West Midlands from the worst effects of the recession in world markets after the First World War. Other industrial regions, which continued to depend upon traditional basic industries such as coal, steel, textiles and shipbuilding, were thrown into lasting decline during the interwar depression. Areas like South Wales, central Scotland and the North of England emerged after 1945 with levels of chronic unemployment persistently above average.

The experience of the West Midlands contrasted sharply with that of the other industrial regions. Although there were severe pockets of economic and social distress in the Black Country, unemployment rates during the interwar years were lower on average in the West Midlands than in most other parts of Britain. Moreover, the region's recovery from the depression was more rapid. The dynamism of the West Midlands economy continued into the postwar period when the region reinforced its position as Britain's industrial heartland. Employment fluctuations during the economic cycles of the 1950s and 1960s were generally much less intense in the West Midlands than in other regions. Protected from structural unemployment, West Midlands skilled manual workers were the highest paid in the UK. The region remained an area of relative labour shortage, attracting black and Asian migrants from Commonwealth countries to occupy many of the less skilled and lower paid jobs.

6.2.2 FROM BOOM TO SLUMP: THE IMPACT OF CORPORATE RESTRUCTURING

From the mid-1960s the former buoyancy of the West Midlands economy began to be undermined. British manufacturing had failed to match the levels of investment and productivity achieved by its major international competitors throughout the 1950s and early 1960s. The continued growth of British industry had been ensured by the buoyancy of world demand during the long postwar boom. But with the slowdown in world growth after 1966 the underlying weaknesses of British industrial performance were dramatically exposed. British manufacturers were ill-prepared for the ensuing period of intense competition for world markets. The competitive weakness of British manufacturing was compounded by an over-valued pound which inhibited British exports while helping foreign competitors to increase progressively their share of the British home market. In these circumstances it is not surprising that the UK's leading manufacturing centre has suffered more than any other region.

By the end of the postwar boom, the jungle of small workshops that had once characterized the West Midlands' industrial landscape had long since ceased to be a dominant feature. The process of capital concentration and merger had created an economy heavily dependent upon a relatively small number of giant companies in a narrow range of industries. By 1976 some 43 per cent of manufacturing employment in Birmingham was accounted for by ten companies while 39 per cent of the city's workforce was employed in the thirty largest plants. In 1977, ten major multinationals – GEC, Lucas, GKN, Tube Investments, Glynwed, Dunlop, BSR, Cadbury-Schweppes, Delta Metals and Imperial Metal Industries – employed an estimated 27 per cent of their combined UK workforce in the West Midlands. From the mid-1960s, faced with a decline in their markets, overcapacity and falling profitability, many of these companies began to rationalize and restructure their productive operations (Gaffikin and Nickson, 1984; Spencer *et al.*, 1986, ch. 5). In many cases this involved diversification away from, and disruption of their links with, the region's traditional industries in their search for new markets and products.

The entrenched industrial conflicts during this period were both a cause and a consequence of this process of industrial restructuring. The growth of large-scale mass production in giant plants, particularly in the car industry, was accompanied by a new mass unionism which greatly increased the bargaining power of trade unions, earning West Midlands workers a reputation for militancy which was quite alien to their earlier tradition (Spencer *et al.* 1986, pp. 26–8). There seems little doubt that this acted as a 'push factor' in the disinvestment from the region by several major West Midlands multinationals while the corporate strategies pursued by some key companies, most prominently British Leyland under the Edwardes Plan, were partly intended to undermine trade-union organization in their West Midlands plants.

The effects of corporate restructuring on the wider regional industrial complex are well illustrated by the experience of the motor industry. Vehicle manufacturing occupies a pivotal position in the West Midlands economy. It is not only the single largest employer in its own right, but also supports substantial numbers of jobs in the metal-based and engineering sectors. It was estimated in 1984 that 60,000 jobs in 4,000 firms within the West Midlands conurbation were directly dependent upon Austin Rover which itself employed a further 17,600 workers (Bessant *et al.*, 1984). The region's major vehicle manufacturers and assemblers, most importantly Austin Rover and Land Rover in Birmingham and Jaguar and Peugeot-Talbot in Coventry, are supported by an industrial substructure of components and equipment suppliers. Among the components firms, the larger and more sophisticated supply car components and sub-assemblies while the smaller firms supply basic castings, forgings and fabrications. Equipment producers supply metal-cutting and -forming machine tools, and mechanical handling and processing equipment.

From the mid-1970s the industry suffered a major contraction of output, capacity and employment. Within the West Midlands conurbation, redundancies announced by the major companies between 1979 and 1985 included 28,000 by British Leyland, 3,000 by Peugeot-Talbot, 8,500 by GKN, 10,000 by Lucas and 4,000 by Dunlop. Up to 1982 the majority of jobs shed were among the major car producers. Since 1982, however, most job losses have been in the supply-base of component and equipment manufacturers which have cut capacity in line with reduced demand from their major customers. The multiplier effects of contraction in the vehicles and engineering sectors are exemplified by the decline of the West Midlands machine-tool industry. In 1980 this industry employed over 10,000 workers in the conurbation. By

1983 the workforce had fallen dramatically by 67 per cent with what remained of the sector becoming increasingly foreign-owned, especially by US multinationals (WMCC EDU, 1983).

6.2.3 INVESTMENT, TECHNOLOGY AND SKILLS

Underlying this deteriorating industrial performance is a long-term record of underinvestment, technological backwardness and a collapse in skill training. Figures for 1983 show that capital investment per employee in West Midlands manufacturing was 21 per cent below the national average, the lowest of all British regions. Underinvestment has been particularly severe among the small and medium-sized, locally owned components producers and engineering subcontractors. Dependence upon a single market for their products has left them vulnerable and lacking in capacity to diversify. The failure of many firms to reinvest profits in more prosperous periods, their resistance to taking in outside share capital and the difficulty in securing loans from financial institutions have meant that many have been starved of investment. Forced to borrow short-term during a period of high interest rates, many went to the wall in the recession of the early 1980s.

Underinvestment in new plant and equipment has been accompanied by a decline of investment in the skilled workforce which has traditionally been one of the region's strongest industrial assets (Lewis and Armstrong, 1986). Between 1978 and 1985 the number of West Midlands engineering apprenticeships fell by 69 per cent. Chronically low levels of investment in research and development (R&D), new equipment and skills has been reflected in the West Midlands' poor rate of adoption of new product and process technologies (Marshall, 1985). Between 1968 and 1983 the number of industrial R&D units in the West Midlands fell from eighty-two to just twenty-seven. On the process side, a survey in 1983 revealed that over 30 per cent of machine tools in West Midlands industry were over 35 years old while only 1 per cent were modern numerically controlled models.

6.2.4 UNEMPLOYMENT AND INCOMES

The most conspicuous consequence of the West Midlands' decline since 1979 has been the dramatic rise in unemployment shown in Figure 6.4. The West Midlands conurbation's unemployment rate climbed from 6 per cent in 1979 to 16 per cent by 1986. The conurbation's inner-urban areas have been worst affected with fifteen local wards in 1986 suffering unemployment rates in excess of 30 per cent including eight wards with male registered unemployment at over 40 per cent. Unemployment among black and Asian workers is, on average, double the white rate and estimated at 80 per cent in some areas. The once highly paid West Midlands manual workers fell from top of the ten-region pay hierarchy in 1972 to eighth a decade later (Low Pay Unit, 1983) while it has been estimated that one in three households in this once prosperous area live on poverty incomes no more than 40 per cent above the minimum state benefit level (West Midlands Enterprise Board, 1986a).

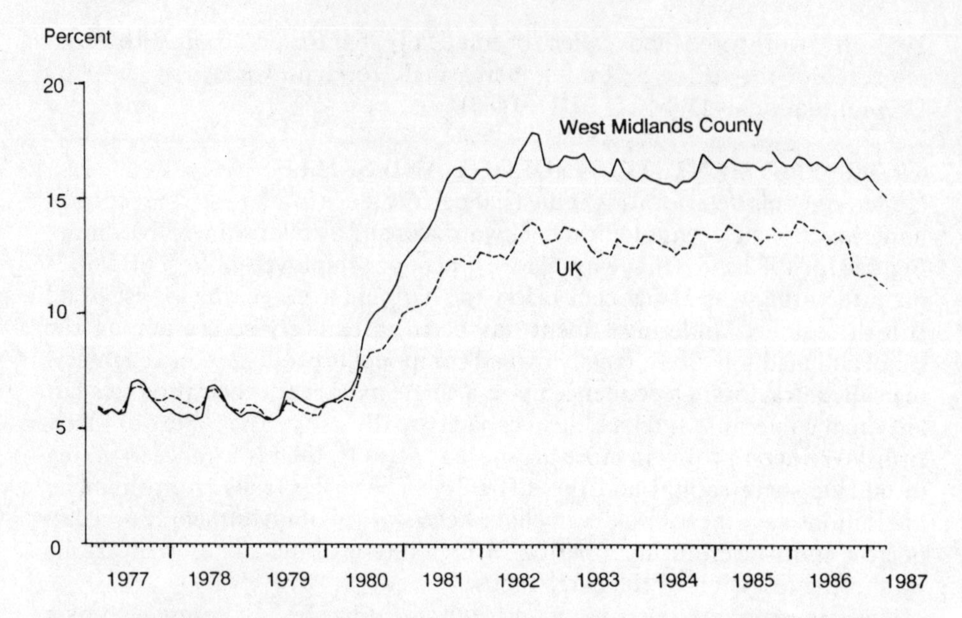

Figure 6.4 Unemployment rates in the West Midlands county and UK, 1977-87. Source: West Midlands Districts Joint Planning and Transportation Data Team.

To conclude this discussion of the West Midlands economic collapse, some appreciation of the scale of the conurbation's problems can be gained by setting them in a wider European context. The European Commission's 'synthetic index' measuring the relative intensity of regional problems in the EC, based on a measure of GDP per head combined with employment levels, ranks the West Midlands conurbation as the twenty-first worst of 131 EC administrative areas (Commission of the European Communities, 1984). This index may understate the scale and intensity of economic decline in the conurbation, since the 131 areas encompass regions with very different types of economic difficulty, ranking declining industrial regions alongside underdeveloped rural regions. Computation of a separate index for industrialized regions alone indicates that since the mid-1970s the West Midlands conurbation has suffered the greatest relative decline of all EC regions (Wabe, 1986).

6.3 CENTRAL-GOVERNMENT REGIONAL AND INDUSTRIAL POLICY

Looking at the critical condition of the West Midlands economy today, it seems incredible that until very recently the region was still officially regarded as one of the most prosperous in the UK. Under the redistributive regional development policies pursued by successive governments with varying intensity since the 1930s, the West Midlands along with the South-East was an industrial 'donor' expected to forgo expansion in favour of the 'assisted areas' in Scotland, Wales and the North. It was not until November 1984 that the West Midlands was itself designated an assisted area for the first time in its history.

6.3.1 REGIONAL POLICY

The effect of past regional-assistance policies on the economic health of the West Midlands has been the subject of much debate (Wood, 1976, ch. 8; Bentley and Mawson, 1985). Under the assisted-area system, West Midlands firms were required to apply to central government for an Industrial Development Certificate granting permission to expand their factory floorspace above a specified threshold. This was intended as an inducement for firms to divert their expansion to the assisted areas where new investment attracted grants and subsidies. This policy operated with varying force throughout the postwar period and was finally suspended in 1982. At its height from the mid-1960s to the early 1970s about 25 per cent of applications in the West Midlands were refused. Government statistics on factory movement indicate that this policy may have been responsible for diverting 39,000 jobs out of the region between 1960 and 1974.

While the precise effects of regional policy on the West Midlands are difficult to quantify, it seems certain that restrictions on new development contributed to the decline in efficency and disruption of intraregional industrial linkages which accompanied the process of corporate restructuring by leading West Midlands manufacturers from the mid-1960s. Relatively few firms relocated to the assisted areas following refusal of permission to expand in the West Midlands. Most took over dilapidated premises in the region or attempted to squeeze additional production from their existing premises. Those large-scale projects which were relocated, most notably the establishment of a new Chrysler car plant in Scotland at the expense of Coventry, contributed to the weakening of linkages between major vehicle producers and the West Midlands' supply-base of motor components and equipment manufacturers.

The long-term, structural nature of the West Midlands' economic decline was only finally formally recognized in November 1984 when a large part of the region, including the whole of the conurbation, was awarded 'intermediate', second-tier assisted area status. It is clear, however, that the gains from being placed on the government's map of regional aid will be much less significant than during the heyday of regional policies in the 1960s and early 1970s. The award of intermediate assisted-area status coincided with the government's announcement of a 42 per cent reduction in the national regional assistance programme over the following four years so that the West Midlands became eligible for support from a much reduced budget.

Prior to the award of assisted-area status in 1984, the present government had responded to the West Midlands' deepening decline with a succession of *ad hoc*, uncoordinated measures. Part of the Black Country district of Dudley was designated an enterprise zone in 1980 and a second zone was subsequently added in Telford. Companies moving into these zones receive a variety of incentives, including exemption from land-use planning controls and local authority rates (property taxes) for ten years. In April 1983 a Coventry Member of Parliament, John Butcher, was appointed as a junior industry minister with special responsibility for the West Midlands. Birmingham Airport was selected in February 1984 to take part in the government's Freeport experiment. The West Midlands has also benefited from a number of industry-specific schemes, such as the short-lived Small Engineering Firms Investment Scheme. Most recently, a large area of the Black Country has been taken out of local authority planning control and placed under a government-appointed urban development corporation charged with promoting redevelopment of derelict industrial sites in the area.

6.3.2 PUBLIC EXPENDITURE AND EC PROGRAMMES

As well as suffering direct restrictions on industrial expansion, the West Midlands has in the past received a low priority in the allocation of government industrial investment and public expenditure. Excluding the £2.6 billion of assistance to British Leyland after 1976, West Midlands firms received a mere £57.1 million of government industrial aid between 1972 and 1983, amounting to just over 1 per cent of total industrial support (Bentley and Mawson, 1984). The only consistent source of central government support for the local economy has been the Urban Programme, launched in 1978, which provides 75 per cent of the costs of economic projects in the conurbation's inner-urban areas. The Urban Programme is, however, predominantly geared to relieving social and housing problems rather than directly promoting economic regeneration. Of the £24.5 million allocated to the Birmingham Inner City Partnership Programme in 1985–6 only £8.1 million was for specific 'economic' projects.

The West Midlands has also fared badly under the EC's various programmes of regional and social assistance since support from a number of the EC's financial instruments, including the European Social Fund, European Investment Bank and European Coal and Steel Community as well as the European Regional Development Fund, is directly or indirectly dependent upon a region being designated for assistance by its own national government. Between 1973 and 1982 the West Midlands was awarded only 1.5 per cent of the total grants and loans made available to English regions from EC sources (Mawson and Gibney, 1984). The West Midlands' position was marginally improved after 1983 when the European Commission granted a special dispensation recognizing the conurbation's inner-urban areas as 'assisted areas' for the purposes of the European Regional Development Fund since it was acknowledged that the UK government's map of assistance blatantly failed to recognize the serious deterioration in the West Midlands' economic status. During 1984 the West Midlands received EC support for a number of infrastructural projects but the level of assistance remained much smaller than that awarded to other UK regions, amounting to only 4.9 per cent of the total £323 million allocated to the UK from the European Regional Development Fund during that year.

6.3.3 NATIONAL ECONOMIC AND INDUSTRIAL POLICIES

Those limited measures that have been taken to support West Midlands industry and employment since 1979, including the award of assisted-area status, pale into insignificance when compared with the negative impact of macro-economic policies which have had damaging consequences for UK manufacturing in general and the West Midlands in particular (Spencer *et al.*, 1986, ch. 6). The abolition of capital exchange controls in 1979 facilitated the ability of West Midlands-based multinationals to transfer investment abroad. Cutbacks and closures in the nationalized industries, as well as the privatization of state holdings, have directly contributed to the loss of employment. At the same time, tight monetary restraint and relatively high interest and exchange rates have inhibited West Midlands exporters, encouraged import penetration, and discouraged the long-term investment required to improve manufacturing productivity and competitiveness. Surveys by the main regional employers' organization during the early 1980s consistently highlighted the recession in demand as the major obstacle to investment and technological modernization (West Midlands Confederation of British Industry, 1983).

It would be misleading, however, to lay the blame for all the West Midlands' economic ills upon the present Conservative government. Earlier attempts by the 1964–70 and 1974–9 Labour governments to intervene in the restructuring of leading companies and sectors were far from wholly successful. The consolidation of British Leyland by the Industrial Reorganisation Corporation and subsequent intervention by the National Enterprise Board had major consequences, not only for the motor vehicle industry itself, but also for the region's component and equipment suppliers. Public funds invested in British Leyland over the past decade have helped to create a restructured and efficient car-production sector – albeit at the cost of massive job losses. But the interlinked component, metal-manufacturing and foundry sectors have been left to 'sink or swim'.

Plant closures and reduced capacity at British Leyland have meant that many smaller companies have seen their traditional major customer disappear practically overnight. The resulting process of intense competition and restructuring among suppliers has not necessarily resulted in the survival of the most efficient producers. In the foundry sector, for example, government assistance schemes between 1975 and 1981 encouraged firms to invest in new productive technologies and capacity, piling up debt at a time of high interest rates and market recession. Paradoxically, therefore, the most modern foundries in the West Midlands have often been those most vulnerable to closure. It is clear that past attempts to intervene in the restructuring of particular sectors have been pursued with scant regard for their effects upon the interlinked industrial structure as a whole while the present government's preference for market-led rationalization has resulted in 'survival of the lucky' as much as of the 'leanest and fittest'.

6.3.4 PROBLEMS OF POLICY COORDINATION

At least part of the explanation for central government's failure to coordinate the impact of its policies and programmes upon the West Midlands lies in the inadequate mechanisms for planned policy coordination by government departments at both national and regional levels (Marshall and Mawson, 1987, pp. 108–9). Outside the assisted areas the Department of Trade and Industry (DTI) does not have a clearly acknowledged geographical role. The department's West Midlands office has been mainly responsible for the regional administration of national programmes, notably the various Support for Innovation schemes. Even programmes with a very strong regional relevance, such as the Small Engineering Firms Investment Scheme, have been managed from outside the region (in this case Swansea in South Wales). Although the DTI's activities clearly have major regional implications, the department has never successfully integrated these considerations within national industrial policy.

As the ministry responsible for overseeing local government, the Department of the Environment (DoE) has a larger and more significant regional presence than the DTI. The DoE has been engaged in a variety of spatially bound programmes of economic development including allocation of resources under the Urban Programme, designation of enterprise zones and urban development corporations, and management of local-authority capital-spending programmes including submission of applications for infrastructure projects to the European Regional Development Fund, as well as overseeing the process of statutory land-use planning. All these programmes have direct regional consequences yet there is no systematic mechanism for coordinating their impact with the DTI's regional and industrial support.

Paradoxically, the appointment just prior to the 1983 general election of an industry minister with special responsibility for the West Midlands seems to have further confused rather than clarified interdepartmental coordination. In practice the minister was only given responsibility for a limited range of DTI activities. Yet he was widely perceived as the minister responsible for *all* matters affecting the region, providing an illusion of coordinated government action which served only to obscure the reality. The appointment was widely seen in the West Midlands as no more than a cosmetic political exercise and the minister was certainly never given the powers and resources to match the scale of the region's problems.

6.4 THE EMERGENCE AND DEVELOPMENT OF LOCAL ECONOMIC INITIATIVES, 1981–6: THE ROLE OF THE WEST MIDLANDS COUNTY COUNCIL

While central government has been slow to recognize, let alone respond to, the West Midlands' economic decline, the region's trade unions and local authorities have long been vocal in calling for major changes in government regional and industrial policies in response to the West Midlands' deteriorating economic performance (WMCC, 1974; West Midlands Economic Planning Council and Planning Authorities Conference, 1979; Trades Union Congress West Midlands Regional Council, 1981). Until the early 1980s the region's local authorities maintained a relatively passive stance towards economic intervention, continuing to see industry and employment as the preserve of national macro-economic and regional policies. Economic initiatives at the local level were mostly confined to provision of new industrial land and premises together with promotion of the region's attractions for inward investment. Since the early 1980s, however, the unprecedented collapse of the regional economy together with the lack of any meaningful response on the part of central government have led many of the West Midlands' local authorities to reassess their role in the local economy (Spencer *et al.*, 1986, ch. 7; Marshall and Mawson, 1987, pp. 112–20). In the case of the West Midlands County Council, a wide range of local initiatives were developed which represented a radical break from previous approaches (Mawson *et al.*, 1984). These initiatives directly confronted the problems of investment, technology, training and business planning as well unemployment and poverty which have been identified as the major causes and consequences of the West Midlands' decline.

The Labour Party took control of the West Midlands County Council in May 1981 with a manifesto commitment to embark upon a radical economic development strategy. The main elements of this strategy had been mapped out prior to the election by various individuals and working parties grouped loosely around the County Labour Party and the Conference of Socialist Planners. Several of those involved had broader national policy interests and were later to play a prominent role in the implementation of the strategy, not least the former Walsall Member of Parliament, Geoff Edge, who became chairman of the county council's newly created Economic Development Committee (EDC), and the late Terry Pitt, who headed the new Economic Development Unit (EDU) until 1984 when he was elected as a Black Country Member of the European Parliament.

Following the county elections, the new economic strategy was given its first detailed public presentation in the chair's statement to the EDC in November 1981 on *Policies and Priorities for Economic Development in the West Midlands* (Edge, 1981). This set out the basis for a new approach to the development and implementation of local economic policies in the conurbation. The principles underlying this approach were very explicitly stated (WMCC EDC, 1984):

'1. The future prosperity of the West Midlands lies in the survival and strengthening of its traditional industries.'
The EDC prioritized support for the conurbation's traditional manufacturing industries. It was acknowledged that although there may be new firms and industries moving into or developing within the West Midlands, these would have only a marginal role in the regeneration of the local economy. The EDC's commitment to the West Midlands' existing manufacturing base was paralleled by an equally strong rejection of arguments for concentrating public support on service industries, so-called 'sunrise' high-technology sectors, or beggar-my-neighbour policies of geographical redistribution.

'2. There is an urgent need to increase the flow of investment funds into the West Midlands.'
The EDC regarded low levels of industrial investment, both past and present, as the key factor underlying the decline of the West Midlands' industrial base. In particular there was concern over the institutional structure of investment in the UK and the lack of regionally based investment agencies which were seen as an important advantage enjoyed by UK manufacturing's chief international competitors such as West Germany.

'3. There is a need to ensure public accountability where public funds are used to support private industrial and commercial activity.'
The EDC argued that public support for industry should be 'an investment rather than a gift', taking the form of shareholdings or loans and not grants. Public investment should be expected to secure a long-term financial return and should be subject to conditions ensuring that companies in receipt of public support acted in the best interests of their workforce and the wider community.

'4. There is a need to invest in human skills as well as buildings and plant.'
The EDC regarded investment in industry as only one half of the formula for economic revival. It was equally important to invest in the region's skilled workforce, not only to ensure that future industrial recovery was not constrained by lack of workers with appropriate skills, but also as a means of upgrading the skills of disadvantaged social groups to enable them to take a full share of economic life.

'5. The Committee recognises that there are within the community a whole series of energies and talents which can be brought together to create employment opportunities.'
Following this principle, the EDC gave special emphasis to cooperative

and community enterprises which provide a means for local people to create their own employment while representing an opportunity for full worker participation in enterprise planning and decision-making.

'6. The Committee recognises that despite the very best efforts of the County Council to create and secure jobs, problems of unemployment, low pay and poverty will all continue on a large scale in the County area.'

This was an important principle of the EDC's approach, recognizing that economic development policies should be concerned not only with employment and investment but also with incomes and living standards.

The centrepiece of the EDC's programme was an explicit 'industrial strategy' whose first priority was to revive investment and employment in West Midlands manufacturing while making all public investment accountable to those who lived and worked in the area. Specific initiatives undertaken included the establishment of a regional investment agency, the West Midlands Enterprise Board; an interest relief scheme for companies seeking long-term loans for investment in new equipment; a scheme to ease the problems of firms struggling in inadequate or burdensome premises; the establishment of a network of local agencies providing finance and support to worker cooperatives; a business-consultancy scheme to assist firms to improve their long-term business strategies; and a variety of initiatives to upgrade the region's technological infrastructure including the establishment of the Warwick University Science Park.

This industrial strategy was paralleled by a 'community strategy' concerned with social opportunities, income distribution and living standards. Measures undertaken included a wide range of skill-training projects; local campaigns to improve awareness of and access to welfare benefits; the establishment of a regional low-pay unit; specific schemes aimed at assisting black- and Asian-owned businesses; and support for community, trade-union and unemployed-workers' centres.

6.5 DEVELOPMENT SINCE 1986: THE ROLE OF THE WEST MIDLANDS ENTERPRISE BOARD

The West Midlands Enterprise Board (WMEB) was the principal agent of the EDC's Industrial Strategy, established by the county council in February 1982 as a regional financial agency providing long-term development capital to indigenous manufacturing firms with over 100 employees (although this size requirement was later reduced to fifty). The WMEB was constituted as a limited company with a majority of county councillors on its board of directors. Investment capital was provided in the form of equity shareholdings and long-term loans, replacing the practice of providing grant aid to industry. All investments were subject to formal planning and investment agreements through which the investee firms were required to make legal commitments covering trade-union recognition and employment practices as well as their commercial development.

By March 1986 the WMEB had approved investments totalling £14.2 million in thirty-three companies. Three of these companies had subsequently failed, while the investments in a further three firms had been realized. The remaining twenty-seven firms in the board investment portfolio employed a combined workforce of 2,635. Each WMEB investment was part of a larger financial injection negotiated with private-sector financial institutions with every £1 invested by the board matched, on average, by £4 of investment from other sources.

The abolition by central government of the West Midlands County Council in March 1986 posed a major threat to the survival and expansion of local economic initiatives in the conurbation. With the exception of Birmingham City Council (1985), by the time of the county council's abolition the local district authorities had committed comparatively meagre resources to economic development in their areas. Moreover, the county council's abolition detracted heavily from the finance that could be raised locally for this function. Most of the financial resources committed by the county council to economic development were raised under a specific legal power which allows local authorities to levy additional local rates up to a specified limit for expenditure on local needs. The removal of the upper tier of local government effectively halved the total sum that could be raised by the conurbation's local authorities under this power.

Notwithstanding the efforts of the local district councils, one of the most important strengths of the former county council's approach to economic development was its inherently *strategic* character. The economic structure and problems of the conurbation are so complex, large-scale and interrelated that any meaningful economic intervention in the West Midlands must be taken on a conurbation-wide basis. The strategic initiatives undertaken by the county council could not be easily replaced by isolated initiatives within individual districts. It was in this context that, following the county council's abolition, the West Midlands Enterprise Board took on a wider role in economic regeneration.

The WMEB was always conceived as one element within a much broader economic strategy. The abolition of the county council in March 1986 and subsequent transfer of control to the seven metropolitan district councils, all but one of which were Labour-controlled, enabled the board to expand its range of economic-development functions, complementing its existing investment activities. This enlarged role involved the establishment of a network of subsidiary companies which kept intact many of the initiatives formerly undertaken by the county council (Edge, 1986). These comprise:

Investment The geographical scope of the WMEB's investment activity has been expanded beyond the conurbation to cover the whole of the West Midlands region. This accompanied the establishment of the West Midlands Regional Unit Trust which the board jointly manages with a London-based merchant bank as an outlet for pension-fund investment in the region's manufacturing industry.

Technology transfer The West Midlands Technology Transfer Centre was established at Aston Science Park in Birmingham in a joint initiative with Aston University. The centre offers consultancy services to manufacturing companies seeking to introduce new product or process technologies, for example on a licensing basis. The WMEB also maintains a shareholding in Warwick University Science Park.

Clothing industry The West Midlands Clothing Resource Centre (discussed in more detailed below) was set up in Sandwell as a focus of support for the expanding local clothing industry. The centre's bureau services include access to computer-aided pattern laying and marking equipment as well as management and operative training.

Cooperative development West Midlands Cooperative Finance works with the local cooperative development agencies to provide support and financial assistance to workers' co-operatives.

Training The West Midlands Training and Community Resource Centre, established in Birmingham, includes an Information Technology Centre for young people and adult-training facilities in fields such as computer numerically controlled machine-tool operation and computer-aided design.

Welfare rights The West Midlands Welfare Rights Agency provides a campaigning resource and advice service to improve access to state benefits.

Research and analysis The WMEB maintains an economic intelligence and strategy team which plays a major role in guiding the development of the board's economic strategy while providing a prime source of information and analysis on the regional economy.

Together these activities enable the WMEB to operate as a strategic, region-wide economic development agency under local democratic control.

6.6 INDUSTRIAL-SECTOR STRATEGY AS A MODEL FOR INTERVENTION

One of the distinctive elements of the county council and WMEB's economic intervention was, and remains, its reliance on a sector-based industrial strategy (Elliott and Marshall, 1989). The choice of this approach has been guided by an analysis both of the structure and development of the West

Midlands economy and of the lessons of previous attempts at industrial intervention pursued by past Labour governments. Given the complex sectoral relationships between the West Midlands' vehicles, engineering, metal-goods and other manufacturing industries, any intervention at the level of the individual enterprise must take into account broader sectoral and intersectoral implications. Past national policies for industrial intervention are judged to have failed precisely because, among other weaknesses, they failed to encompass and plan for these broader sectoral and regional consequences.

The sector strategy developed by the WMEB is founded on an appreciation of the dynamic interlinkages between local firms and industries as well as their relationships with trade unions, central and local government and other agencies. Sectors, in this sense, are groups of firms within an industry which may – not unproblematically – be characterized by common inputs, outputs, productive technologies or external influences. The approach recognizes that sectors are subject to a variety of economy-wide forces, such as the structures and strategies of multinational companies and government industrial or economic policies. Hence the concept of 'sectors' refers to much more than industrial activities defined for statistical classification purposes. Sectors are conceived of as the total ensemble of economic, social and political interests grouped around a particular industrial focus.

The WMEB's sector-based approach to industrial intervention is seen as 'strategic' in a variety of ways. It is strategic in geographical scale since it recognizes that economic processes and relationships transcend local or regional boundaries and must be confronted in complementary ways at different levels. The approach signifies the opposite of 'piecemeal', implying that strategic sectoral intervention must take into account the combined effects of a range of particular interventions to address interrelated issues within and across industrial sectors. This does not mean that intervention must be founded upon a comprehensive sectoral analysis and pursued via a predetermined plan. Rather, intervention and analysis should proceed simultaneously, the one guiding the other, through an iterative process of investigation, intervention and evaluation within overall policy parameters.

This conceptual framework has a number of practical advantages (Elliott, 1986):

Coherence The approach offers the potential for assessing the impact of a combination of coordinated policy instruments in related fields such as direct investment, skill training and technology transfer.

Consensus Sectors are industrial categories which have real meanings for managers, trade unions, employers' organizations, training agencies and governments. Policies and initiatives directed at particular sectors offer the possibility for mobilizing these interests around common projects.

Analysis The identification of sectoral linkages offers the basis for developing a dynamic analysis of the economy, highlighting targets where intervention can have the strongest strategic impact.

Advocacy Analysis identifies not only the potential but also the limitations of economic intervention at the local or regional levels. Sectoral analysis, together with evaluation of the practical experience of sectoral intervention, indicates the extent to which local initiatives might be more widely imitated or need to be supplemented by complementary forms of intervention on national or international scales. This provides a strong basis for effective advocacy of economic and industrial policies that should be adopted by other agencies or by higher levels of government.

The practical application of this approach is best illustrated through examples of the WMEB and former county council's work with particular industries discussed below. The foundry industry exemplifies the contribution of direct investment to intervention in the restructuring of a traditional West Midlands metal-based sector while the motor and components industry illustrates how direct intervention has been extended by policy advocacy in alliance with other groups. The WMEB's intervention in the clothing industry shows how the sector-based approach has assisted the development of a range of varied initiatives addressing interrelated issues within this sector. Finally, the WMEB's technology-transfer initiative illustrates one way in which sectoral analysis has guided intervention in an area of broad, cross-sectoral significance.

6.6.1 THE FOUNDRY INDUSTRY: DIRECT INVESTMENT IN RESTRUCTURING

The West Midlands foundry industry typifies many of the problems faced by the region's metal-based sectors. The region dominates national production in the industry, accounting for 37 per cent of all UK foundry employment in 1983. Foundries provide primary and intermediate products for a wide range of industrial customers led by vehicles and engineering applications. Employment in the West Midlands sector is heavily concentrated in medium-sized firms, typically employing 50–500 workers. Many foundries rely on a small number of customers, often locally based. Since the late 1970s the contraction of foundries' major markets has led to a dramatic decline in demand for castings. Between 1979 and 1986 output of ferrous foundries fell by 61 per cent and of non-ferrous foundries by 24 per cent. In the West Midlands this was reflected in employment decline of 43 per cent between 1978 and 1984. Despite many closures and two national rationalization schemes, the fragmented nature of foundry ownership together with continued decline in industrial demand has maintained chronic overcapacity in the foundry sector. This results in periodic phases of suicidal price-cutting as a short-term expediency which undermines attempts by some of the more farsighted manufacturers to introduce new products and processes and enter fresh markets.

The launch of the WMEB in 1982 therefore coincided with a period of dramatic change in the foundry sector. Perhaps not surprisingly, the sector accounted for three of the first four investment proposals considered by the board. Typical of these was Butler Foundries, a Black Country manufacturer producing castings for the automotive and mechanical engineering markets. Despite strong exports and modern production techniques, the company had suffered the effects of escalating interest rates shortly after investing in new equipment during 1978. High debt charges had militated against further expansion of the business and the company's long-term future was in doubt. Butlers had been unable to identify any alternative source of long-term development capital.

In December 1982 the WMEB invested £500,000 of share capital to restructure Butlers' balance sheet and fund expansion. The investment was accompanied by insistence that Butlers systematically develop its long-term business strategy through improvements in financial control and marketing. This was enforced by the WMEB through rigorous quarterly monitoring and the appointment of a nominee director. The board maintains the right under certain circumstances to convert its minority shareholding into a large majority, providing a further sanction against under-performance by the company.

In the event, Butlers has fulfilled the potential envisaged in 1982. The workforce has increased from 120 to 160 while annual turnover has almost doubled to £5 million. The company has retained a strong customer spread and achieves profits substantially above the average for the sector. The WMEB made a further small investment in 1984 to assist the development of new testing facilities and the company is now funding this programme from internally generated funds.

The WMEB currently has four investments in the foundry sector and continues to receive a steady flow of investment enquiries from foundries. It is clear that most institutional investors still regard the sector as a poor prospect and are predisposed against investment. Without the board's intervention, Butlers and other foundries would doubtless not have survived.

The WMEB's foundry investments have been supported by other initiatives towards the sector. In 1982 the county council undertook a major study of the West Midlands foundry sector (WMCC EDU, 1982) culminating in a conference on the industry's performance and prospects in May 1983. Arising from this conference, a joint export marketing scheme was set up involving two foundries and a machining company in a collaborative venture to spread costs of entering the North American market.

6.6.2 MOTOR VEHICLES AND COMPONENTS: POLICY ADVOCACY AND ALLIANCES

The motor and component industry is of leading strategic significance within the West Midlands economy, employing an estimated 145,000 people in the region. The industry is dominated by leading companies operating on a national, European or even global basis. This clearly presents difficulties for a regionally based economic development agency seeking to intervene effectively in the sector. Although the WMEB has provided direct support to a number of small and medium-sized components manufacturers, the main thrust of the WMEB's intervention has been to build broader alliances with other groups on the basis of the board's industrial research and policy advocacy.

The WMEB's work in the sector has been founded upon a series of research studies examining the relationship between the Rover Group and its component suppliers. These studies have revealed the extent to which investment in modern manufacturing technologies and the development of aggressive purchasing policies by Rover have weakened the components sector, leaving it comparatively technologically backward yet unable to invest in complementary technologies. These trends have been compounded by the government-backed establishment of UK productive facilities by two major Japanese car and components producers, Nissan and Honda. This poses a major threat both to Rover's domestic market share and to the already weakened linkages between Rover and its longstanding UK components suppliers.

The main form of direct intervention by the WMEB has been through capital investment. By mid-1987 the board had invested in seven motor components firms, a vehicle spares and parts distributor and a number of firms supplying productive equipment, such as robotic systems, to the components sector. Direct investment has been supplemented by training courses in electronics, robotics, pneumatics and hydraulics to equip engineers with the skills necessary for the introduction of new technologies.

Drawing on research and analysis, as well as this experience of intervention in the sector, the WMEB's advocacy work on the motor and components industries has been undertaken within a broader policy framework consisting of five main elements (WMEB, 1986b):

- the need for policies to be developed within an international perspective of restructuring in the sector, focusing on the corporate strategies of the major multinational producers;
- the need for a strategy towards the West Midlands and UK vehicle and components industries to be based upon a publicly owned UK motor vehicle manufacturer producing the full range of vehicles;
- the need to encourage greater cooperation between manufacturers and component suppliers, particularly in the design and development of new vehicles;

- the need for national investment, training and technology policies towards the sector which explicitly take account of the importance of motor vehicle and component manufacture to the West Midlands economy;
- the need to support and build alliances with trade unions, companies, employers' organizations and other interested bodies where they share similar policy perspectives to those of the WMEB.

It has been recognized that the WMEB's role can be more effective if joined by other organizations sharing similar perspectives towards the industry. Hence, the board has embarked upon a longer-term campaign to build alliances with other local and regional authorities in the UK and Europe as well as with the trade union movement.

The WMEB has been instrumental in forming two UK and European local authority working groups. The UK Motor Industry Local Authority Network is undertaking a programme of area studies covering the main UK vehicle manufacturing regions. It is anticipated that these will generate practical proposals for local intervention in the industry, for example through training initiatives. The UK Network is complemented by a European working group of local authorities in Italy, France, West Germany and Belgium as well as the UK. The group is directing its energies towards influencing the EEC to improve coordination of its regional, social, economic and technological programmes in relation to the motor industry and to explicitly take account of the specific actual or potential impact of these programmes on European vehicle-producing regions (Council of European Municipalities and Regions *et al.*, 1985). In addition, the group is campaigning for a new form of special EEC assistance to regions suffering employment decline due to technological change and restructuring in the industry. Both the UK and European working groups aim to articulate a 'fourth interest', representing the views of regions and communities affected by restructuring of the motor industry alongside those of management, trade unions and national government. Clearly there are tensions inherent in any collaborative efforts, particularly on an international scale, between regions whose industries are in mutual competition. This has limited the scope of the European group which is mainly concerned with the development of planned measures to address socio-economic consequences of regional industrial *decline*, as has been achieved in the cases of the European coal, steel and textiles industries, rather than intervention in international competitive relations.

Alliances with employers and trade unions are more difficult to establish than with other local and regional authorities because the overlap of policies and interests is neither as strong nor as permanent and because the different representative organizational structures do not mesh so easily. Employers seldom share all the elements of the WMEB's policy perspective and alliances with this group therefore tend to be transitory and based around a single issue on which the companies' commercial interests coincide with the WMEB's strategic orientation towards the sector. One example currently being explored is the possibility of developing an interface between aftermarket distributive and retail chains and West Midlands manufacturers of car spares and replacement parts.

Alliances with trade unions are less short-lived but inter-union differences and the problem of separating immediate industrial relations issues from a longer-term strategy for the industry require careful negotiation. The board has recently established a West Midlands Auto Industry Trade Union Forum (1987) which brings together trade-union officials and local-authority representatives. This is a potentially powerful campaigning group, capable of minimizing the divisions which have hitherto impeded a coherent and united trade-union and community view on the future of the motor and components sector.

6.6.3 THE CLOTHING INDUSTRY: INTEGRATED SECTORAL INITIATIVES

Unlike the foundry and vehicle sectors, the clothing industry has not traditionally been of major significance to the West Midlands. There are fewer than twenty major clothing manufacturers in the region. However, since the mid-1970s there has been a dramatic growth in the number of smaller clothing firms in the West Midlands conurbation, virtually amounting to a whole new local industry. A study commissioned by the former county council identified 500 of these small firms, mostly in the conurbation's inner-urban areas, employing 7,000 workers directly plus an estimated 5,000 homeworkers (Leigh and North, 1983).

The emergence of these firms was originally stimulated by redundancies among Asian workers in the metal-based sectors. Given their disadvantaged position in the labour market, many used their access to small amounts of capital and a pool of cheap female labour within the Asian community to start up in an industry with almost non-existent barriers to entry. Many of the newer owners have neither managerial experience nor relevant manual skills and often have no prior involvement in the industry. Hence, while the few long-established West Midlands clothing manufacturers tend to focus on high-quality tailored goods, the newer, small firms are concentrated in the lower-quality-product areas. These firms are characterized by low profitability, a high degree of sub-contracting work and inefficient production techniques. These features ensure generally low, and sometimes illegal, pay levels together with poor working conditions (West Midlands Low Pay Unit, 1984). There is a marked lack of formal operative training and reliance on outdated equipment which in turn leads to low productivity, excessive material wastage and poor product standards.

It was clear from the outset that any programme of intervention in the sector had not only to assist firms through investment and management training and advice, but also to support the workforce by tackling the low wages and poor conditions, especially among homeworkers, which are endemic to the industry. Initiatives were developed to address each of these problems. Direct investments were made by the WMEB in some of the larger, high-volume, established clothing businesses. A business-advice scheme, aimed partly at Asian-owned firms, together with a clothing-design project were also developed. Training courses for operatives, managers and senior technical staff were set up in collaboration with a local technical college, the main government training agency, the employers' organization and professional association. Measures to improve pay and conditions in the industry involved working closely with the West Midlands Low Pay Unit, an action-based research project and a campaign highlighting the conditions suffered by homeworkers in the industry.

The tension implicit in a strategy which sought to strengthen the interests of both capital and labour was always accepted. But it was also recognized that any intervention which simply sought to strengthen one side of the industrial divide would ultimately destabilize the industry to the benefit of neither group. The overall aim was to upgrade the industry *as a whole* from a low-wage, low-investment and low-productivity sector to one where higher levels of investment and productivity could support improved wages and working conditions. The disparate range of initiatives listed above clearly could not achieve this in isolation. There was a need to ensure coordination and consolidation of these initiatives as well as to develop links with other local clothing initiatives being developed elsewhere in the UK.

The West Midlands Clothing Resource Centre was established by the WMEB in 1986, providing a physical and organizational focus for initiatives towards the sector. In contrast to the fashion centres set up by several local authorities in other parts of the UK to address problems of marketing, the West Midlands centre was aimed directly at intervention in production. The centre houses computer-aided pattern-laying and -marking facilities to which local clothing firms subscribe on a bureau basis as a means of improving production efficiency. Other types of business assistance, such as management and operative training, are provided around this core activity. Ongoing access to these services is conditional on employers making a commitment to improve pay and conditions to recognized standards. In the longer term, those firms with sufficient potential will be led upmarket, away from dependence upon sub-contracting and sales agencies towards integrated production and direct marketing.

Links between the West Midlands Centre and other UK local initiatives have been developed through the formation of a joint working and campaigning group, Local Action for Clothing and Textiles. The group coordinates, supports and informs local authority initiatives while advocating national policies to provide targeted support for the clothing and textiles industries, building upon the experiences at local level.

6.6.4 TECHNOLOGY TRANSFER: CROSS-SECTORAL INTERVENTION

The West Midlands Technology Transfer Centre, developed jointly by Aston University and the WMEB at Aston Science Park, is one of a range of complementary technology-oriented initiatives. Others include training courses in new technology-based skills, the WMEB's continuing involvement in the Warwick University Science Park and specific investments in a number of firms producing or introducing advanced manufacturing technologies such as robotics. The broad objectives of the WMEB's technology-based interventions are:

- to encourage use of new technology where it enhances competitiveness or develops new linkages and new areas of production in the regional economy;
- to do this in ways which encourage participation by the workforce, build existing and new skills and open up novel areas of employment (WMEB, 1986c).

The Technology Transfer Centre aims to make a specific contribution to these objectives by encouraging the diffusion of new technologies between different users and producers, strengthening linkages between sectors and subsectors of West Midlands industry. Unlike many other UK local initiatives to promote technological innovation, the West Midlands Centre aims to support the technological upgrading of *existing*, indigenous manufacturing sectors rather than foster the development of new, high-technology industries. It is argued that the potential for development of new industries in advanced technology fields, as well as their contribution to the regeneration of the regional industrial structure, can be maximized if these activities are closely linked to the modernization of existing manufacturing industry.

The centre provides consultancy services to assess the technological needs and potential of client firms in the context of their wider business development. Assistance is provided in drawing up specific projects which may include the adoption of new productive techniques or introduction of new products, for example on a licensing basis. The centre is also able to assist in the implementation of projects through access to Aston University's technical expertise as well as the training and investment services of the WMEB. Prior to the centre's establishment, a regional Innovation Campaign was conducted through a series of seminars for industrial managers coupled with follow-up surveys.

The centre has adopted a sector-based approach, working with selected local firms on the basis of their actual or potential intra- and inter-sectoral relationships. The complex, interlinked structure of the West Midlands' engineering and metal-based industries necessitates a strategic perspective of technology transfer capable of addressing interrelated sectoral issues. By working with groups of firms in related industrial and technological fields it is possible to identify opportunities for developing linkages and collaborative relationships between the centre's client firms.

Two initial priority sectors have been identified for the centre, namely motor components and machine tools. The motor components sector has been prioritized since, as well as being the single most important West Midlands industry, the sector is characterized by strong inter-company relationships between vehicle producers, first-tier component suppliers, second-tier suppliers and materials and equipment producers. Partly because of these customer–supplier relationships, the sector is undergoing rapid changes as new technological requirements, for example in computer-aided design, are passed downstream from the vehicle producers through the various tiers of component suppliers. Hence there are strong opportunities for developing and strengthening intra-sectoral linkages on the basis of technological collaboration in vehicle design and production.

The machine-tools industry is also a key regional industry, located at the core of all metal-based manufacturing processes. Several of the Technology Transfer Centre's contact firms in other industries are known to be seeking new tooling or modifications to existing tools. The centre's work in the machine-tools sector can help foster links between West Midlands suppliers and users while encouraging suppliers to explore the wider market for any new product innovation piloted in the region. In addition to the machine-tool industry's contribution to the technological upgrading of customer sectors, the machine-tool industry is itself undergoing rapid technological changes. Several West Midlands manufacturers have restructured their product range towards the expanding markets for computer-numerically-controlled equipment through licensing arrangements with overseas producers. Some of these arrangements have involved little more than licensed assembly or distribution with no meaningful technology transfer, for example in software control systems. This is an area where some of the smaller local producers may benefit from external expertise.

6.7 CONCLUSION: LOCAL ECONOMIC INTERVENTION IN LONG-TERM PERSPECTIVE

The West Midlands' recent economic slump is the latest phase in a long-term process of industrial change and development spanning the region's 200-year industrial history. One of the most striking characteristics of the West Midlands' historical pattern of industrial development has been its *continuity* as the region's core metal-based industries have evolved and developed, re-emerging in each fresh long wave of economic development in a metamorphosed and rejuvenated form. Unlike the 'branch plant' economies of many other European industrial regions, the West Midlands has maintained a unique capacity for self-sustaining indigenous development. Many of the key industries and companies that make up the West Midlands' present industrial structure can be traced back to family-owned nineteenth-century metalworking businesses while half the West Midlands' top twenty multinational enterprises have their headquarters in the region.

While it seems unlikely that the West Midlands will regain the levels of manufacturing activity enjoyed in the past, it would be premature to write off the region's traditional industrial strengths. Although service industries may continue to grow steadily and there may be new manufacturing sectors developing in the region, the direct contribution of these activities to the region's economic regeneration will be relatively marginal. The West Midlands' economic development will continue to be dependent upon the interlinked structure of vehicles, engineering and metal-based sectors which, despite major losses of employment and capacity, remain the dominant feature of the regional industrial landscape. Recent advances in productive organization and technologies offer the potential for a long-term regeneration of industrial competitiveness involving strong inter- and intra-sectoral industrial linkages. However, experience suggests that these types of industrial linkages will not develop spontaneously. The market-led process of industrial restructuring has already weakened the intra-regional linkages that have in the past been a major strength of the West Midlands' industrial structure. In many companies and sectors, short-term survival strategies have been pursued at the expense of the long-term investment in fixed capital, skills and new product and process technologies required to renew the West Midlands' capacity for indigenous growth.

The West Midlands' key manufacturing industries occupy a strategic position, not only within the regional economy but also from a national perspective. The West Midlands has historically been the manufacturing powerhouse of the world's first industrial nation. In many respects, the postwar decline of the UK economy can be equated with Britain's decline as an international manufacturing nation which is in turn closely connected with the decline of the industrial heartland of the West Midlands. The past development and future prospects of the region are inseparably linked with the internal foundations of the British economy and its external international relationships.

The single most important factor in the decline of West Midlands manufacturing has been underinvestment in the new products, productive techniques and skills required to preserve the region's competitive position in world markets. The problem of investment is not new. On the contrary, it is deeply rooted in the socio-economic, political and geographical division between industrial production and Britain's financial and commercial role which is as old as British industrial capitalism itself. Britain today remains divided between two competing visions of the basis for future prosperity. The first regards a reconstructed modern manufacturing base as the key to restoring employment levels and economic growth, not only in the manufacturing sector itself but also in supporting service industries. The second sees

Britain's continuing role as a leading international financial and commercial services centre, pivoting on the role of the City of London, as the vehicle by which Britain will 'pay its way' in the world.

In all their essentials, these two visions have remained remarkably unchanged during the past two centuries. For most of Britain's industrial history they have coexisted because the international financial and commercial interests of Britain's ancient ruling order of aristocratic, financier and mercantile classes have coincided with the aspirations of industrial capital. Britain's primacy as the world's first industrial nation was in large part a consequence of British manufacturing industry's ability to link into the pre-existing international trading network established by the preceding mercantile order. But at each subsequent historical crisis and turning point in the ensuing course of British economic development this social and political consensus has broken down in protracted conflict over the basis of Britain's international economic role. During the first Kondratieff downturn of the 1830s and 1840s it was the factory- and millowners of Manchester, whose cotton textiles industry had been the leading sector of the Industrial Revolution, that led the free trade movement against the protectionist aristocratic landowners. Following the Great Depression of the late nineteenth century it was the Birmingham-based social imperialists who sought to redefine Britain's international industrial role in opposition to the overseas investments pursued by the London-based financiers at the expense of domestic manufacturing.

In this historical context, it can be seen that the recent resurgence of local and regional economic initiatives in the UK is not without precedent. The leading industrial regions of the UK economy have provided the basis for regional social and political movements which have, at critical moments, contested the ensuing course of national and even international economic development. The key factor in the overall course of British economic development has been the unique social character of British capitalism in the division between industry and finance capital which has itself entailed a spatial division between the provincial industrial regions of the North and Midlands and the metropolitan centre of finance and commerce. Over the past two decades the British economy has once again been undergoing a critical period of change which once again has been the subject of intense political conflict in which the continuing regional socio-economic divide has been a marked feature.

It would be a mistake to regard the regional economic strategies developed in the West Midlands and elsewhere since 1981 as purely *regional* strategies of self-help. Interventionist local and regional industrial strategies have been consciously developed as *demonstrations* of the kind of policies which need to be pursued at national levels of government, as 'paradigms for a planned economy' (Benington, 1986). The industrial-sector strategy pursued by the WMEB and former county council provides a practical demonstration, first, of the continuing importance of manufacturing activity in the national economy; second, of the need for a coherent national industrial strategy to secure the revival of the UK manufacturing base; and third, of the role of democratically accountable regional development agencies within such a national strategy. Experience and analysis suggest that traditional forms of arm's-length economic management are an insufficient response to the problems of UK and West Midlands industry. While manufacturing industry would undoubtedly benefit from changes in macro-economic policies, including planned reflation of demand and more favourable interest and exchange rates, such measures will not in themselves solve the structural weaknesses of the UK manufacturing sector (Marshall, 1986). Problems of investment, training, technology and business planning will need to be confronted through direct industrial and regional intervention in key companies and sectors.

Over the past few years, local examples of economic intervention by Labour authorities in the West Midlands and elsewhere have won growing national recognition. The alternative economic programme on which Labour contested the 1987 general election explicitly acknowledged the contribution of local authorities, regional development agencies and enterprise boards to a national programme of investment in jobs, skills and industry in partnership with new national institutions for industrial investment and planning (Labour Party, 1987, p. 6). This is an important advance on previous models for economic planning, recognizing that industrial intervention should not be the sole prerogative of central government but must necessarily involve a major role for local and regional economic agencies (Batkin, 1987).

No one involved in the development of local, regional and national economic strategies underestimates the enormity of this project. The Thatcher government was re-elected in 1987 for a third term of office and there is clearly no immediate prospect of any change in central government policy towards the industrial regions. It is worth noting, however, that while the Conservatives won 88 per cent of the parliamentary seats contested in the south of Britain, their share of seats was 63 per cent in the Midlands, 40 per cent in the north of England, 21 per cent in Wales and only 14 per cent in Scotland (*The Economist*, 1987). Britain remains as regionally divided politically as it is socio-economically. While the immediate prospects for regional and local economic initiatives remain uncertain, they will doubtless continue to influence national policy debate. Regionalism will remain a major factor in British economic, social and political life for the remainder of the century.

REFERENCES

Batkin, A. (1987), 'The impact of local authorities on Labour Party economic policy', *Local Economy*, vol. 2, no. 1, pp. 14–24.

Benington, J. (1986), 'Local economic strategies: paradigms for a planned economy?', *Local Economy*, no. 1, pp. 7–24.

Bentley, G., and Mawson, J. (1984), *Industrial Policy 1972–1983: Government Expenditure and Assistance to Industry in the West Midlands*, ESRC Inner City in Context Research Programme, West Midlands Study Working Paper No. 6, Birmingham: Joint Centre for Regional, Urban and Local Government Studies, University of Birmingham.

Bentley, G., and Mawson, J. (1985), *The Industrial Development Certificate and the Decline of the West Midlands: Much Ado About Nothing*, ESRC Inner City in Context Research Programme, West Midlands Study Working Paper No. 15, Birmingham: Joint Centre for Regional, Urban and Local Government Studies, University of Birmingham.

Bessant, J., Jones, D., Lamming, R., and Pollard, A. (1984), *The West Midlands Automobile Components Industry*, West Midlands County Council Economic Development Unit Sector Report No. 4, Birmingham: West Midlands County Council.

Birmingham City Council (1985), *An Economic Strategy for Birmingham 1985–6*, Birmingham: City Council.

Boddy, M. (1984), 'Local economic and employment strategies', in M. Boddy and C. Fudge (eds), *Local Socialism*, London: Macmillan.

Briggs, A. (1968), *Victorian Cities*, Harmondsworth: Penguin.

Commission of the European Communities (1984), *The Regions of Europe: Second Periodic Report on the Social and Economic Situation and Development of the Regions of the Community*, Brussels: European Commission.

Council of European Municipalities and Regions/Birmingham City Council/West Midlands County Council (1985), *The European Car Industry – A Strategy for the Future*, Report of Second European Conference of Local and Regional Authorities in Motor Manufacturing Areas, Birmingham, 29 October–1 November.

Crouzet, F. (1982), *The Victorian Economy*, London: Methuen.

Department of Employment (1987), 'Historical Supplement No. 1', *Employment Gazette*, vol. 95, no. 2.

The Economist (1987), 'How the regions voted', *The Economist*, 20 June, pp. 31–3.

Edge, G. (1981), *Policies and Priorities for Economic Development in the West Midlands*, Birmingham: West Midlands County Council Economic Development Unit.

Edge, G. (1986), *Priorities for Economic Regeneration in the West Midlands: The Future Role of the West Midlands Enterprise Board Ltd. Group of Companies*, Birmingham: West Midlands Enterprise Board.

Elliott, D. (1986), 'The sector approach to industrial research: fiddling while Rome burns?', in *West Midlands County Council Economic Development Committee Economic Review No. 4: Research for Economic Development*, Birmingham: West Midlands County Council.

Elliott, D., and Marshall, M. (1989), 'Sector strategy in the West Midlands', in P. Hirst and J. Zeitlin (eds), *Reversing Industrial Decline? Industrial Structure and Industrial Policy in Britain and Her Competitors*, Oxford: Berg.

Financial Times (1981), 'Survey: Birmingham and the West Midlands', *Financial Times*, 10 September.

Gaffikin, F., and Nickson, A. (1984), *Jobs Crisis and the Multinationals: The Case of the West Midlands*, Birmingham: Birmingham Trade Union Group for World Development/Birmingham Trade Union Resource Centre.

Gamble, A. (1981), *Britain in Decline*, London: Macmillan.

Hobsbawm, E. J. (1969), *Industry and Empire*, Harmondsworth: Penguin.

Ingham, G. (1984), *Capitalism Divided? The City and Industry in British Social Development*, London: Macmillan.

Labour Party (1987), *Labour Manifesto: Britain Will Win*, London: Labour Party.

Lee, C. H. (1979), *British Regional Employment Statistics, 1841–1971*, Cambridge: Cambridge University Press.

Leigh, R., and North, D. (1983), *The Clothing Sector in the West Midlands*, West Midlands County Council Economic Development Unit Sector Report No. 3, Birmingham: West Midlands County Council.

Lewis, J., and Armstrong, K. (1986), 'Skill shortages and recruitment problems in West Midlands engineering industry', *National Westminster Bank Quarterly Review*, November.

Low Pay Unit (1983), 'Poverty wages in the West Midlands', *Low Pay Review*, no. 15, London: Low Pay Unit.

Marshall, M. (1985), 'Technological change and local economic strategy in the West Midlands', *Regional Studies*, vol. 19, no. 6, pp. 570–8.

Marshall, M. (1986), 'National economic forecasting at local level: applications of the Cambridge model in the West Midlands', *Local Economy*, no. 3, pp. 49–60.

Marshall, M. (1987), *Long Waves of Regional Development*, London: Macmillan.

Marshall, M., and Mawson, J. (1987), 'The West Midlands', in P. Damesick and P. Wood (eds), *Regional Problems, Problem Regions and Public Policy in the UK*, Oxford: Oxford University Press.

Martin, R. (1986), 'Thatcherism and Britain's industrial landscape', in R. Martin and B. Rowthorn (eds), *The Geography of De-Industrialisation*, London: Macmillan.

Mawson, J., and Gibney, J. (1984), *Memorandum of Evidence to the House of Lords Select Committee on the European Communities, 1984*, Birmingham: Centre for Urban and Regional Studies, University of Birmingham.

Mawson, J., Jepson, D., and Marshall, M. (1984), 'Economic regeneration in the West Midlands: the role of the county council', *Local Government Policy Making*, November, pp. 61–72.

Spencer, K., Taylor, A., Smith, B., Mawson, J., Flynn, N., and Batley, R. (1986), *Crisis in the Industrial Heartland: A Study of the West Midlands*, Oxford: Oxford University Press.

Trades Union Congress West Midlands Regional Council (1981), *Our Future: A Planned Programme of Economic and Social Advance*, Birmingham: West Midlands TUC.

Wabe, J. S. (1986), 'The regional impact of de-industrialisation in the European Community', *Regional Studies*, vol. 20, no. 1, pp. 23–36.

West Midlands Auto Industry Forum (1987), *The Road Ahead: A Future for the West Midlands Motor Industry*, Birmingham: West Midlands Auto Industry Forum.

West Midlands Confederation of British Industry (1983), *Manufacturing in the West Midlands: Problems and Prospects*, Birmingham: West Midlands CBI.

West Midlands County Council (1974), *A Time For Action*, Birmingham: West Midlands County Council.

West Midlands County Council Economic Development Committee (1984), *Action in the Local Economy*, Birmingham: West Midlands County Council.

West Midlands County Council Economic Development Unit (1982), *The Foundry Industry in the West Midlands*, West Midlands County Council Economic Development Unit Sector Report No. 1, Birmingham: West Midlands County Council.

West Midlands County Council Economic Development Unit (1983), *The Machine Tool Industry in the West Midlands*, West Midlands County Council Economic Development Unit Sector Report No. 6, Birmingham: West Midlands County Council.

West Midlands Economic Planning Council and the West Midlands Planning Authorities Conference (1979), *A Developing Strategy for the West Midlands: The Regional Economy – Problems and Proposals*, Birmingham: West Midlands Regional Study.

West Midlands Enterprise Board (1986a), *Poverty and Family Income Trends in the West Midlands*, Birmingham: West Midlands Enterprise Board.

West Midlands Enterprise Board (1986b), *Policies and Priorities Towards the West Midlands Motor Vehicle and Component Industries*, Birmingham: West Midlands Enterprise Board.

West Midlands Enterprise Board (1986c), *Policies and Priorities Towards Technological Change in West Midlands Industry*, Birmingham: West Midlands Enterprise Board.

West Midlands Low Pay Unit (1984), *Below the Minimum: Low Wages in the Clothing Trade*, Birmingham: West Midlands Low Pay Unit.

Wood, P. (1976), *Industrial Britain: The West Midlands*, Newton Abbot: David & Charles.

CHAPTER 7

Priority to local economic development: industrial restructuring and local development responses in the Ruhr area – the case of Dortmund

G. Hennings and K. R. Kunzmann

7.1 ECONOMIC DECLINE IN DORTMUND AND THE RUHRGEBIET

The city of Dortmund is the most important urban centre in the south-east of the Ruhrgebiet (Map 7.1). The legally approved state development plan (*Landesentwicklungsplan*) of North Rhine-Westphalia has assigned Dortmund the functional role of high-order centre (*Oberzentrum*) in the agglomeration core (*Ballungskern*). With about 580,000 inhabitants in 1985, it ranks eighth largest city of the Federal Republic of Germany. Within its catchment area live more than 2 million people. Since 1970, when the city reached a population figure of 645,000, the number of inhabitants has been continuously declining – by 10 per cent between 1970 and 1985. The most recent population forecast for Dortmund is of around 500,000 inhabitants in the year 2000.

The population development of Dortmund parallels almost exactly the demographic decline of the Ruhrgebiet. The Ruhrgebiet, still one of the biggest industrial agglomerations in Western Europe, has lost about 500,000 people or 9 per cent since 1970. According to a recent forecast the population of the Ruhrgebiet will further decline from 5.1 million to 4.55 million in the year 2000 (Schönebeck and Wegener, 1984).

To a great extent the economic structure and the production pattern of Dortmund and the Ruhrgebiet were shaped in the period of early industrialization. Based on the exploitation of the largest hard-coal deposits of Europe, the once rural Ruhrgebiet was transformed into the most important European centre for mining and for iron and steel industries with considerable forward and backward linkages. A variety of manufacturing industries (steel construction, mechanical engineering, iron and sheet-metal working, chemical engineering) contributed to a highly interlaced regional economy which, however, was very much dominated by the coal and iron industries.

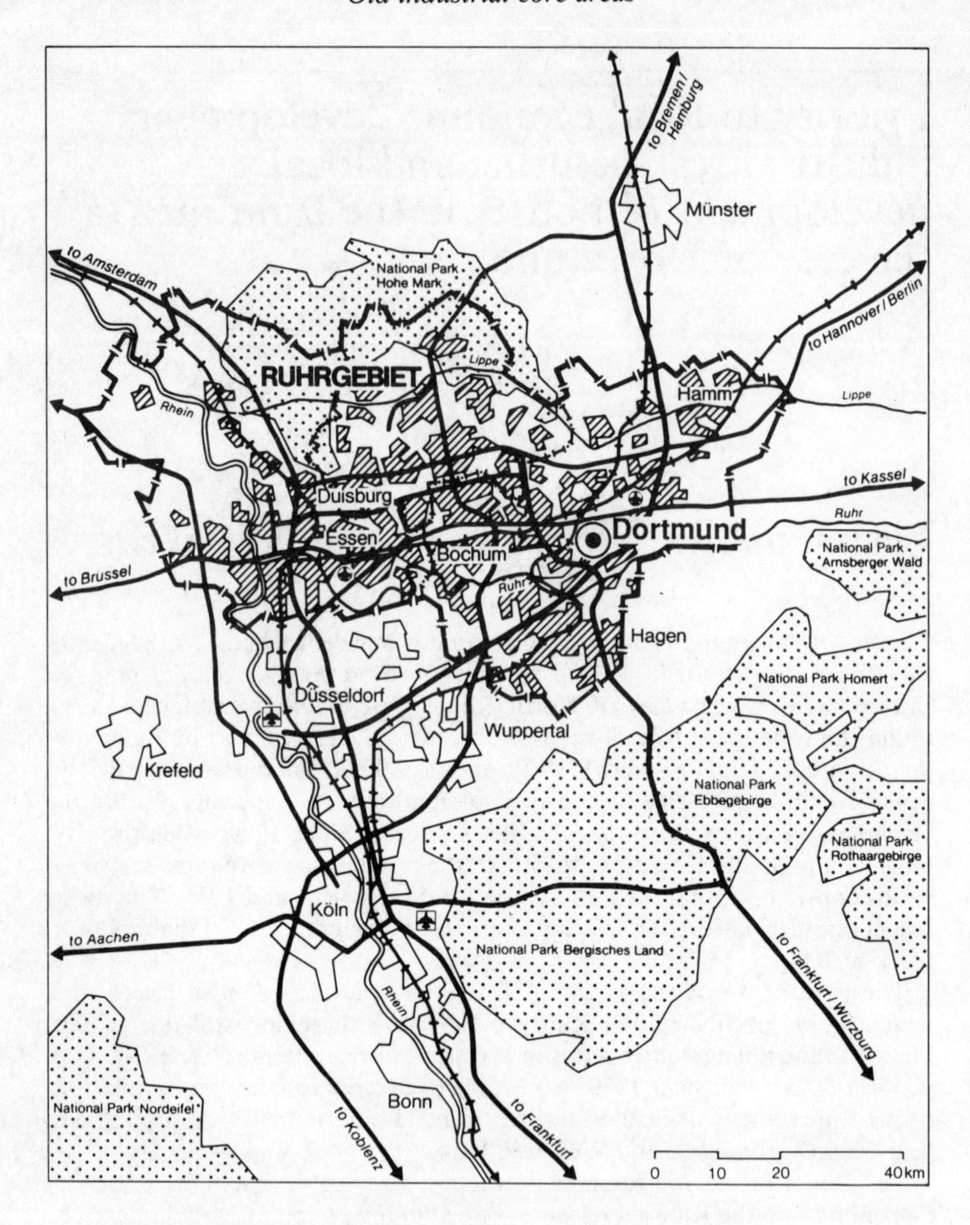

Map 7.1 The Ruhrgebeit: regional context and settlement structure.

Coal, of course, was also the regional basis for making the Ruhrgebiet the most important centre of energy production (Clemens and Tengler, 1983).

It is the structural decline of this old industrial complex which is the predominant reason for the economic decline of the region. The so-called coal crisis has been a permanent crisis since the beginning of the 1960s. In spite of all measures of adaptation and modernization the Ruhr coalmining industry is no longer competitive in either the world coal or the national energy market due to geological conditions. The adjustment process reduced pitcoal output from 124 million tons (1956) to 61.2 million tons (1984). At the same time the number of jobs fell from 393,831 to 130,000 in 1984, the number of active pits from 140 to 24 (KVR 1987).

The German coal industry can only survive with high state and federal subsidies granted under the national *Kohlevorrangpolitik* ('policy of priority for coal energy') which aims at maintaining coalmining as a national energy reserve. In spite of considerable subsidies of DM 5–9 billion in recent years (according to different sources: Miegel, 1987) the coal industry in 1987 ran into a new acute crisis. There are rumours that the number of jobs in the coalmining sector will have to be cut by a further 30,000 in the coming years.

Because of the exhaustion of the coal deposits coalmining has been gradually moving to the northern fringe of the Ruhrgebiet. In the big cities of the southern Ruhrgebiet - Essen, Bochum and Dortmund - all coalmines have now been closed. The last pit in Dortmund was closed in spring 1987, marking the end of more than 150 years of coalmining in Dortmund.

The structural problems of the Ruhrgebiet have further increased since the beginning of the steel crisis in the 1970s. The first city to be hit hard was Dortmund. In 1979, the Hoesch Company, Dortmund's biggest employer, announced the closure of one of three big steelworks. Within a period of three years 13,000 steelworkers were dismissed. From 1974 to 1985 employment in iron and steel production of the Ruhrgebiet decreased by 38 per cent, from 344,000 to 212,000. In 1987 the big steel companies of the Ruhrgebiet announced a further 20,000 dismissals up to 1990.

The restructuring process of the Ruhrgebiet is further hampered by a number of other structural problems:

- The decline of the coal and iron sectors in the Ruhrgebiet has considerably affected most of the other manufacturing industries, which heavily rely on sub-contracts from the two once-dominating sectors (Eckey and Schwickert, 1982). In terms of employment, there is virtually no manufacturing growth industry in the Ruhrgebiet.

- The big industrial enterprises of the Ruhrgebiet realize their diversification strategies often outside the region. They justify their investments at other locations with certain locational disadvantages, like high energy costs, high costs of environmental control, relatively high wages, lack of manpower with managerial qualifications, bureaucratic behaviour of the local administration and relatively expensive industrial land (Clemens and Tengler, 1983).

- Especially in the 1960s, when overall growth rates and industrial mobility were high, many chances to locate new and modern industries in the Ruhrgebiet were missed. Many studies registered innovation deficits as the most important structural problem of the Ruhrgebiet (Arras, Müller and Eckerle, 1984), while new modern industrial growth centres have emerged in southern Germany. High-tech industries have formed new industrial complexes in the regions around Frankfurt, Stuttgart and Munich; they are not willing to invest in the Ruhrgebiet.
- The opportunities for the Ruhrgebiet to attract national and international enterprises are rare. This is mainly due to the negative image of the Ruhrgebiet as that of a typical smokestack region, where – according to expressed prejudices – cultural opportunities and recreational values are low. Although the reality of today's Ruhrgebiet is different, this image is very difficult to change.

Given such contextual conditions it is not surprising that the overall number of jobs in Dortmund decreased from 278,000 to 238,000, by 14 per cent, between 1970 and 1985. While in the same period employment in the Ruhrgebiet declined from 2.2 millions to 2 million, or 9.3 per cent, employment in the Federal Republic decreased only by 3.9 per cent. This, above all, shows the dwindling importance of the Ruhrgebiet and of Dortmund for the economy of the Federal Republic as a whole.

The loss of employment opportunities in the whole country is a result of the job losses in the producing sector (mining, manufacturing and construction). Producing-sector employment in the Federal Republic decreased by 20 per cent from 1970 to 1985, in the Ruhrgebiet by 31 per cent and in Dortmund by 42 per cent. Consequently, employment development in the Ruhrgebiet, must be interpreted in the light of the deindustrialization taking place in the Federal Republic. This process affects traditional industrial areas overproportionately. While employment in the producing sector declines, service-sector employment has grown continuously since 1970: in the Federal Republic by 21 per cent, in the Ruhrgebiet by 21 per cent, and in Dortmund by 15 per cent. The whole country, including the Ruhrgebiet and Dortmund, is changing gradually into a service-sector economy and a service-sector society (Table 7.1).

In 1985 the 57 per cent share of the service sector in the Ruhrgebiet was slightly higher than that of the Federal Republic with its 53 per cent. At the same time the service sector in Dortmund accounted for 65 per cent of overall employment, whereas the share of manufacturing employment, which was as high as 50 per cent in 1970, is now down to 34 per cent. The transformation process from an industrial to a service-sector-based economy in Dortmund is mainly due to job gains in science, education, health, private business services as well as in public and semi-public organizations. Also banking and insurance increasingly contribute to the changing economic base of Dortmund.

However, regional employment growth in the service industries had not been high enough to compensate for all the job losses in the producing sector. In April 1987 unemployment rates were high: 8.5 per cent in the Federal Republic, 15 per cent in the Ruhrgebiet and 17 per cent in Dortmund.

The gloomy figure indicates that the prospects of Dortmund are not bright. Coal and steel industries will have to dismiss more blue- and white-collar workers, and the city's budget is burdened by increasing social welfare expenditures for a steadily growing number of persons falling under the

Table 7.1 Employment development 1970 to 1985 in Dortmund, the Ruhrgebiet and the Federal Republic of Germany as a whole

	1970		*1985*		*Variation*
	absolute	*%*	*absolute*	*%*	*1970 to 1985 %*
			DORTMUND		
Agriculture, forestry	2,270	0.8	820	0.3	−64.0
Producing sector	139,880	50.3	81,496	34.2	−42.0
Services	136,150	48.9	156,512	65.5	+14.6
Total	278,300	100.0	238,828	100.0	−14.2
			RUHRGEBIET		
Agriculture, forestry	39,700	1.8	21,063	1.0	−46.9
Producing sector	1,222,800	55.4	837,745	41.8	−31.5
Services	943,600	42.8	1,143,088	57.1	+21.1
Total	2,206,100	100.0	2,001,896	100.0	−9.3
			FEDERAL REPUBLIC OF GERMANY		
Agriculture, forestry	2,262,000	8.5	1,390,000	5.4	−38.6
Producing sector	12,987,000	48.9	10,461,000	41.0	−19.5
Services	11,311,000	42.6	13,680,000	53.6	+20.9
Total	26,560,000	100.0	25,531,000	100.0	−3.9

Sources: Employment figures for Dortmund and the Ruhrgebiet 1970: Komunalverband Ruhrgebiet (Hrsg.), *Beschäftigte im Ruhrgebiet*, Arbeitshefte Ruhrgebiet, Essen, 1980, Results of the 'Arbeitsstättenzählung 1970'.

Employment figures for Dortmund and the Ruhrgebiet 1985: *Beschäftigtenanalyse des Landesamtes für Datenverarbeitung und Statistik des Landes Nordrhein-Westfalen*, Düsseldorf, 1986.

Employment figures for the Federal Republic of Germany 1970 and 1985: Statistisches Bundesamt (Hrsg.), *Statistisches Jahrbuch 1986 für die Bundesrepublik Deutschland. Erwerbstätige nach Wirtschaftsbereichen und Stellung im Beruf.*

poverty line. Although the city government's early efforts to limit public expenditure and to apply for regional and federal grants have been quite successful, the financial resources of Dortmund are continuously shrinking.

The present conservative federal government in Bonn refuses, mainly for political reasons, to develop specific policies for its traditional industrial regions. The state government of North Rhine-Westphalia, for financial reasons, cannot afford a new economic development programme for the Ruhrgebiet, as it undertook in 1966, in 1970 and in 1979. As a consequence the cities of the Ruhrgebiet are left very much on their own.

In an early response to the increasing structural problems Dortmund's city managers started in the early 1980s manifold new activities aiming at a reversal of the negative economic trend. Since then, local economic development has received the highest political priority in Dortmund. By 1987 Dortmund was regarded as the most active and most successful town in the Ruhrgebiet in restructuring and modernizing its local economy.

7.2 ACTORS AND INSTITUTIONS INVOLVED IN LOCAL ECONOMIC DEVELOPMENT POLICIES

To understand the context of local economic policies and the actors involved it is essential to know that the principle of local self-government as an expression of civic freedom plays a decisive role in Germany. It has a long and great tradition and is guaranteed by the *Grundgesetz* (constitution). Accordingly, the local governments have the right to regulate their own communal affairs – within the framework of state and federal laws. These includes among others local economic development, local transport, culture and urban development planning. In all such fields local authorities are subject only to legal control by regional or state levels.

7.2.1 PARTICIPANTS AND EXECUTING AGENCIES OF LOCAL ECONOMIC DEVELOPMENT

Three groups of institutions or actors are involved in the economic development process in Dortmund:

(i) various departments of the local governments preparing, planning, programming, implementing and monitoring policies, measures and projects aiming at local economic development:

(ii) public and semi-public institutions executing general public tasks or representing private interests;

(iii) political parties and committees engaged in the formal political decision-making process, and in setting policies, guidelines and targets of local economic development.

In the first group, the Amt für Wirtschafts- und Strukturförderung (Department of Local Economic Development) plays a dominant role in the formulation, implementation and monitoring of policies and measures to restructure the city's economy. The department's duties are among others to advise local investors, to attract new enterprises, to maintain close contacts with all actors of the local economy, to develop and to market local industrial land and to promote Dortmund as an attractive location for new industries.

In the context of its usual urban planning business the Stadtplanungsamt (Urban Planning Department) is responsible for all aspects of planning, zoning, control and physical monitoring of local industrial areas. This also includes mediation and conciliation of spatial conflicts between competing local interests.

Other departments involved are the Amt für Liegenschaftswesen (Department of Municipal Properties) and the Planungsstab für Stadtentwicklung (Urban Policy Unit). The latter, once an influential policy unit attached to the Oberstadtdirektor (Chief City Manager) has meanwhile lost most of its importance.

The second group includes both the Industrie- und Handelskammer (Chamber of Industry and Commerce) and the Handwerkskammer (Chamber of Trades and Crafts). They traditionally represent the interests of local industries and trades. As Träger öffentlicher Belange (Body of Public Concern) both chambers are officially involved in legal land-use planning and zoning procedures.

Both chambers have established their own professional training centres which are of inter-regional importance. In addition, Dortmund's Chamber of Industry and Commerce is actively engaged in industrial extension services, in advising start-ups and in the transfer of high technologies. It also participates in the management of the new Technologiezentrum (Technology Centre Dortmund) at the university's premises.

Other institutions involved in local economic development policies are the Landesentwicklungsgesellschaft (State Development Corporation) which manages the Grundstücksfonds Ruhr (Real Estate Fund of the Ruhr) established for the recycling of industrial land, the local Arbeitsamt (labour office) which is responsible for so-called *Arbeitsbeschaffungsmassnahmen* (job creation measures) and for many activities in the field of continuing education.

Other local actors which belong to this category are the representatives of the local trade unions and those of the university and the polytechnic.

The third group concerns two political parties which have always been represented in Dortmund City Council: the Social Democrats (SPD), which (since the Second World War) have held the majority in the council (1984: 55.4 per cent) and the Christian Democratic Union (CDU) (30.1 per cent). Following the 1984 election a new ecological party, Die Grünen, joined the local council, winning a considerable 10.7 per cent of all votes.

Political decisions in the city council concerning economic restructuring are prepared in mainly two formal political subcommittees, the Wirtschaftsförderungsausschuss (Committee for Economic Development) and the Ausschuss für Stadtentwicklung und Planung (Committee for Urban Development and Planning). Both committees formulate the political guidelines in their respective areas of concern and control the administration's performance and follow-up activities.

7.2.2 INTERACTION OF ACTORS INVOLVED IN LOCAL ECONOMIC DEVELOPMENT

Traditionally the various participants and executing agencies at the local level and a few institutions at the regional and state level form a dense network of formal and informal interactions. The Ministerium für Landes- und Stadtentwicklung (Ministry of Country and Town Development), the Ministerium für Wirtschaft, Mittelstand und Verkehr (Ministry of Economy, Medium-Scale Industries, and Transport) at the state level and the Bezirksregierung (District Authority) as well as the Kommunalverband Ruhr (KVR) (Intercommunal Association of Ruhr Municipalities) at the regional level, are important institutions involved in one way or another in local economic restructuring.

According to the general philosophy of the federal system and its institutional set-up and responsibilities, all top-down and bottom-up bargaining and decision-making processes follow a counter-current principle. This established administrative system favours policy innovation at the various levels, avoids a cumbrous top-down decision-making machinery, and allows considerable public participation and involvement at the local level. However, informal personal links upwards through political parties, unions or employers' associations play a continuously decisive role in preparing and concerting actions at the respective policy levels. The dense but flexible network of the established administrative system is a guarantee of competent and efficient execution of innovative policies and programmes. It usually obviates the creation of additional new agencies and institutions, if new problems and tasks evolve.

7.3 INTRODUCING A NEW LOCAL ECONOMIC DEVELOPMENT POLICY

It was the decision of the Hoesch Company in 1981 to close down one of its three steelworks, which originally caused the formulation of a new local economic development policy for Dortmund. This entrepreneurial decision, which involved almost halving the enterprise's labour force of 23,000, provoked considerable uneasiness among the local community.

The city's lord mayor reacted to the new situation by convening a so-called 'Hoesch Conference', to which he invited representatives of federal and state ministries, influential local industrialists and businessmen, as well as members of the state parliament and local government. The purpose of this conference was to have a forum for discussing Dortmund's prospects and those of the eastern Ruhrgebiet. The conference's participants' assessment of the city's prospects were rather negative. Obviously this gloomy outlook convinced the responsible opinion leaders of the region to strive for a new economic basis for the city and its immediate hinterland.

One important outcome of the Hoesch Conference was the formal decision of Dortmund City Council in 1982 to give local economic development top priority among the many public tasks. This decision included the allocation of a considerable city budget for the Department of Local Economic Development. Its budget line rose from DM 4.2 million in 1980 to DM 35.7 million in 1983. The number of staff was also considerably increased. Another effect of this decision was the formulation of a politically binding policy paper on 'local economic development' and of operational guidelines for the Local Economic Development Department (Stadt Dortmund, 1983).

Overall aim of the new local economic policy was to consolidate Dortmund's function and position as an *Oberzentrum* (high-order centre) and to convert it into a future-oriented regional manufacturing and services centre. In order to reach this goal and to leave the coal and steel structure behind, two objectives were proclaimed:

- accelerating technological change and transforming the city into a modern centre of technology and service industries; and
- improving environmental conditions and the quality of residential areas and leisure facilities to render the city more attractive to skilled labour.

Development on the basis of these policy objectives comprised three partial-strategy elements: a rather defensive strategy 'Dortmund Today', a more preventive strategy 'Dortmund Tomorrow', and an offensive strategy 'Dortmund the Day after Tomorrow'.

The 'Dortmund Today' element aimed at consolidating the existing economic structure and at securing the local steel basis. It also formulated measures to retain local enterprises threatening to leave Dortmund for more attractive locations.

The 'Dortmund Tomorrow' element initiated measures supporting indigenous firms and enterprises in their efforts to adopt new technologies and to speed up the innovation process. Among the measures formulated, the linking of economic activities to research and development facilities of the university and the polytechnic played an eminent role.

The 'Dortmund the Day after Tomorrow' element aimed at attracting to the city new high-tech firms in the micro-electronic, communication and biotechnology sectors. The university and other local research and development facilities were supposed to take a key role in this future-oriented strategy element.

For the implementation of this comprehensive strategy to local economic development a variety of policies have been initiated, concerning an innovation-oriented local economic policy; employment initiatives; industrial land regeneration; and restructuring the local steel industry. We will deal with each of them in more detail. Other initiatives which are of a more complementary character are better utilization of grants and funds provided by European, federal and state programmes; improving local infrastructure and environmental conditions; intensifying and expanding the advisory services of the city's Department of Local Economic Development; and establishing an advisory network for guiding and monitoring the restructuring process.

It is important to state that all these initiatives explicitly excluded the establishment of new institutions for local and economic development. Instead, they aimed at reviewing, redefining and changing conservative attitudes of the existing economic development machinery in the city.

7.3.1 INNOVATION-ORIENTED LOCAL ECONOMIC POLICY

Currently most local government activities and measures to promote Dortmund's economic restructuring focus on a consistent innovation policy. This high-tech-oriented policy aims at:

- accelerating the innovation process of indigenous local firms and small and medium-size enterprises;
- shaping a new image of the town as an attractive location of research and development;
- encouraging the establishment of new enterprises and the creation of jobs in high-tech industries;
- attracting small and medium-size high-tech enterprises from other regions in the Federal Republic to Dortmund.

Particular achievements of efforts to implement this innovation oriented local economic policy are described below.

Attracting public research and development institutes to Dortmund

Through intensive lobbying at state and federal level the university and the local government succeeded in achieving newly created research and development facilities which have been or will be located at Dortmund (e.g. the Fraunhofer Institute of Transport Technology and Goods Distribution, the Institute of Automatization and Robotics, an experimental track of Germany's first fully automated suspended passenger transit system and the Centre of Expert Systems).

Creating agencies for local and regional technology and innovation transfer

Transfer centres have been established within the university, the polytechnic and within the local chambers of industry and commerce and of trade and crafts. A technology centre has been built. Zoning and landscaping plans have been initiated to transform the university's surroundings into a huge modern research and technology park.

Promoting technology-oriented new firms
Both the city and the local savings bank have launched own programmes for entrepreneurial start-ups. The Local Economic Development Department, the local chambers of industry and commerce and of trading and crafts, do offer client-tailored and advisory services to founders of new firms.

Locating high-tech firms
A major aim has also been that of attracting and financially supporting the location of high-tech firms and enterprises in Dortmund and aiding individual local enterprises to implement their innovation plans.

The University of Dortmund, with its strong engineering and science faculties (e.g. the departments of informatics, chemical engineering, mechanical engineering and electronics) and its 18,000 students holds a key position in this innovation process. Since the early 1980s personal contacts between chairholders and scientists of the new university, founded in 1968, local political leaders and union leaders, and representatives of the local business community have considerably intensified.

The most spectacular result of this new spirit of cooperation, however, is the Technology Centre Dortmund opened in June 1985. The impulse to establish this first technology centre in the Ruhrgebiet came not only from the university but also from officials from the Department of Local Economic Development. Within a few weeks the idea was taken up by the leading city officials, the chamber of industry and commerce as well as the university's board of governors. In the spring of 1984 financial backing was approved given by the city and the state Ministry of Economy, Medium Scale Industries, and Transport.

The fully booked centre now hosts R&D activities carried out by local and out-of-town enterprises in cooperation with scientists of the university. Major fields of cooperation are material flows, logistics, material technologies and material-handling systems. Also R&D projects in the fields or information and software development as well as business-consultancy firms have found a new home in the technology centre. Among the consultancy firms in the centre is one which offers venture capital to promising projects. A year after its opening pressing demand for additional space encouraged the centre to plan for considerable expansion.

Finance for the centre came from three sources: from the City of Dortmund; the Ministry of Economy, Medium-Scale Industries and Transport of North Rhine-Westphalia, and from the European Regional Fund. The technology centre is managed by a GmbH (a partnership with limited liability), in which the City of Dortmund, the local chamber of industry and commerce, a few local banks, including the powerful local Stadtsparkasse (savings bank), and a consortium of local enterprises are partners. The centre has become a prominent example of successful public–private partnership in Dortmund.

Such efforts of local economic development are complemented by two environmental planning projects of the city's urban planning department. In the immediate neighbourhood of the university 18 hectares of industrial area are being developed for high-tech firms and enterprises. Enterprises willing to locate in these 'new technology areas' have to accept and to meet high ecological and architectural criteria. The second project concerns the open space between the university's scattered campus facilities, which will be transformed into a 'landscape park'. Ecological considerations play a dominant role: a sector of the park, for example, will be made available for an eco-tech pilot project experimenting with perma-cultures. The consistent linking of high-tech activities with landscaping projects which accentuate ecological factors has already improved the image of Dortmund (Map 7.2).

In 1986 the first high-tech firms were established on the campus. Since that time there has been an increasing demand for developed land at this location. The first start-ups from the technology centre have meanwhile started to construct their own buildings on adjacent grounds. Successful start-ups are in the fields of sensor technologies, software development and technology consultancy. New enterprises which relocated from outside Dortmund to the attractive 'new-technology areas' are in the fields of chip production, biotechnology and electrical engineering.

7.3.2 POLICY FOR EMPLOYMENT INITIATIVES

Continuously high unemployment (17 per cent in 1987) and not unrealistic fears that its rate would increase further caused the city government to embark also on a new policy field previously unknown in Dortmund: job creation measures and unconventional employment initiatives. Encouraged by local politicians and based on a concept, designed by the local government's Urban Policy Unit, steps were undertaken to support the local informal economic sector. The respective initiatives additionally profited from the fact that Dortmund recently qualified as a 'development area' for the European Social and Regional Funds.

Actions taken by the local government in this policy area included:

- free management consultancy services to initiatives in the manufacturing sector. The Department of Local Economic Development assists such initiatives to find appropriate management consultants;
- financial support through funds from the European Regional and Social Funds, and through own financial means. A small budget line was established for such activities;
- mediation of 'ABM measures' to initiatives in the social and in the cultural sectors (ABM is a more than DM2 billion per year federal-

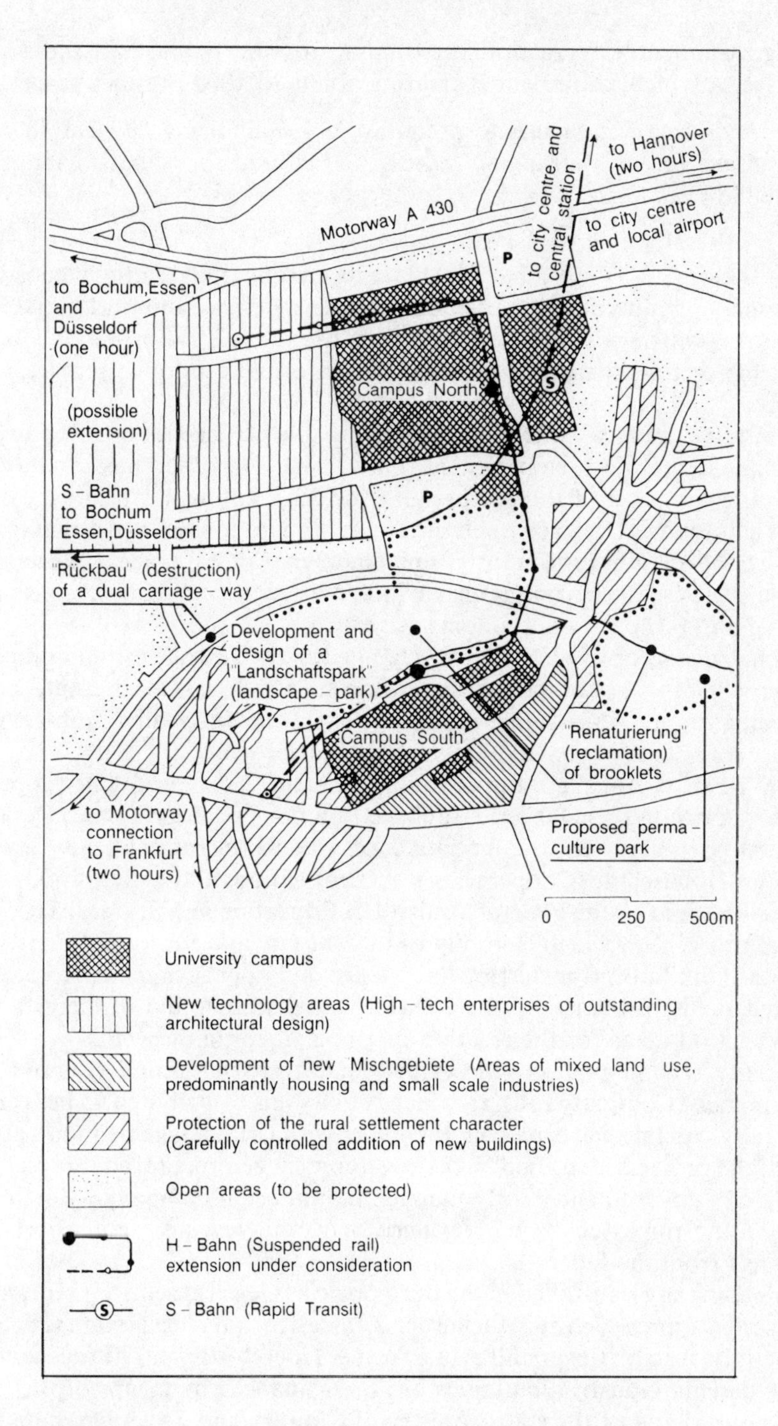

Map 7.2 The high-technology precincts of the University of Dortmund.

government job-creation programme for the public and the private sector which, at present is creating about 100,000 jobs per year);

- provision of appropriate premises for small firms and employment initiatives, by arranging access to unused or derelict industrial buildings.

These actions, however, have not yet been integrated into a consistent programme, mainly because of communication deficits and traditional fears of contacts which exist between the actors, the administration and the formal economic sector on the one side and the informal economic sector on the other.

Obviously the local unions have considerable difficulties in accepting the informal sector and employment initiatives. They emphasize the self-exploiting character of this sector and consider it an inappropriate means to create employment or to compensate for redundant jobs in the formal sector. But also the chambers of industry and commerce and of trade and crafts are still suspicious of the unorganized informal economic sector, considered threatening to the formal economy.

Such views seem to have slightly changed in the recent past and prepared the ground for a comprehensive project, which is, at present, under implementation. This rather complex but very innovative job-creation project consists of three fully interrelated sub-projects.

The first, a so-called Entwicklungszentrum Dortmund (Development Centre Dortmund), has been established by the Local Economic Development Department, on the initiative of the Kooperationsstelle Gewerkschaften-Hochschule (Cooperation Centre of Unions and University), a pilot agency financed by the Federal Ministry of Education and Science. The function of this centre, which is jointly run by the city and the local unions, is to act as a think tank, transferring knowledge and appropriate technologies in selected fields, primarily in the environmental sector, and to identify promising action areas for the creation of employment initiatives.

For the second project, the Arbeit und Umwelt GmbH (Labour and Environment Company Ltd), to which the city has contributed 75 per cent of the equity capital, has been commissioned with the management and execution of the projects identified by the development centre. Staff of Hoesch and of the city government jointly manage this unusual company. Most of its eighty-four employees were formerly unemployed and got short-term contracts from the federal ABM programme.

This company is installed in the third project, Gewerbezentrum Huckarder Strasse (Enterprise Centre, Huckarder Strasse), an industrial estate which has been established on the premises of a former Hoesch works. The regeneration of the derelict industrial buildings has been financed by means of the 'Real Estate Fund' and of the state Ministry of Country and Town Development. Production space in this regenerated industrial estate is also leased to private local enterprises. This contributes to reduce the overall operating costs of the 'enterprise centre'.

From its inception, this complex project has been implemented with considerable enthusiasm by its initiators and with full political support. Recent experience suggests, however, that continuous efforts and visible success are necessary to overcome the day-to-day difficulties and bottlenecks which now threaten to frustate the participating parties beyond their idealism in this unusual and innovative project.

7.3.3 INDUSTRIAL-LAND-REGENERATION POLICY

Increasing environmental awareness, and scarce land reserves suitable for industrial development within the city's boundary, have rendered it more and more difficult for the city government to provide appropriate land for industrial firms in attractive locations. As a consequence of this environmentally justified blockage of open-land reserves, the local government had to shift its focus to the reuse of derelict industrial land.

In the past the city has had some experience with the regeneration of industrial land, when looking for new uses of derelict coal mines in the area, but a consistent long-term and prospective policy to regenerate all derelict industrial land in Dortmund has never been approached. One reason for this inactivity was the land blockage exercised by the powerful regional coal and steel industries. Only the obvious decline of these industries in the early 1980s changed their presumption and speculative attitudes.

A preparatory step for the formulation of a new local industrial-land policy was a survey of land reserves. It classified a total of 254 hectares of derelict land in the city as being suitable for industrial purposes. Theoretically, this considerable land reserve is sufficient to meet the local demand for more than a decade.

However, only a small proportion of these 254 hectares of unused industrial land could be regenerated. It is either the low motivation of the landowners to sell the land at reasonable (low) prices, or zoning restrictions which render the regenerations so difficult, or the extremely high costs to regenerate the land for any productive use. Since 1982, however, 82 hectares of derelict land have been acquired by means of the Real Estate Fund Ruhr, a DM 500 million fund established in 1980 by the government of North Rhine-Westphalia for the regeneration of derelict industrial land. Almost 90 per cent of this land (72.6 hectares) has now been rezoned as industrial land.

The biggest single project was the regeneration of one of the former local Hoesch steelworks, a 56.4 hectares plot of land at an extremely good location near the city port and the city centre with easy connections to the regional motorway network. The very expensive regeneration measures included the demolition of obsolete premises, the total redevelopment of the area, its internal infrastructure and utilities, the improvement of access roads and, wherever necessary, the decontamination of polluted land.

In 1987 the first firms and enterprises commenced construction of their own buildings and started production. By 1989 about five new firms had been located there. Given the high local demand for industrial land at attractive locations, there is no doubt that the city will find sufficient clients and investors for the new industrial land offered.

Besides this project, the local government has also started activities to improve those industrial areas that have been run down and lost their former attractions. Through comprehensive measures such as consolidation of land, improvement of infrastructure or landscaping measures, for which so far neither a state nor a federal programme exists, Dortmund intends to stabilize existing industrial areas and to meet the qualitative expectations of new investors.

A pilot project in this field was the modernization of the industrial area near Dortmund's port, which started in 1980. Measures to rejuvenate the area included among others the replacement of obsolete public utilities, the improvement of the internal road network, the repair of the canal's banks, the deepening of the bottom of the canal and landscaping measures to create a better physical outdoor environment for workers and employees.

Additional administrative efforts have been undertaken to arrange land for such enterprises in the area which depend on waterway transport. Incentives for relocation have been given to those firms which no longer require the access to the canal. There are two innovative elements in this undertaking. First, whereas the administrative co-ordination of all city government's departments being involved in the project was carried out by the Economic Development Department, the coordination of the measures and the contact to the existing enterprises was entrusted to the Port Authority, the Dortmunder Hafen AG. This organizational arrangement was unusual for Dortmund, but it turned out to be very effective, as it bundled professional competence and considerably shortened time periods for substantial decisions.

Second, financial means necessary for regenerating the area were primarily provided by the local government. Through thoughtful design and programming of the necessary measures, some additional financial means could be acquired from a variety of public-sector programmes aiming at projects with a quite different character. This innovative communal exercise in 'multi-budgeting' proved to be very successful. It was possible because of the organizational arrangements made for the project.

Resulting from positive experience made so far, the city administration has embarked on a second quite similar comprehensive project.

7.3.4 POLICY OF RESTRUCTURING THE LOCAL STEEL INDUSTRY

In 1981 the Hoesch Company, the biggest employer in Dortmund and one of the biggest steel enterprises in the Federal Republic of Germany, was almost bankrupt. The reasons were:

- the enterprise was faced by enormous sales difficulties as a result of worldwide overproduction;
- its production was poorly diversified; although operating a number of subsidiaries, Hoesch was – compared with other steel enterprises in Germany – hardly diversified; even in the late 1970s the production still focused on raw steel;
- parts of the production lines were rather outdated;
- the production in Dortmund was scattered over three separate locations, a fact which led to comparably high production costs;
- a rather inefficient waterway connection and an outdated city harbour made transportation costs high and rendered Dortmund's steel production less competitive compared with steel locations along the Rhine;
- the quota of borrowed capital was too high.

To overcome the crisis Hoesch initiated a number of measures. First, the former overcapacity of production has been scaled down, and employment has been reduced. One of the three local steel works was closed; further production lines were shut down. Second, serious efforts to modernize its steel production have been undertaken. The production policy has been changed. It now focuses primarily on sheet metal for motor industries. Third, the diversification of production has been accelerated. This was done mainly by acquisition of firms and enterprises offering future-oriented products. It is now the entrepreneurial aim by the early 1990s to limit the turnover of the steel sector to 25 per cent (1986: 50 per cent) of the enterprise's total turnover.

To implement such modernization measures Hoesch profited from a number of public programmes. Negotiations with the state and federal governments for financial support happened at the highest political levels. Local politicians and senior officials, local members of the parliaments in Düsseldorf, Bonn and Brussels (European Community) and local union leaders lobbied intensively and successfully for the enterprise, to receive grants from not less than nineteen public programmes. As a consequence, and within an extremely short time, the enterprise succeeded in regaining profitability.

In autumn 1986, however, new clouds appeared on the horizon. In the course of its continuing restructuring efforts, Hoesch announced a further dismissal of 4,000 employees, including 2,000 in Dortmund. Now it is openly speculated that the company may abandon steel production by the year 2000, and convert gradually to a modern diversified enterprise of mechanical engineering.

Dortmund's own assistance to Hoesch had primarily a complementary character. Apart from preparing adequate zoning plans and granting the necessary building permits, the city government supported the enterprise in other ways. To reduce the enterprise's transportation costs, efforts have been made to modernize the Dortmund–Ems waterways and the inland harbour of the city and to construct two major access roads. To reduce Hoesch's burden of high operating costs, the city has taken over some of the enterprise's privately run social facilities.

During the last decade the importance of Hoesch for the local economy has considerably declined. This fact is demonstrated by the enterprise's recent willingness to sell its derelict land reserves to the city. In the past, coal and steel industries in the Ruhrgebiet, have continuously blocked their land reserves. By that they more than once discouraged the location of new industries in the region. Changing economic conditions have now forced enterprises like Hoesch to sell their land. This in turn opens up new potentials for the city of Dortmund to redevelop derelict industrial land.

7.4 ASSESSMENT OF THE NEW LOCAL ECONOMIC DEVELOPMENT POLICIES

Efforts to intensify activities for accelerated restructuring of Dortmund's traditional economy are relatively recent. As late as 1981 leading actors involved in local economic development became aware of the shortcomings of a scarcely diversified local economy. In that year the first contours of the city's new economic development policy became noticeable. It is still too early for any formal and quantitative assessment of this policy (e.g. concerning the number of new jobs created). However, there are already a number of indications which justify an anticipation of success for this policy. Since 1983, local actors have changed their attitudes towards local economic development policies; more innovative approaches to local policies and urban management have been adopted; the indigenous development of local resources and potentials has been given preference; local economic development and local urban policy are increasingly becoming integrated; local actors better concert their involvement in urban economic development; new local communication networks have evolved; and the image of the city is progressively improving.

Although such indicators are just qualitative signs of success, they are simultaneously key factors of success as well as indispensable prerequisites for the successful mid- and long-term restructuring of a traditional industrial city. In the following section they will be discussed in more detail.

7.4.1 KEY FACTORS OF SUCCESS

The fact that the city of Dortmund has started numerous new activities and initiated innovative projects is already a success *per se*. But only the successful implementation of these projects will lead to the creation of new jobs, to a diversified economic structure, the reduction of unemployment or to the hoped-for economic upswing.

The key factors of success, which are described below, are simultaneously the target and instrument of efforts to restructure a traditional industrial region.

Changed attitudes of local actors towards local economic development policies

Throughout the 1970s and similar to all other traditional industrial regions, political and administrative structures in Dortmund were closely linked to the locally dominating coal and steel industries. It was not until the obvious decline of local coalmining and the crisis at the local steel enterprise Hoesch that the public consciousness came to accept that traditional structures were declining. It was the gloomy economic situation which ultimately forced the local government to fight seriously for the restructuring of the local economic base. At the beginning, all modernization activities proposed were again largely linked to the traditional coal and steel sector. Step by step, however, these activities were paralleled by combined efforts aiming at the innovation of other local industries; the allocation of new high-tech firms and enterprises; the transfer between university and local/regional structures; the qualitative expansion of the local cultural sector; and the ecological regeneration of the local environment.

To a certain extent this change happened because a new and younger generation of local actors had evolved, which could no longer rest on its postwar-reconstruction laurels.

Innovative approaches to local policies and urban management

The new awareness of local problems and changing attitudes have increasingly encouraged local actors to be much more open to innovative and integrative economic and urban development approaches. Because of this change, innovative projects in policy areas, such as employment initiatives, land development, and promotion of high-tech activities, have at last found the necessary local support for implementation. These new innovative approaches to local economic development are characterized by two features, namely a considerable expansion of instruments available for local economic development, and a change of target groups of local economic policies. Undoubtedly, these two features of the new economic development policy are a key to the successful restructuring and modernization of the local economy.

Preference given to the indigenous development of local resources and potentials

Since the early 1960s state programmes have been drawn up aiming at the structural modernization of the Ruhrgebiet. The initiative for such programmes was always taken from above. This may have been one of the reasons that they had a rather limited economic impact at the local level and did not result in the modernization of the local economic base.

It was not until they became aware of the serious situation of the city that local actors developed their own initiatives and set new priorities. It was the better knowledge of local conditions and problems which evoked the formulation of a new economic policy. This policy was based primarily on the development of endogenous potentials of the city. Undoubtedly the constitutional autonomy of local government in Germany facilitates such endogenous development.

Integrated local economic and local urban policies

The growing integration of local economic and local urban policies is manifested by close coordination of an increasing number of actions initiated by the respective departments of the city. Such an integration does play an eminent role in the implementation of innovative approaches to local economic development. From the consistent linking of economic and of physical and social measures both policy areas benefit in two ways.

(i) Appropriate local economic development policies contribute to the rehabilitation of the city's physical environment under the ecologically oriented urban-regeneration policy. This policy, in turn, creates the indispensable qualitative preconditions for the new local economic development policy.

(ii) Financial means from different federal, state and local budget lines and programmes can be bundled for implementing particular projects. Such bundling allows a much more effective utilization of limited funds available for urban restructuring. In this way they can be utilized much more effectively. The financial support of the Ministry of Country and Town Development of North Rhine-Westphalia, for example, has played a considerably bigger role in implementing the new local economic development policy than financial means of the state Ministry of Economy, Medium Scale Industries, and Transport.

Such linking from which both departments, the Department of Local Economic Development and the Urban Planning Department, mutually benefit requires competent and innovative urban managers with strategic trans-departmental thinking, bargaining skills and a good knowledge of multi-budgeting techniques. The better- and the earlier-informed the local decision-making machinery is, the more likely it can successfully initiate and implement innovative policies.

Local actors' willingness to act in concert

An essential factor in the city's local economic development process has been the continuous and open communication among local actors. Traditionally, capital, labour and local government in Dortmund have had relatively close relations. During the last few decades, no really militant conflicts between local unions and the industries have occurred. This fact undoubtedly contributed much towards the concerting of efforts to restructure the local economic base. The relationship between capital and labour is characterized by extensive bargaining for compromises, with the local government as a mediator.

Evolution of new local communication networks

Innovative local economic development policies cannot be implemented successfully unless new networks of actions, actors and personal relations do evolve. In Dortmund this proves to be essential. Whereas the traditional network of personal relations centred around representatives of the traditional coal and steel community and the Social Democratic Party, a new network of opinion leaders of a younger generation is evolving and gaining ground. The new local actors which are involved in innovative local economic development policies are the representatives of the local chambers and of the local savings bank, are management consultants, owners and managers of medium-size firms, active council members of major political parties represented in the city council, as well as scientists and research managers of the university and the polytechnic. They have created new opportunities and circles for discussion facilitating the exchange of information, informal decision-making, and the coordination of actions.

Parallel to this network a second overlapping network is evolving. This network links younger unionists, a few politicians, courageous administrators, critical scientists and representatives of ecologically oriented action groups together.

The influence of both networks on the city's long-term economic development is growing. Thus it can be assumed that in future these new networks will exceed even their present significance.

Improvement of the city's regional image

Continuous economic decline and a slow improvement of environmental conditions on the one hand, and the changing value system of German society as a whole on the other, have contributed much to the negative image of the Ruhrgebiet and its urban centres.

Consequent marketing policies emphasizing the innovative activities of the city do slowly change this negative image within and outside the city boundaries. At present Dortmund is considered to be more innovative and more sympathetic to the private economic sector than other cities in the traditional industrial agglomeration. This new image will certainly have a positive impact on future local economic development. At present, the city seems to be more successful than other cities in the region in receiving financial support or in attracting new firms.

7.4.2 CONCLUSIONS

For the time being the success of Dortmund's new local economic development policy lies primarily in the fact that the city has started systematically to break up traditional structures. It can be anticipated that the manifold endeavours to initiate political, social, economic, ecological and environmental innovations at the local level will be successful in the long run.

If the local actors in Dortmund have the strength and persistence to pursue the new policy approach continuously, and if they do not discontinue the innovation process started, the city's prospects will, in the long run, become brighter.

This optimism is reflected in Table 7.2, which sums up the qualitative interim assessment of particular sub-programmes of the new local economic development policy. This assessment is based primarily on evaluations and judgements made by local actors involved, by local entrepreneurs interviewed, and on our own assessments. The local impacts of the new policy are differentiated according to their effects on modernizing the local industrial base, respectively according to their impact on the local labour market. The impacts of the various sub-programmes are additionally evaluated as to their short-, and their possible medium- and long-term effects. This overall benevolent assessment of possible positive impacts of Dortmund's new economic development policy should not be overvalued. In the coming years the still dominating industrial sectors will continue to reduce their number of employees. Consequently unemployment figures in Dortmund will rather increase than decrease and will further worsen the overall economic situation of the city. It is still difficult to predict when the bottom of the valley will be reached.

The initiatives for the future economic development of Dortmund taken so far by the local actors are a sound and solid basis for the future local economy. Compared with other cities in the region, which suffer from similar problems, Dortmund seems to have reacted just in time. It has taken the right measures for restructuring its local economy. However, it will be essential that present innovative efforts endure and that successful action, projects and programmes are replicated and complemented. Nevertheless, it will be indispensable for upper tiers of government, primarily for the federal government, to support this process of endogenous economic development. Spatial redistribution policies have more than ever a role in the restructuring process of traditional industrial regions.

The whole bundle of necessary actions to restructure a local economy also requires the continuous support of many actors (Table 7.3). In the Federal Republic of Germany, it is not a top-down approach but a process whereby individual actors share the responsibility as well as the financial or managerial burden, while one, predominantly local, institution, in full accord with others, acts as an agency taking the lead.

The role individual actors have in initiating and implementing projects in the process of restructuring the local economy is dependent on their vested interests, their professional experience, respectively their functions and responsibilities. Whether or to what extent they engage in supporting and/or implementing projects, programmes and policies, aiming at contributing to local economic development, does, however, also depend on individual

Table 7.2 Interim assessment of local economic development policies

Policy/Programme	Impact on modernizing the local industrial base			Impact on local labour markets			Comments
	Short-term	Medium-term	Long-term	Short-term	Medium-term	Long-term	
1 Locating technology-oriented new enterprises	+	+	+	+	(++)	(++)	
2 Helping existing enterprises innovate:							
Products	+	+	+	0	+	+	Focusing on indigenous potential for economic development
Processes	+	+	+	–	×	×	
3 Locating public research facilities	+	++	++	+	+	++	Spill-over effects on local labour markets
4 Technology centre (linking university research to private enterprises)	+	++	++	+	+	+	Small number of additive jobs with high qualifications
5 Science and technology park	(+)	(+)	(++)	(+)	(+)	(+)	Interrelated to item 1
6 Aiding start-ups	+	++	++	+	+	(+)	Probability of failure
7 Supporting voluntary employment initiatives	?	(+)	(+)	+	?	?	Indirect impact on employment through training
8 Providing industrial land:							Removing barriers for locating firms, keeping land available for potential demand
Developing virgin land	+	+	+	+	+	+	
Recycling derelict land	+	+	+	++	+	+	
9 Providing subsidies for steel industry to aid the process of modernization	+	+	?	––	×	?	Helping to prevent additional shut-down, stabilizing employment on lower levels
10 Improving public infrastructure:							Short-term employment effects through construction
Roads, freeways	?	?	?	++	?	?	
Canal and harbour	+	+	+	++	?	?	

Note: + net increase 0 no impact – net decrease ? not available × stabilizing ( ) uncertainty.

The assessment is based on qualitative judgement, reflecting the still short period of programme implementation. Medium-term and long-term effects represent presumable *ex ante* estimations.

Table 7.3 Actors at local and state levels involved in local economic development policies in Dortmund

Economic development programme/policy	*Local government*				*Politicians and political committees*			*Public and semi-public institutions*					*Regional*		*State ministries*			*Others*		
	Department of Local Economic Development	*Urban Planning Department*	*Municipal Property Department*	*Urban Policy Unit*	*Key politicians*	*Committee for Economic Development*	*Committee for Urban Planning and Development*	*Chamber of Industry and Commerce*	*Chamber of Trades*	*Trade Supervisory Authority*	*Land Development Corporation*	*Labour Office*	*District authority*	*Intercommunal Association of Ruhr Municipalities*	*Ministry of Economy, Medium Scale Industries and Transportation*	*Ministry of Labour and Social Affairs*	*Ministry of Country and Town Development*	*Local trade unions*	*Hoesch-Werke AG*	*University*
1 Locating technology-oriented new enterprises	■	●	●	-	●	●	○	●	-	○	-	○	○	-	●	-	○	-	-	○
2 Helping existing enterprises innovate	●	-	-	-	○	○	-	■	●	○	-	●	-	-	●	○	○	○	○	■
3 Locating public research facilities	●	○	●	-	■	○	○	●	-	-	-	-	○	-	○	-	○	-	-	■
4 Technology centre (linking university research to private enterprises)	■	-	-	-	●	●	-	■	-	-	-	-	-	-	●	-	-	-	-	■
5 Science and technology park	●	■	○	-	●	●	●	●	-	-	-	-	○	○	○	-	●	-	-	●
6 Aiding start-ups	●	○	○	-	○	○	○	■	■	○	-	○	-	-	●	-	-	-	-	○
7 Supporting voluntary employment initiatives	●	○	○	○	■	●	○	-	-	○	-	●	-	-	○	●	-	●	○	○
8 Providing industrial land:																				
Developing virgin land	■	●	●	○	●	●	●	○	○	○	-	-	●	○	○	-	●	-	-	-
Recycling derelict land	■	■	●	○	●	●	●	○	○	○	●	-	○	○	○	-	●	-	●	-
9 Improving public infrastructure roads, freeways, canal and harbour	●	■	●	○	●	●	●	■	○	○	-	-	●	○	○	-	●	-	●	-
10 Providing subsidies for steel industry to aid the process of modernization	○	-	○	-	●	○	○	●	-	-	-	●	-	-	●	●	-	●	■	-

■ lead ● highly involved ○ supportive or involved - not involved personalities in the institutions concerned. Individual commitment to certain activities and identification with projects and policies seem to be essential prerequisites to shorten time periods for decision-making, to take over responsibility for decisions and for efforts to communicate commitment to the other actors involved.

Without a certain mutual understanding and respect of public concerns and private interests the necessary timely concentration and coordination of activities will not take place.

ACKNOWLEDGEMENTS

The article is based on the findings of a research project for the OECD, which was commissioned by the Bundesministerium für Raumordnung, Bauwesen und Städtebau (Federal German Ministry of Spatial Planning, Housing and Urban Development) in 1985. Eberhard von Einem from the Institut für Stadtforschung (IfS), Berlin, and Rainer Kahnert from the Institut für Raumplanung (IRPUD), University of Dortmund, cooperated in that project.

REFERENCES

Arras, E. H., Müller, K., and Eckerle, K. (1984), 'Innovations- und Diversifikationshemmnisse im Ruhrgebiet. Empfehlungen zum Abbau' unpublished TS, Basel: PROGNOS AG.

Clemens, Reinhard, and Tengler, Hermann (1983), *Standortprobleme von Industrieunternehmen in Ballungsräumen. Eine empirische Untersuchung im IHK-Bezirk Dortmund unter besonderer Berücksichtigung der Unternehmensgrösse*, Göttingen: Otto Schwartz.

Eckey, F. F., and Schwickert, J. (1982), *Analyse der sektoralen Entwicklung im Ruhrgebiet*. Essen: Strukturberichterstattung Ruhrgebiet, Kommunalverband Ruhrgebiet.

KVR (ed.) (1986), *Ruhr-Report*, Essen: Kommunalverband Ruhrgebiet.

Miegel, M. (1987), *Kurswechsel in der Kohlepolitik?*, Stuttgart: Verlag BONN AKTUELL.

Schönebeck, Claus and Wegener, Michael (1984), 'Wirtschaftsentwicklung und Raumstruktur – Gesamträumliche und kleinräumige Auswirkungen der Stahlkrise im Raum Dortmund', in Wilhelm Gryczan (ed.), *Zukünfte alter Industrieregionen*, Dortmund: Institut für Raumplanung.

Stadt Dortmund, Amt für Wirtschafts- und Strukturförderung (1983), Beschluss des Rates vom 17.02.1983 zum Konzept für die Wirtschafts- und Strukturförderung, Dortmund.

CHAPTER 8

Regional restructuring in French-speaking Europe: the examples of the Swiss watch-making area and the French Montpellier region

Denis Maillat

8.1 ARE LOCAL COMPANIES OR IS LARGE-SCALE INNOVATION BEHIND LOCAL RESTRUCTURING?

For about fifteen years, the economic geography of the developed countries has been undergoing a number of changes. Some dynamic new regions have emerged while previously rich regions are declining. In general it is the industrial regions which have recorded the least-favourable results. At the same time some new regions which have been able to monopolize new technologies have emerged. More recently we have witnessed the development of a new category of regions with so-called 'spontaneous' or 'diffuse' industrialization whose dynamism is based on very small enterprises.

These changes in the spatial dynamic, accompanied by a transformation of the distribution of activities, are manifesting themselves in most industrialized countries. In the United States the trend is moving from the east coast towards the west and south. In Italy the three leading provinces, which employed a quarter of the industrial workforce in 1971, accounted for only one-fifth of industrial employment in 1981; in Switzerland, traditionally touristic regions like Ticino or traditionally rural ones like Fribourg are renewing the structure of their activities; in France, the eleven regions that were best-placed between 1968 and 1975 notched up less favourable performances between 1975 and 1981 than the ten worst-off regions. Regions such as Provence-Côte-d'Azur and Languedoc-Roussillon in particular are doing better than those with old industrial traditions (Lorraine, Nord-Pas-de-Calais, or even Ile-de-France). Up to now the phenomenon has been marked enough for some authors to speak of spatial reversal and new territorial links (Aydalot, 1984; Courlet, 1986). Even if one does not go as far as that, it is nevertheless important to enquire as to the reasons for the success of these new regions.

Of the new regions, those which are most frequently mentioned have been able either to build their success on the new technologies or to establish a form of 'spontaneous' or 'diffuse' industrialization.

The regions with new 'spontaneous' or 'diffuse' industrialization based on small enterprises are to be found all over Europe. These small and medium-sized enterprises (SMEs) often fill the niches left vacant by the large companies (small electric household appliances, small-scale engineering, etc.) or those from which they have ousted the large companies, such as clothing and shoes. They may be very specialized and oriented towards small series or made-to-measure production.

This form of industrialization is due to the initiatives of local entrepreneurs and is tailored to the skills of the local workforce. It does not therefore result from the decentralization of large companies' activities. The dynamism of these regions stems from the creation of a network of interdependent relations between firms, known as an integrated territorial system of SMEs. Within this system companies maintain links (whether formalized or not) ranging from financial participation through supply contracts to ordinary family ties or temporary informal agreements (Courlet, 1986).

As for the new-technology-oriented regions, these are characterized by the physical proximity of research and training institutions and production units. By way of example, one may cite the Cambridge region with its technology park, or the Grenoble region with the Meylan ZIRST. In these areas firms are strongly oriented towards research and design activities. They are generally connected with the work of a university or a research centre and employ many scientists, technicians, and specialists. In principle, these firms do not devote themselves to mass production. The development of this type of region is largely due to local initiative (private or public). Their management of innovation is highly territorialized. They provide a territorial base for scientific and technological applications and they rely on the effects of proximity to stimulate the process of innovation (OECD, 1986).

There are, of course, other types of region that are favourable for the development of new activities. We do not intend to review them all. What is important here is to note that the emergence of the new regions cannot be explained simply in terms of decentralization stemming from large companies or central regions. It is more the result of local entrepreneurial ability and the tapping of local potential. In other words, local firms may be regarded as a source of social and technical innovation. A new development model is emerging which emphasizes the milieu. We shall endeavour to show that this model may be generalized and applied to other regions, including traditional ones.

8.2 A NEW DEVELOPMENT MODEL: THE ROLE OF THE MILIEU

Tapping local potential and highlighting the role of local entrepreneurial talent result in an original regional development model. This model shows that a region's fate is not necessarily dictated by its natural resources nor by the attractive force it exerts over the branches of large companies, but that it depends on its ability to innovate, its creativity and will to act. If these intangible skills are not to disappear, they must be nurtured and renewed. This can be achieved by creating a territorial innovation dynamic (Perrin, 1986).

The territorial innovation dynamic is a function of the characteristics of the milieu (sum of all the economic, political, social and cultural relations). Through multiple communications and trade networks (commercial and non-commercial, formal or informal), the milieux generate specific territorial skills and solidarity among local firms and individuals. In this way competition, supplies, various forms of assistance, colleagues etc. eventually form a more or less fertile set of vectors capable of either enriching or stifling the will to create or the reality of creation (Arocena *et al.*, 1983). Local milieux function like incubators of innovation. The enterprise is therefore not regarded as an isolated innovating agent; it is part of the milieu which has made it act. Consequently, territories' past, their organization, collective behaviour and the consensus which structures them are major components of innovation (Aydalot, 1986). Rather than considering innovation as a product of the milieu, it could, according to conventional models, be seen in terms of companies' patterns of location or technology dissemination models. The approach centred on companies - particularly large companies - implies investigating the factors that determine where its various branches are located. It can, for example, be shown that units incorporating high technology have patterns that require very specific location conditions which few regions can offer. The technology-oriented approach - technology being regarded as exogenous - prompts investigation of the modalities of technology transfer and its reception by the various regions, but does not highlight the process of innovation within the region. In fact, only the milieux-based approach exphasizes the territorial innovation process and the ability of each milieu to articulate the form of technical progress suited to its structures and characteristics. In addition, it stresses that the milieu is more than a mere juxtaposition of the production units; it is a complex of commercial and non-commercial proximate relations involving all regional factors.

If innovation is a product of the milieu, of the latter's inventiveness, each milieu may be regarded as capable of generating a specific innovation process. Seen in this light, the most important factor is not the appearance of a new technology but the decision to set in motion the territorial innovation process.

8.3 SMEs AND THE TERRITORIAL INNOVATION PROCESS

The renewed importance of the SMEs enables the territorial innovation dynamic to be seen in a new perspective. The SMEs are re-emerging as some of the most active agents of innovation. Furthermore, at the territorial level, the SMEs do not operate in isolation from one another: they are part of a network. This network which they supply and which provides them with multifaceted services in the areas of technology, training, management, marketing and financing is essential if they are to function smoothly. The network decides their innovative abilities by the skills it puts in contact with each other (Planque, 1987).

For a long time large companies were ascribed a prime role in regional development owing to the inductive effects they were likely to generate. Analysts readily contrasted the big, innovative companies with the SMEs, which were regarded as sub-contractors possessing little innovative ability.

The economic crisis revealed the deficiencies of large companies and the (sometimes exaggerated) virtues of the SMEs were highlighted. Now it is held that in large companies the entrepreneurial spirit and the progress of innovation are restrained by administrative complications and the cumbersomeness of corporate structures. Consequently, they often react slowly to changes in the environment. Compared with the structural complexity and sluggish decision-making of large corporations, SMEs are in many respects better able to adjust to technological progress and fluctuations in demand. Their flexibility is further enhanced by robotization, which makes it possible to manufacture in small series at a reasonable cost. These phenomena explain why, since the early 1970s, most new enterprises have been small, both in industrial sectors and in services. Their emergence is generally not due to their belonging to a sub-contractors' network dominated by the large companies. Some of these SMEs, which are particularly dynamic, are developing independently. The specific characteristics of the products or services they offer make them more partners than sub-contractors (Maillat, 1986).

True, SMEs cannot systematically be contrasted with big companies. The latter remain determining factors in the development and functioning of production systems. However, from the point of view of a territorial production system, SMEs play a strategic role. Their decision-making centres are attached to such a territorial system. In particular, they maintain more relations (whether commercial or non-commercial) with the other firms of the milieu than do businesses owned by large companies.

The dynamic of territorial development thus depends on local entrepreneurial ability and not on the spatial strategy of the large companies. Indeed, the large company only rarely acts as a function of the milieu in which it is located. The SME, on the other hand, derives its efficiency from the ability of all the social factors to create an environment favourable to its development.

8.4 THE MILIEU AT THE CROSSROADS OF A DUAL LOGIC

Placing the emphasis on the milieu does not mean a region is inward-looking. On the contrary, the milieu is at the crossroads of a horizontal (territorial) and a vertical (functional) development logic. Although innovation depends on the specific nature and intensity of territorial relations (horizontal logic), that does not mean there is no need for links with the outside world (vertical logic). But by emphasizing proximate relations, the effects of solidarity, corporate networks, the role of local factors etc., the new development model teaches us that it is through an adequate milieu that a region can acquire a certain degree of autonomy for its innovation strategy (Perrin, 1986; Mifsud, 1987; Maillat, 1987). We shall illustrate this range of problems with two examples: the Jura Arc (Arc Jurassien), a region with an industrial tradition which has suffered the effects of the economic recession and the introduction of the new technologies, and the Languedoc-Roussillon region, which for a long time was regarded as relatively undeveloped but which, despite the recession and thanks to its ability to innovate, is emerging and attracting attention. In both cases the nature of the milieu plays an important role.

8.5 THE JURA ARC, THE ROLE OF THE MILIEU AND KNOW-HOW

The surveys undertaken by GREMI (European Research Group on Innovative Milieux) have revealed that at least some industrial regions possess the resources necessary to revitalize their production systems (Aydalot, 1986). These traditional industrial fabrics innovate by direct continuity, as for them innovation is connected to what exists (shift from mechanical skills towards electronic skills, link-up of mechanical and electronic technologies (robots), use of the traditional market as a base for diversification etc.). All in all, in this case the success of innovation is based on elements of continuity with the experience acquired from the milieu. In these regions the territorial systems conceal 'traditional' resources and skills which may be recombined with new elements. These skills are crystallized in the form of know-how. This know-how comprises all the practical and intellectual skills needed to master technologies within the production system. It belongs to the milieu.

Thanks to these skills that have been built up over time, such regions have resources which enable them to relaunch the territorial innovation dynamic. It is obvious that this is only possible if the new technologies are compatible with the local technical culture, as it is a question of achieving the appropriation of novelty by the milieu (Maillat, 1987). The recent history of the Jura Arc is a good example of such a development.

With an industrial tradition (more than 52 per cent of employment is in the secondary sector) centred mainly on the watchmaking and machinery industries, this region lost 34,000 inhabitants between 1970 and 1985, a decrease of 5 per cent. Today it has a population of 671,000. Over the same period industrial employment declined by 30 per cent, which is equivalent to a loss of more than 50,000 jobs. It was obviously the watchmaking industry that was mainly responsible for this decline.

These figures show just how vulnerable the production system in the Jura

Arc was at the beginning of the 1970s. It was composed of a strongly integrated complex of watchmaking, machine-tool and precision-engineering industries. The arrival of electronics and tougher international competition were radically to alter this production system. The machine-tool industry transformed its products by introducing numerical control systems and modernizing its workshops; the precision-engineering industry did the same, but it was undisputably the watchmaking industry that underwent the most profound changes. New technology eliminated some of the segments of the traditional watchmaking industry. This in turn caused the closure of a number of companies, mergers, changes of product lines and new management and marketing techniques.

Thus all the activities present in the Jura Arc were among those to which electronics was to cause radical changes, both to products and to manufacturing processes. The immediate consequences were a reduction in jobs, one of the highest unemployment rates in Switzerland (2–3 per cent) and substantial emigration (mainly by the foreign population). But these changes were also a manifestation of the region's ability to adjust, which took the form of the adoption of new production techniques, new forms of management and the appearance of new products (Crevoisier, Rudolf and Vasserot, 1987).

Traditional expertise played a key role in this transformation. Accustomed to having a technical and intellectual command of micro-mechanics within the production system, the workforce had no difficulty adjusting to micro-electronics. Most of the time it was the mechanics who attended further training courses. Young people leaving the training system continued to be integrated into the workshops as before. The training of electronics engineers and their incorporation into companies was carried out at the same time, without the mechanics being stripped of their qualifications or of their influence over job contents. In this way it was possible – thanks to the mechanic's broad professionalism – to avoid creating electronics specialists who would have deskilled and destroyed the sense of responsibility inherent in mechanical work by breaking the links between workers, machinery and materials.

In the watchmaking industry the change was introduced mainly by a number of firms which already had a certain amount of electronic expertise in research work (the quartz watch was invented in Switzerland). The small enterprises specializing in the traditional watchmaking parts (hands, faces etc.) either had to close down or produce other articles.

In fact, it was the precision-engineering professions that bore the brunt of the region's restructuring, but which at the same time were its vectors. Today there are fewer watchmaking mechanics than in the past, but their expertise was reusable in the other precision engineering industries, so that these professions have not vanished. Their skills were maintained at the technical, organizational and social levels. Moreover, it was undoubtedly in order to defend this strong professional culture that the new technologies were adpoted (Crevoisier, 1987).

In the final analysis it was the defence of an indentity that was at stake. As has been emphasized, there always has been a strong technical culture in the Jura Arc. It is important *per se*, but even more so as a fundamental representation of the identity of the milieu. Measures were taken very early on to ensure the survival of this milieu. Paradoxically, however, in a first stage no trust was placed in local enterprises.

8.6 THE QUEST FOR OUTSIDE ENTERPRISES

With the economic crisis and the ensuing decline in employment, there was a loss of confidence in local firms' ability to maintain or create new jobs. Thus the reaction of the local (cantonal) political authorities was unanimous: to look to the outside world and to attract into the region enterprises or branches of companies located outside the region. Representatives responsible for economic promotion were appointed, whose task involved (and still involves) prospecting outside the region, particularly abroad, with a view to persuading companies to set up in the region.

Regional companies are usually mistrustful of this type of action. In the event, though, they did not put up much resistance in view of the very disturbing situation of the regional economy. After the skirmishes that inevitably occur when such measures are envisaged, a genuine local consensus was finally reached in support of the political authorities' initiatives.

Initially the aim was reindustrialization on all fronts, but gradually the search for outside companies became more discriminating (the main target was firms using new technologies compatible with regional expertise); however, it expanded to take in services.

While the cantons were trying to attract outside firms to their territory, the confederation (central government) was participating in the policy of restructuring the economy by adopting a series of legislative measures, in particular a law of 1978 designed to encourage innovation and diversification projects. The regions of the Jura Arc were the main beneficiaries of these measures. The idea was to encourage the companies of these regions to develop their products, to manufacture or market new ones, to adopt new production processes, to modify their marketing policy and to create new distribution channels. In addition, it was decided to grant subsidies to regional information services in the field of innovation (consultancy). As various surveys have shown, the overall result of these cantonal and federal measures has been positive since it is estimated that between 8,000 and 10,000 jobs were created or maintained over the period 1978–86.

8.7 MASTERING TECHNOLOGY

The policy of looking to outside enterprises helped to create a new climate of confidence in the region, but it was soon realized that it was not sufficient to revitalize the region's economy in a durable way. Gradually, the idea of creating a micro-technology centre in the region took shape. The presence of a university at Neuchâtel certainly played a crucial role in this. Indeed, that university decided to create a micro-technology institute as early as 1982. Furthermore, a number of research laboratories already existed in the region, especially the CEH (Electronics and Watchmaking Centre) and the LSRH (Swiss Watchmaking Research Laboratory). But in order to take up the new challenge and to change mentalities it was necessary to find a new structure and more substantial resources, hence the idea to locate a national research centre at Neuchâtel. This was the CSEM (Swiss Electronics and Microelectronics Centre SA), which was set up thanks to the perseverance of the local authorities. Rather than aiming at an industry (watchmaking), as had been done until then, measures were targeted at a technology, in this case micro-technology. This encompasses micro-mechanics, micro-electronics and opto-electronics, and is closely linked to information technology. It creates microprocessors, sensors, optical fibres, robots and other small, reliable, high-performance components or devices on which depend both the improvement of industrial processes and products and the development of information processing and transmission systems.

The remit of the CSEM, with its 200 personnel, is to develop and perfect state-of-the-art expertise in the various micro-technology disciplines. Its aim is to anticipate the needs of industry and to make available in good time the knowledge and processes it needs. In addition, the CSEM has to offer universities a framework for efficient cooperation. To do so, it makes its skills and facilities available to the public teaching and research institutions. It is also involved in disseminating the results of university research throughout the economy.

Although located at Neuchâtel, this institution was conceived with the whole country in mind, in order to 'take over and adapt the results of basic research carried out abroad to the needs of Swiss industry'. It was thus not intended, *a priori*, to ensure the dissemination of new technologies at the regional level. True, it is very obvious that some local enterprises in the region collaborate actively with the CSEM, but its other customers come from the rest of Switzerland or abroad. However, the location of the CSEM at Neuchâtel has had a considerable effect on the region's image. It has mobilized and channelled energies towards a project that has made it possible to reorganize and boost the research carried out in the region. Today efforts are being made to increase this centre's territorial impact by exploiting the effects of proximity. In order to achieve this objective, adequate structures have to be put in place, particularly as the region's industrial fabric is essentially composed of SMEs which have no opportunity to dialogue with the CSEM or to entrust it with financially worthwhile research contracts. The aim is to lower the barriers to entry, thereby exploiting the synergetic effects with local enterprises.

8.8 INNOVATION IN LOCAL ENTERPRISES AND THE CREATION OF ENTERPRISES

The latest analyses clearly reveal that the Jura Arc has not succeeded in compensating for the jobs that were shed over the past fifteen years. However, in the area of new technologies, the number of jobs has grown faster than in the other parts of the country. In addition, a certain dynamism is manifesting itself with regard to company innovation and the creation of new enterprises.

Out of a sample of 320 firms, it was found that 252 (79 per cent) had innovated, 55 (22 per cent) of which in products, 91 (36 per cent) in processes and 106 (42 per cent) in products and processes (Maillat and Vasserot, 1986).

Product innovation, which enables the progress of innovative enterprises to be pinpointed, takes place in three different types of enterprise: the enterprise that innovates by imitation (75 per cent of cases), the enterprise that innovates by diversification (15 per cent of cases) and that which innovates by bifurcation (10 per cent of cases). The survey showed that 75 per cent of enterprises develop new products on the basis of known technologies, consisting of improvements and updated versions of existing products. These enterprises adjust to the technical development of the industry or the product, depending on competition and the market. Firms that innovate by diversification develop new products from their expertise. They help to transform the region's production structure. As for firms that innovate by bifurcation, they introduce so-called high-technology products. As can be seen the majority of product innovations do not involve high technology; this is proof that in an industrial region one should not have too restrictive a vision of innovation. The important thing is that the territorial innovation process should be set in motion.

We do not have any exact figures on the creation of new enterprises. However, a recent survey, carried out on a significant sample over the period 1981–7, shows that 75 per cent of new enterprises are of local origin and 25 per cent came to the region following the action taken by the economic promotion services. Most of these new enterprises belong to the electronics and machinery industries. Foreign firms provide more jobs than local ones do, supplying about 50 per cent of the jobs created, whereas they account for only 25 per cent of new enterprises. On the other hand, local firms employ more skilled labour than foreign companies. Thus 9 per cent of the workforce work on R&D in local firms whereas the percentage is only 2.6 per cent in foreign outfits. The information on innovation in regional enterprises and on the creation of new enterprises clearly shows that the region has regained its internal creativity.

All in all, the Jura Arc is behaving like many other industrial regions which, after pursuing a policy of attraction, discover or rediscover the need for a more territorial dynamic. As is emphasized by Mifsud (1987):

the establishment of outside companies may be desirable, or even indispensable, to inject additional dynamism at the local level. However, they only become genuine vectors of development if they are integrated into the local economic structure and are in harmony with the initiatives and achievements taking place within it. The extent and importance of the latter attest to the degree of dynamism of the local economy, and to its ability to weather change and to innovate.

8.9 LANGUEDOC-ROUSSILLON – THE CREATION OF A TERRITORIAL ECONOMIC SYSTEM

In a typology based both on recent economic history and an analysis of economic and demographic structures, Laget (1984) distinguishes, in the case of France, four groups of regions: those close to the Ile de France, which are called 'overspill' regions since they function thanks to the capital's diffusion effects; the western and central regions, so-called transitional regions, which are characterized by a relative growth in industrial employment and services but still have a substantial proportion of agricultural employment; and the regions of industrial restructuring of the east and north. In the latter it is as if the decline of the large traditional industries were going to last for as many decades as were necessary to establish them; finally, the pressure regions (Provence-Côte-d'Azur, Rhône-Alpes, Midi-Pyrénées, Aquitaine, Languedoc-Roussillon) which, while having high unemployment rates, are creating new jobs and new enterprises and are inventing new development strategies. In this category the example of Languedoc-Roussillon is particularly interesting. The low growth rates it had last century, the absence of polluting industries, and the availability of quality space on the edge of the Mediterranean enable it to envisage an unconstrained approach to the information revolution. This relatively underindustrialized wine and vine-growing region is developing towards modern industrial and tertiary activities, i.e. towards a post-industrial type of society. In particular, it attaches great importance to research and technology.

Languedoc-Roussillon is in fact a region of contrasts. The conclusions may be diametrically opposite, depending on the analysis carried out and the indicators used. Thus the region is characterized by one of the highest job-creation rates, but at the same time it has one of the highest unemployment rates. Yet its economic structure is being transformed, and a process of territorial innovation is being established.

8.10 THE RESTRUCTURING OF ACTIVITIES

In order to understand the mechanisms at work it is necessary to go back to the 1950s.

Infrastructural investments have been substantial, with the construction of the canal linking the Lower Rhône with the Languedoc, and in the early 1960s with the interministerial coastal development mission. These investments were the result of public and private decisions which relayed or linked up with one another. Thus, for example, the development of tourism resulted from private investment opportunities and a public action, since it was the state which bought the plots of land and created the infrastructure of the resorts.

Productive private investments developed rapidly in the 1960s. The landmark was the decision of IBM-France to locate in Montpellier in 1965, with a plant designed to produce the company's biggest mainframes for the whole of Europe. It has become the region's largest industrial enterprise (2,700 employees) and attracts a great deal of sub-contracting (Verlaque, 1987). Other companies, too, have located in the region. Heliotropism is not extraneous to this phenomenon of attractiveness, but one should not underestimate the arrival of the repatriates from North Africa (some 100,000 people between 1960 and 1964), a population that was industrious and often endowed with resources (Laguet, 1987; Marchesnay, 1985).

As of the 1960s, the region's progressive sectors developed favourably and faster than at the national level (thus employment in electrical and electronic engineering soared from 300 in 1964 to 6,000 in 1982) and at the same time the recession-hit sectors (the consumer-goods industry or sectors using traditional labour, extracting industries) were also contracting more rapidly.

These movements caused a restructuring of activities and generated a dynamism that was subsequently confirmed. Thus over the period 1975–82, the number of jobs in the Languedoc-Roussillon region increased by almost 10 per cent, four times faster than that of national employment. The slowdowns observed in recent years do not invalidate this general trend (Fornairon, 1987).

Over this period seven industrial sectors and ten service sectors, comprising more than half of regional employment, enjoyed a faster growth rate than at the national level. The industrial sectors included timber, furnishing, electric and electronic engineering, printing, the press, publishing and non-ferrous-metals industries. The tertiary sector included mainly commercial activities: the non-food wholesale trade, transport, posts and telecommunications, financial services and above all commercial services for companies, which grew faster than in the rest of France. As for traditional industries such as leather/shoes, fuels and solid minerals and food industries, these declined more rapidly than at the national level. Behind these relative rises and falls lies the most significant phenomenon of this period, the creation of new links between industrial and tertiary activities and new economic growth logics. In the 1970s, inductive activities sprung up in many sectors, unlike the situation of the 1960s when a fairly clear-cut distinction could be drawn between inductive jobs and induced jobs. At that time industry was regarded as being the opposite of services, since the latter had always been presented as dependent, induced activities. In fact, many industrial sectors gradually found themselves to be dependent on the liveliness and dynamism of certain tertiary activities. Services to companies in particular appear to be inductive services with extra-regional markets. They exercise an inductive influence on the structure of the economy as a whole. A new system, characterized by constant interrelations between service activities and industrial activities, has thus gradually been established (Bel, 1985; Layet, 1985; Marchesnay, 1985).

8.11 COMMAND OF TECHNOLOGY

This system is completed by a network of relations with the universities and research centres. Indeed, in this region the mesh of university institutions and research centres is both very dense and very old. The present networks that are being created around the notion of 'technology park' are incorporating and coordinating existing bodies and institutions, which are expanding rapidly. Here again heliotropism is facilitating the trend, as senior executives are particularly sensitive to the quality of life and the environment. But at the base there exists a local scientific potential. For example, the four industries in which Montpellier excels use their acquired knowledge and experience and enhance its value:

(a) the bio-medical and pharmaceutical industry is founded on a tradition of medical teaching and research in the region;

(b) the 'agropolis' industry comprises a remarkable training and research potential composed of public institutions (School of Agriculture, National Agronomic Research Institute, Mediterranean Agronomic Institute) and semi-public bodies (National Agricultural Machinery Centre, Centre for Agronomic Research for Hot Regions, French Marine Development Institute);

(c) the information technology and robotics industries use the twofold opportunity afforded by the presence of IBM at Montpellier and the National Southern University Data Processing Centre;

(d) the 'new media' which, although linked to the information technology and robotics industries, rely on the Montpellier local authority, which has banked heavily on telecommunications.

Basic research and development research are in many respects more developed than in most other French regions: they appear to have reached the size necessary to give rise to induced activities and new location by companies (Laget, 1986).

Furthermore, it is probable that the research carried out in this region in the human, economic and social sciences is not extraneous to the population's behaviour, which is both critical and innovative. The exceptional wealth of human resources in this region, where vocational qualification indicators, graduate rates and cultural levels are remarkable and sometimes two to three times better than those of the underprivileged regions, is at one and the same time a great opportunity and a development challenge.

8.12 HUMAN RESOURCES

The 2 million people in Languedoc-Roussillon constitute a human capital which increases annually by 1.1 per cent. The accumulation of this human capital is the result of immigration on the one hand and of vocational training on the other.

The demographic trends of Languedoc-Roussillon can be explained mainly by migratory phenomena and not by natural excess. Immigration helps to regenerate an ageing population. Indeed, contrary to the received view, immigrants are not only retired persons or job-seekers, they are distributed over all age groups. Often they are Languedocians returning to their region after an exile of several years, but each of the French regions contributes its share to the increase in the regional population. These 'new Languedocians' are motivated by the possibility of practising a profession or landing a job. This immigration rejuvenates not only Languedoc-Roussillon but also transforms its professional structure, since the highest immigration rates are for the higher professions and managerial staff. These regular arrivals exert an inductive influence on the local population. They have helped to create, to use Laget's expression, a 'socio-economy of mixing and selection'. In this type of region which, it should not be forgotten, is described as a pressure region, there exist maximum forces which induce selection. Several hundred people compete for ten new jobs (whereas in other regions there is only one). By degrees, and for decades, many jobs have thus been held by over-qualified workers as a result of this selection process.

Not only do immigrants have a high qualification rate but the region itself has a good training system which enables it to train qualified individuals. A Languedocian, for example, is three times more likely to go to university than the young people of the regions least well off in this respect.

In Languedoc-Roussillon perhaps more than elsewhere it is obvious that human resources have become one of the important development variables. Today all the indications are that the prime investment is tending to be made in human resources and the rate of increase in highly-skilled jobs in the structure of regional qualifications allows one to posit a reversal of the coefficient of capital to the benefit of acquired knowledge and at the expense of machines.

8.13 THE CREATION OF ENTERPRISES AND THE EMERGENCE OF LOCAL ENTREPRENEURS

This importance of human resources is manifesting itself notably in the creation of enterprises which is increasingly resulting from a labour–capital combination, not a capital–labour one.

In Languedoc-Roussillon as elsewhere the large company is not the only explanation for the trend, particularly with regard to the number of jobs created. Today small units play an essential role. In this respect Languedoc-Roussillon appears to be a leading region. It is estimated that since 1980 the creation of enterprises (measured in terms of the number of enterprises per 1,000 employees) is 50 per cent higher than at the national level (Laget, 1987). In this region the development of the SME was to take place in two stages (Marchesnay, 1985).

In a first stage, that is, between 1965 and 1975, the stock of local enterprises suffered the consequences of industrial restructuring. Although some firms succeeded in adjusting, a considerable number were to get into difficulties and vanish. This was the case particularly for those belonging to traditional sectors (agribusiness, textiles, steelmaking etc.). On the other hand, during this period extra-regional entrepreneurs arrived who created SMEs, invariably sub-contracting following the establishment of national or multinational companies: IBM Montpellier in particular.

The second stage, which began in 1975, was that of the new Languedocians. Many of these new arrivals created businesses: one head of household in five and one spouse in four either created their own job or maintained their self-employed status (Fornairon, 1987). But gradually the region began to generate its own entrepreneurs. They mostly came from the 'modernist' classes belonging to previous migratory flows, but also included native Languedocians who refused to become expatriates. They are just as active in services as industrial activities.

To a certain extent, and this is characteristic of a pressure region, 'the unemployment problem and the difficulty of solving it cause a certain number of job-seekers to set up their own businesses' (Fornairon, 1987).

It is true that although many SMEs are created many die as well. Yet according to one writer, Naro, these SMEs are 'a revitalizing force for the regional economy in that they maintain – among each other as well as with larger companies – a whole complex of trade flows whose inductive effects can but be positive for the creation of a regional production system'.

In fact, although the Languedoc-Roussillon region has borrowed a lot from outside it has been gradually moving towards the creation of a territorial production system. As elsewhere, the latter is characterized by ever greater interactions between the various regional components, and between the latter and extra-regional factors. The result is a specific territorial dynamic which enables the region to steer its development more effectively.

This region derives its dynamism from these contrasts. Because it is a pressure region it has to generate more jobs than the others; because its industrialization has been lagging behind it has to enter the post-industrial age. It should therefore no longer create external 'industrial-milieu' economies but external economies based on services and intelligence. Now its milieu conceals skills which can allow it to restructure its production system. In this region the renewal and enrichment of human resources has been remarkable. The opportunities arising from the scientific milieu can be developed thanks to a body of research sufficient to lead to synergy and creations. In addition, service activities for companies have been developing remarkably. Thus little by little the components of the territorial production system have been put in place. Moreover, the process that leads to the quest for a regional identity is indisputably at work. Finally, at the political level - which is not the least important after all - politicians, especially the youngest of them, have realized that innovation needed a territorial base.

8.14 CONCLUSION

In Languedoc-Roussillon, as in the Jura Arc, a considerable amount of borrowing from outside the region has taken place. But in each case revitalization requires the restructuring of a territorial production system and a milieu favourable to creation.

Thus it appears that the important thing for a region is to set up a full and coherent territorial production system. Full, because it should comprise manufacturing and service activities organized from the upstream end (research and development) towards the downstream end (the market. Coherent, because it has to be able to generate specific regional skills and cooperation phenomena (Maillat, 1987).

Once this has been done the territorial innovation process can be envisaged. Indeed, the latter's dynamism presupposes the collaboration and connection of complementary functions: basic research, applied research, development, preparation of prototypes, industrial investment, putting into production, marketing and tailoring production to the market (Planque, 1987).

The mastery of the upstream end is essential to produce and develop specific scientific and technical skills. To the extent that innovation presupposes a command of new scientific knowledge, it can no longer be the preserve of enterprises (especially where SMEs are concerned). It also relies on the territorial scientific set-up - the relations that are created between research institutions, the training system and the production system. The interaction between these different elements creates cooperation phenomena which enrich the milieu. This cooperation makes it possible to attract from outside and to develop inside the region the scientific and technical knowledge suited to the milieu.

Moreover, if it is to be able to detect market trends and stimulate its research system by development, a region has to strengthen the downstream end of its production system. This implies the development in the region of service activities such as marketing, market research and export advisory services. This downstream extension seems to be easier in Languedoc-Roussillon than in the Jura Arc, an industrial region which has always known how to organize the upstream part of its production system better than the downstream part.

Thus by organizing its production system on a territorial basis a region is able to develop its own specific creativity. As a result, a milieu may be formed in which all the regional elements have a role to play. This is the approach by which a region can gain or regain its innovational independence.

REFERENCES

Arocena, J., Bernoux, P., Minguet, G., Paul-Cavallier, M., and Richard, P. (1983), 'La création d'entreprise, un enjeu local', *La documentation française*, Notes et études documentaires, no. 4709–10.

Aydalot, P. (1984), 'Recherche de nouveaux dynamismes spatiaux', in P. Aydalot (ed.), *Crise et espace*, Paris: Economica.

Aydalot, P. (1986), *Milieux innovateurs en Europe*, Paris: GREMI.

Bel, P. (1985), 'Evolutions sectorielles de l'emploi des qualifications et adaptation formation emploi en Languedoc-Roussillon', *Revue de l'économie méridionale*, vol. 33, nos 130–1, pp. 43–56.

Courlet, C. (1986), 'Industrialisation et développement: analyse traditionnelle et mouvements récents', *Economie et Humanisme*, no. 289.

Crevoisier, O. (1987), *Le processus d'appropriation de l'électronique dans le milieu de l'Arc jurassien*, mimeograph, Neuchâtel.

Crevoisier, O., Jenny, A., Rudolf, J.-Ph., and Vasserot, J.-Y. (1987), *Innovations et nouvelles entreprises dans l'Arc jurassien*, Dossiers IRER 13, Neuchâtel.

Fornoiron, J. D. (1987), 'La mutation des activités et des emplois', *Revue de l'économique méridionale*, vol. 35, no. 137, pp. 83–94.

Laget, M. (1984), 'Peut-on parler de système économique régional?' Colloquium on 'Systémique et économétrie appliquées à la modélisation régionale et spatiale', Poitiers.

Laget, M. (1985), 'Quelques hypothèses pour comprendre l'évolution de l'emploi et de la démographie en Languedoc-Roussillon', *Revue d'économie méridionale*, vol. 33, nos 130–1, pp. 9–19.

Laget, M. (1987), 'Le système économique du Languedoc-Roussillon', *Revue de l'économie méridionale*, vo. 35, no. 137, pp. 31–56.

Laget, M. *et al.* (1986), *Transfert de technologie et développement régional*, Montpellier: Centre régional de la Productivité et des Etudes Economiques.

Maillat, D. (1986), 'Les PME innovatrices et la revitalisation des régions', *Revue d'Economie Régionale et Urbaine*, no. 5, pp. 688–93.

Marchesnay, M. (1985), 'Petite entreprise, services et région', *Revue de l'économie régionale*, vol. 33, no. 132, pp. 79–86.

Mifsud, P. (1987), *Milieux urbains et développement local*, Saint-Etienne: CREUSET.

OECD (1984), *Rapport analytique sur la recherche, la technologie et la politique régionale*, Paris: OECD.

OECD (1986), *Restructuration régionale*, Paris: OECD.

Perrin, J.-C. (1986), *Technologies nouvelles et synergies locales*, Notes de Recherches no. 67, Aix-en-Provence: CER.

Planque, B. (1987), *PME innovatrices et potentiel territorial d'information et de compétences*, Aix-en-Provence: CER.

Planque, B., and Py, B. (1986), 'La dynamique de l'insertion des PME innovatrices', *Revue d'Economie régionale et urbaine*, no. 5, pp. 587–607.

Raveyre, M.-F., and Saglio, J. (1984), 'Les systèmes industriels localisés: éléments pour une analyse sociologique des ensembles de PME', *Sociologie du travail*, no. 2.

Stöhr, W. B. (1986), 'Regional innovation complexes', *Papers of the Regional Science Association*, vol. 59, pp. 29–44.

Verlaque, C. (1987), *Le Languedoc-Roussillon*, Paris: PUF.

PART III

DIVERGENT LOCAL DEVELOPMENT PATTERNS IN EASTERN EUROPE: CASES FROM TWO COMECON COUNTRIES

CHAPTER 9

Learning experiences of local development in a centrally planned economy: the example of the Poznań region, Poland*

Bohdan Gruchman

9.1 INTRODUCTION

Local development in the centrally planned economy that existed in Poland after the Second World War has been shaped by major decisions at the centre. It is there that development funds have been allocated to sectors and regions, major locational decisions reached, and general planning targets determined. These decisions together with 'rules of the game' concerning behaviour of enterprises, budgetary units, and administration constituted the framework within which local development could take place. This framework has not been uniformly tight during the entire postwar period. The interference of central authorities into local development was particularly strong during the first half of the 1950s, and in the second half of the 1960s and 1970s. But even during these years there was room for local development initiatives, albeit restricted to minor areas. Besides that, the way central development policies were executed in particular domains was determined by local specificities and allowed for some local initiatives. And then there were longer periods of marked loosening up of central control and management, with the first half of the 1980s representing the beginning of a systemic change towards even more local autonomy.[1]

The sub-national authorities which enjoyed the relatively broadest scope of autonomy were those at the province (*voivodeship*) level. Until 1975 the country was divided into seventeen provinces. They constituted a vital link between the lower-level subdivisions of the territory (counties and communes) and the central government. The provincial governments supervised the sectors of the economy directly serving the local population, e.g. retail trade, local transport, health services, education, but also small-scale industry, handicrafts, agriculture and local construction companies. They were responsible for establishing development plans and balancing the local budgets, and allocated investment funds to lower-level units. They were also responsible for locational decisions with respect to numerous investment

* This chapter refers to the conditions just before the recently initiated drastic reforms towards a market economy and multi-party democracy in Poland.

projects of lesser importance and cooperated with central authorities in locational decisions for major projects.

In 1975 an administrative reform was undertaken: the seventeen provinces were broken down into forty-nine units (also named *voivodeships*) and simultaneously counties were abolished in order to simplify and shorten administrative communication lines from the centre to the communes. The economic power of the new *voivodeships* has diminished not only owing to their smaller size, but also because of the transfer of a great number of enterprises under the direct control of the centre. In the early 1980s, with the beginning of the economic reform, it was decided to strengthen the *voivodeship* authorities once more and to increase their sphere of direct control. The income base of the local budget has been broadened and more decision power granted in economic and social matters.

9.2 PRESENTATION OF THE POZNAŃ REGION AND ITS DEVELOPMENT PROBLEMS

The province of Poznań in western Poland, together with the city of Poznań, represented in its pre-1975 borders 8.6 per cent of the total area of the country and was inhabited by 2.4 million people (1960), i.e. by 8.2 per cent of the total population of Poland. In a relatively short time after 1945 it managed to overcome considerable war damages and reach its prewar level of the economy. It was then a predominantly agricultural region with relatively little industry. The only major industrial concentration was in the city of Poznań. Industrial establishments in the city employed 77,600 people (1960). Their production per inhabitant was 72 per cent above the national level whereas the same indicator for the province without Poznań was 31 per cent below the national average.[2]

The province of Poznań supported the largest number of statutory towns among all *voivodeships* of the country – ninety-six in all (1960). These towns were mostly small in size: half of their number had fewer than 5,000 inhabitants, eighteen were in the 5–10,000 group, sixteen in the 10–20,000 group and only five towns had a population of more than 20,000. Most towns functioned as supply and service centres for agriculture in the surrounding vicinity.

This structure of the regional economy did not guarantee sound development for the future. With progress in mechanization and modernization of agriculture this sector could not support the growing population. Migrants from rural areas had difficulties in finding work in towns owing to their weak industrial base. This prompted the regional authorities to formulate a programme of restructuring the regional economy. The main provision of this programme was speeding up industrialization of the region. As the region was already relatively well-equipped with food-processing industries, restructuring meant an accelerated development of other branches of industry which could be offered favourable locational conditions.

The region supplied skilled labour from numerous vocational schools. It already had well-developed infrastructural facilities, particularly roads and railways. The latter offered good links to other regions of the country and to other countries as well. The numerous towns relatively well-equipped with communal facilities could be attractive for many urban-oriented industries. The region could also be considered promising for market-oriented branches owing to relatively high income in agriculture. Finally, labour in the Poznań region was widely believed to be disciplined, industrious and efficient.

Restructuring through accelerated industrialization also had a distinct geographical dimension.[3] It was planned to speed it up, particularly outside the region's capital. The city of Poznań, which loomed large as an industrial centre over the rest of the *voivodeship*, should stop developing quantitatively in this field and, instead, modernize its existing industry and make it more efficient. It should even help the surrounding smaller towns to grow faster by transferring to them some enterprises hitherto operating in the city. Lastly, regional authorities decided to pay particular attention to the industrialization of the eastern parts of the province, where the level of economic and also social development was the lowest in comparison to the rest of the region.

This programme of regional restructuring was worked out by a special scientific council attached to the *voivodeship* government. It was composed of scholars from different academic and research institutions in Poznań, representatives of major enterprises, banks, professional organizations and social and cultural institutions. The council solicited wider support for its development through regular publication of its studies and findings discussed at period conferences. It did not enjoy any administrative power yet its proposals were accepted by the region's authorities as official strategy in its long-term development plan and its consecutive five-year plans.

Implementation of the adopted development strategy required concurrence of the central government as the latter assigned the bulk of investment funds to sectors and decided the location of major industrial projects. However, regional authorities were instrumental in preparing locational decisions of the central government with respect to sites within the *voivodeship*. Apart from that, they had the final say with respect to location of minor projects. They could also finance minor local projects from their own budget and help local population to execute self-help projects (e.g. in melioration, road construction, building of schools, health centres, fire-brigade centres). In the following, characteristics of two types of local development will be given: mainly externally (centrally) induced, and mainly internally (locally) induced development.

9.3 EXTERNALLY INDUCED REGIONAL INDUSTRIAL DEVELOPMENT: LOCATION OF NEW ESTABLISHMENTS

The authorities of Poznań *voivodeship* actively solicited the location of new industrial establishments on their territory. They cooperated with industrial boards which prepared locational decisions in each manufacturing branch for the government (and particularly for the central Planning Commission) in drawing new industrial projects to the *voivodeship*, and suggested the location sites within the region in accordance with the above-mentioned development strategy. Their task was facilitated during the 1960s and the early 1970s by an increased industralization drive of the Polish economy which resulted in a large number of new establishments being build around the country.

It is during that period that a number of new manufacturing enterprises were built in towns of the province, particularly in those surrounding the city of Poznań at a distance of 30–50 kilometres. The new establishments included a large shoe factory in Gniezno (47,000 inhabitants in 1965), a big foundry in Śrem (12,000 inhabitants), metal factories in Kościan (17,000), Środa (14,000), and Oborniki Wielkopolskie (9,000). Industrial works established further from Poznań included an aluminium works in Konin (26,000), a construction-materials factory in Słupca (6,000), sanitary-outfits and abrasive-materials factories in Koło (12,000) and an electric bulb factory in Piła (38,000). All the examples cited are relatively large-scale concerns for which the final locational decisions were approved by the central government. In addition, regional authorities located in many towns new small-scale establishments, mostly in food-processing, light industry, and metal and plastics.[4]

Some of the factories mentioned above were originally to be constructed in Poznań, either in order to substitute for an existing obsolete factory or as an entirely new establishment attracted to the city. Examples of the former are the metal factories in Środa and in Kościan. They operated in Poznań for many years and needed new facilities on another site. In accordance with their restructuring policy the regional authorities did not allow them to remain in Poznań and assigned them new locations in nearby towns situated some 35–45 kilometres south of Poznań. It is from this direction that most of the daily commuters to work in Poznań industry came. Apart from manpower supply, good site conditions for expansion, urban amenities offered by county seats, and last but not least the vicinity of Poznań were locational factors in favour of the towns chosen. Despite protests from employees of existing factories in Poznań, who appealed to the central government and even to the political authorities, the decisions to build the new establishments outside Poznań were upheld. Thus both towns hitherto possessing a weak industrial base obtained enterprises which substantially altered their economic fate (see the case study of Kościan).

It was relatively easier to reach locational decisions on industrial projects entirely new to the region, as on the location of a new foundry in Śrem. As a branch plant of the Cegielski Metal Works in Poznań it would have been easier for the parent factory to have it constructed in Poznań. Again, following the development policy formulated by regional authorities, the central government accepted the location of the foundry in Śrem where it totally upset the economic structure and growth rate of the hitherto sleepy town.

In all of the above-mentioned locational decisions locally available manpower was a major factor in favour of the towns selected. The manpower reserves on the local labour markets were indicated by the relatively low occupational rates of males and females of working age, the growing number of people commuting to work in Poznań, and outmigration to other regions. However, when the new industrial facility started to operate and reached its full capacity it encountered difficulties in recruiting the full manpower required. Typical examples of this are the Polania shoe factory in Gniezno with a staff of over 3,500 and the H. Cegielski foundry in Śrem with a staff of over 3,000. Both factories succeeded in mobilizing sufficient resources for filling the vacant places by means of extensive housing construction and recruitment reaching far out to other towns and surrounding villages. Smaller enterprises which could not afford this would seek full employment through competitive wages and salaries. They also employed more women. Nowadays, with even stronger competition in the local labour market, many enterprises put labour-saving very high on the list of reasons for innovations.

Ties of the new establishments to the local and regional economy developed to a different degree from one enterprise to another. The H. Cegielski foundry in Śrem shows a strong orientation to the regional market as it delivers a considerable part of its production to the parent factory in Poznań and to other machine-building factories and railway-equipment works in the region. The shoe factory in Gniezno is partly supplied by the local tannery, but the bulk of its material comes from other regions and from abroad. In return its production is being delivered to stores around the country. The same is true of the electric bulb factory in Piła, which is the main national supplier for home consumption and for exports. The metal factory in Środa, which produces mainly metal elements for tyres, supplies part of them to the Stomil tyre factory in Poznań, whereas the factory in Oborniki, producer of construction elements, sells them primarily to construction companies in the Poznań region and in neighbouring regions.

Although internal demand was the prime reason for the construction of new capacities mentioned above, some of the factories, once established on the internal market, started to export part of their output. The most successful until now have been the foundry in Śrem (parts of ship engines) and the bulb factory in Piła (bulbs).

Several new industrial establishments have profited from economies of juxtaposition or were instrumental in creating them. This was primarily the case in Kościan and Koło where, because of the new industrial projects, 'industrial parks' have been opened up and the new sites provided with necessary technical infrastructure. Other enterprises subsequently located there saved considerable investment funds.

However, the strongest impact of the new industrial investment projects was felt in the overall development of the particular towns (measured by the

Table 9.1

Towns	*Total population*			*Persons per room*		
	1960	*1979*	*1984*	*1960*	*1978*	*1984*
Gniezno	44,200	61,100	67,400	1.58	1.12	1.05
Oborniki	8,100	12,800	13,900	1.68	1.13	1.04
Śrem	10,400	21,400	24,200	1.61	1.17	1.04
Środa	12,700	17,100	18,400	1.70	1.16	1.00
Kościan	15,700	21,200	22,300	1.61	1.12	0.95

Source: *Rocznik Statystyczny Miast 1985* (Statistical Yearbook of Towns 1985), Warsaw, GUS, Statystyka Polski, 1986.

increase of total population) and in housing construction (measured by a drop in the number of persons per room). Table 9.1 contains data for the towns mentioned above.

Whereas during the period 1960–84 the population of the city of Poznań increased by only 42 per cent, in all towns included in the table the rate of growth was higher: in Śrem it amounted to 133 per cent, in Oborniki to 72 per cent and in Gniezno to 52 per cent. Despite these high rates of population growth the housing conditions of the population improved substantially owing to the accelerated construction of dwellings both in the cooperative and state sectors and in the private one. Towns with new industrial projects could count on higher subsidies from the *voivodeship* level for cooperative construction and also on additional housing funds attached directly to the industrial project. Thus, in the above-mentioned towns entirely new suburbs were built. In addition, private-housing construction also accelerated in such towns owing to more income accruing to the local population.

The new industrial enterprises also contributed to the construction and expansion of some communal facilities. Thus, they helped to construct streets (such as in Kościan), kindergartens and educational facilities (Gniezno, Kościan), parks and recreational facilities (Śrem) and several others. However, activities of new enterprises in these areas were not sufficient. The state of communal and social facilities did not directly affect production; hence, the management of industrial enterprises economized on them as far as possible. Funds which the towns obtained from the *voivodeship* budget for communal projects and also for educational and health facilities were also very modest. Therefore, with very limited investment funds of local origin the towns were not able to meet the growing needs in many communal, social and cultural areas. Multipliers generated by industrial investment did not operate in a horizontal manner the way they should.

Finally, one should mention another side-effect of the rapid industralization of some towns in the Poznań region: the deterioration of the environment caused directly by industry or by a rapid overall growth of population not matched by proportional investments in environmental protection. Industrial and communal waste water could not be treated by the existing purification plants and had to be dumped into nearby rivers partly untreated. In Oborniki and Środa untreated sewage amounted to four-fifths of the total (1984). Air pollution by industry generally did not exceed acceptable standards although it has increased in comparison to the pre-investment period. However, in Śrem the foundry emitted a considerably increased amount of dust into the air: it was three times higher than the average per square kilometre for all towns in the country (1984). Gases emitted in Śrem also exceeded the national average per town.[5] Local communities started to organize themselves against these consequences of industrialization. In Śrem the local Society of Friends of Śrem has pressurized the foundry's management to install protective devices in the factory and has initiated popular action to increase the area of parks and green areas in the town. A similar society in Kościan, which is also very active in the cultural field, has undertaken comparable activities.

9.4 INTERNALLY INDUCED LOCAL INDUSTRIAL DEVELOPMENT: EXPANSION OF EXISTING ENTERPRISES

Industrialization through construction of new enterprises was capital-intensive. Exercised on a large scale, it caused investment funds to become scarce and, therefore, the central and regional authorities could not afford to locate new industrial projects in too many towns of the country, the more so since scale economies reduced the overall number of producing units needed to meet market demand. This applied also to the province of Poznań which, as already mentioned, had the highest number of towns of all Polish provinces.

These towns were seats of numerous small-scale enterprises. Most of them were processing agricultural produce from their surroundings – dairies, flour mills, fruit- and potato-processing plants, sugar mills etc. There were also small machine plants and repairs shops, brickyards and other factories producing construction material, furniture workshops and other light-industry plants. Most of them were former private enterprises, nationalized in the 1950s or taken over by cooperatives. They were mostly supervised by local authorities and could not count on sizeable investment funds for expansion. Strong competition from large-scale enterprises aggravated their situation. Their future seemed to be bleak.

Yet, during the 1960s and 1970s some of the small enterprises found products with an expanding market and gradually evolved to become relatively large units with a production noteworthy in their branch. In the process of expansion some of them branched out from the parent enterprise and all changed their supervising agency from a local to a central one, making access to central modernization funds easier. Their growth reached such a rate and scale that it also decisively influenced the overall growth of their home town. Some of them even created new units as spin-offs. Examples of such enterprises are the loudspeaker factory in Września (15,500 inhabitants in 1965), surgical-instruments plant in Nowy Tomyśl (5,300) and an enterprise producing coupling devices in Ostrzeszów (8,300).

All of them were initially small units struggling with material-supply difficulties in search of a production profile which would guarantee them a secure future. After a longer period of trial and error during which the staff gained valuable experience they found products with high market potential. The prime agents in this search were groups of ambitious engineers engaged in the factories and supported by the management. When the factories proved their ability to master complex production processes and to launch products with high growth potential the local and, later on, central authorities were more willing to grant them subsidies for expansion, the more so since the funds needed were much smaller than those necessary to start an entirely new plant. New development impulses appeared when the enterprises obtained lucrative export contracts. In order to maintain and even expand their exports to markets in the West and East they have constantly had to innovate.

Today, the Tonsil loudspeaker factory in Września employs approximately 3,000 people, of which over 10 per cent are engineering staff. Besides Tonsil there exists another enterprise, which is a spin-off of the former: a R&D unit producing low-power electrical machines. The surgical plant in Nowy Tomyśl employs over 1,000 workers with a similar share of engineers, and the coupling devices plant in Ostrzeszów about 1,400 persons.

As already mentioned the prime agents in the expansion process were engineers working in the enterprises concerned. However, they would not have been successful without the backing of the management and the operation of some special local factors. In all three enterprises, innovative engineers were particularly strongly backed and assisted by the management. Innovations meant changes in production processes, risks, additional costs and a need for retraining. Since, under the price system existing until recently in Poland, prices did not adequately favour such changes, it was necessary for the management to regard innovations from a broader, longer-term point of view, beyond the immediate troubles.

There were also special local factors at work in each case. Cooperation among enterprises in these towns has been good, which made it easier for each of them to overcome many difficulties. They helped each other in repair works, sometimes in construction, and (in emergencies) with supplies of certain materials. There usually existed a gentlemen's agreement among them not to undercut each other on the labour market. In Ostrzeszów such cooperation has even acquired a formal structure of a 'directors' club' under the prodding of local authorities.

All three towns are strong centres of vocational training. Close cooperation exists between trade schools and the enterprises in matters of training profile, exchange of staff and mutual inspiration. This has been particularly true of Września, where over 2,600 students attended professional schools and an additional 500 pupils were at the general secondary school (during the school year 1984/5). It is particularly thanks to this factor that in the *voivodeship* of Poznań there is a common view that in Wrzesnia innovations 'hang in the air'.[6] The vocational school attendance is also very high in Ostrzeszów (2,200 pupils) and Nowy Tomyśl (1,900 pupils), particularly when the small overall number of the towns' inhabitants is taken into account.

Last but not least, all three towns mentioned are relatively good places to live. Housing conditions are generally better then elsewhere. In addition, industrial enterprises were treated with priority by local authorities in allocation of new housing for their cadres. The towns mentioned combine urban amenities with a country environment. Air and water pollution is low. All of them offer favourable recreational opportunities (lakes and extensive wooded areas in the vicinity) and at the same time are only a short distance to the cultural facilities of Poznań (Września and Nowy Tomyśl) and Kalisz-Ostrow (Ostrzeszów).

Naturally, the development of the enterprises we have discussed, gradual but steady, was not without costs and efforts. In the process of expansion new production halls and other facilities had to be constructed and new equipment acquired. However, the investment effort had been extended over a longer time period than would be needed in the case of a new factory.

With this gradual growth, the towns' populations grew considerably, as Table 9.2 shows.

Table 9.2 Population growth

Towns	*1960*	*1979*	*1984*	*1979/1960*	*1984/1960*
	Population			*in %*	
Września	13,900	23,100	25,200	166	181
Nowy Tomyśl	5,100	9,800	11,800	192	231
Ostrzeszów	6,900	10,900	12,300	158	178

Source: *Rocznik Statystyczny Miast 1985*, GUS Statystyka Polski, Warsaw, 1986.

It turns out that in the end over the entire period 1960–84 the towns with internally induced industrialization enjoyed a higher overall rate of population growth than the ones with development impulses from outside (with the exception of Śrem, where the population in 1984 was 233 per cent of that in 1960).

Despite such high growth rates local labour supply did not suffice to meet the demand of the expanding enterprises. In consequence, the number of commuters increased considerably and at the end of 1983 reached approximately 3,500 in Września, 3,200 in Ostrzeszów, and 2,500 in Nowy Tomyśl.[7] Most of the commuters, however, came from villages and small towns close to the destination towns. Thus the expansion of the existing plants brought income multiplier effects not only to the towns in question but also the surrounding rural areas.

The rapid development of Września, Nowy Tomyśl and Ostrzeszów carried with it problems and tensions similar to those encountered by towns with new industrial projects. However, owing to the already mentioned fact that growth was gradual and stretched over a longer period the local means supported by assistance from enterprises and the *voivodeship* budget were able to meet the evolving needs more adequately and could solve them at relatively lower cost.

9.5 CONCLUSIONS

Looking back at the entire industrialization process in the *voivodeship* of Poznań during the last two decades or more, it can be said that the original development-policy aims of the regional authorities were generally fulfilled. The industrialization process in the areas outside the city of Poznań was faster than in Poznań itself.

Over the period 1960–73 industrial employment in the *voivodeship* outside Poznań increased from 120,600 to 219,000, i.e. by 82 per cent, whereas industrial employment in the city of Poznań during the same period rose from 71,600 to 93,000, i.e. only by 30 per cent. This growth in employment together with progress in labour productivity contributed to an increase in overall production in the *voivodeship* outside Poznań by 216 per cent and in the city of Poznań by 169 per cent.

Later on growth rates of industrial employment in the *voivodeship* (within 1975 borders) slowed down, more in Poznań than in the surrounding towns, while production continued to grow. Then came the crisis of the early 1980s with a considerable drop in production and employment. Industry in areas around Poznań has recovered faster than in Poznań city: employment there in 1984 was only 6 per cent lower than in 1979, while that of Poznań industry is still 16 per cent behind the pre-crisis level. Thus, in the period of expansion industrial development in the region surrounding Poznań was markedly faster than in the city of Poznań, and during the crisis more resilient and able to recover at a faster rate. Since 1960 the economic structure of the region was substantially changed: from a predominantly agricultural one to a region with a substantial share of industry in its overall production and employment.

The towns surrounding Poznań obtained a stronger economic base both from the new industrial projects and from the expansion of existing enterprises. The new projects represented development induced 'from above' through locational decisions of the central authorities in cooperation with the regional ones. The expansion of existing enterprises could be called 'from below', because it was initiated from within the particular enterprises with backing from local authorities. Both approaches had advantages and disadvantages, and these will have to be taken into account when charting future courses of development.

1. Location of large-scale plants caused rapid structural changes in the towns concerned. It also created many tensions and negative side-effects (environmental deterioration). Locally induced industrial development was gradual and stretched over a longer period of time. In the end, the last mode of industrial development brought a faster development of the entire towns than in the case of externally induced growth.

2. New industrial plants brought with them from outside new technologies hitherto unknown to the towns. Once installed they have retained their innovative character only for a relatively short time. The staff of some new enterprises had difficulties at first in mastering the technologies introduced and then in developing them further. Those difficulties were particularly felt where an R&D unit was absent. On the other hand, the locally initiated industrial expansion came about owing to systematic innovation which originated from local engineering staff very active in this field.

3. New large-scale industrial enterprises were established primarily in order to meet internal market demand. For a longer period their production was entirely absorbed within the region or the country. Relatively recently some of them started to export. The locally initiated industrial development, on the other hand, launched export production very early in its expansion. In most cases exports became the dominant factor of further growth.

4. New large-scale industrial projects were usually accompanied by additional investment funds for housing and infrastructure either within the same project or granted separately to the local budget by regional authorities. Locally induced industrial expansion did not offer much additional funding in this field. Local authorities had to economize restricted funds at their disposal and to organize assistance from existing enterprises or to advocate self-help projects.

5. Local labour markets showed signs of bifurcation in the case of new large-scale projects. Training for the latter had to be organized at relatively short notice, often in similar enterprises in other towns. New expertise had to be imported from outside, sometimes on a relatively large scale. Often the original estimates of surplus labour in a given town proved to be exaggerated, forcing the new enterprises to recruit people from distant places or to buy them up from existing enterprises. These effects did not occur or were much less pronounced in towns with gradual, locally induced development.

6. Two kinds of horizontal organizational forms could be observed, generated by the two different types of industrialization. In the case of towns with new large-scale plants, local citizen groups mobilized themselves in efforts to protect the environment or to pressure enterprises and local authorities to action in certain areas. In towns with locally induced expansion it was the enterprises themselves that entered into cooperation with each

other and with local authorities in order to facilitate their expansion. In both cases there were also citizen groups which were particularly active in the cultural field, especially in popularizing local history and cultural heritage. They wanted them to be preserved in a fast-growing and changing town.

The case of the Poznań region is to a great extent representative of the entire country. Similarities stem from the unified systemic conditions of development. On the other hand, specific features of the Poznań case have their roots mainly in the attitudes and tradition of the region's inhabitants. They are known throughout Poland as resourceful, hardworking and well-organized. The Polish case is also to a certain extent typical for most other 'socialist' countries of Eastern Europe. The local development in the period analysed took place within a central planning system which dominated in all of them.[8] That system is being substantially revised in Poland and in other countries. The changes under way should create better conditions for local development than before.[9]

The following case studies analyse towns with development induced pre-

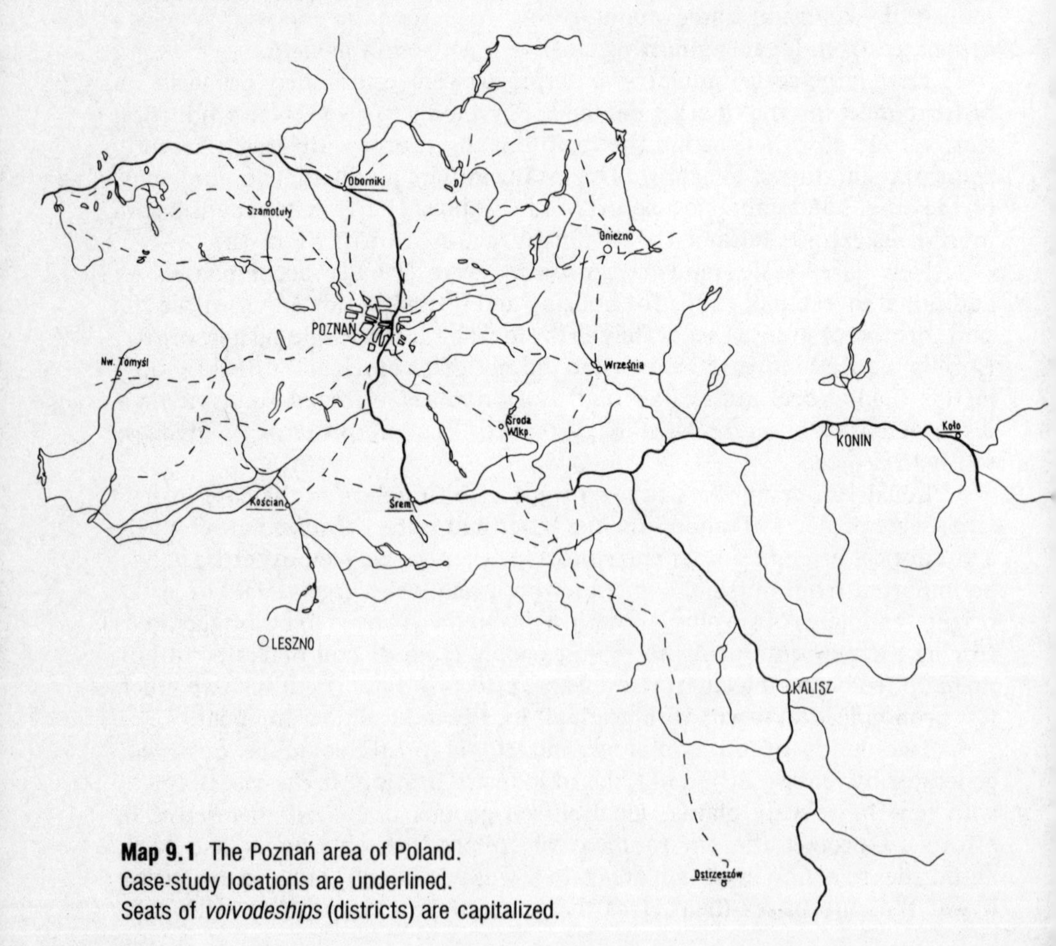

Map 9.1 The Poznań area of Poland.
Case-study locations are underlined.
Seats of *voivodeships* (districts) are capitalized.

dominantly from outside (Śrem and Koło), from inside (Ostrzeszów) and a combination of both (Kościan and Oborniki).[10]

9.6 ŚREM: INDUSTRIAL DEVELOPMENT INDUCED FROM OUTSIDE

The town of Śrem is about 40 kilometres south of Poznań. As a former county town it developed an economic base which served the surrounding agricultural region: a flour mill, dairy, grain extract and coffee factory, sawmill, agricultural-machinery factory, cement works and a few other establishments producing for local needs (such as a furniture factory, small weaving mill and a press). These industrial establishments together with retail trade, handicrafts, social and administrative facilities supported a population of 9,600 (1955). The town grew slowly to reach 11,800 inhabitants ten years later. Local industry employed about 1,300 persons.

A drastic change began in 1963 when the central authorities accepted the regional administration's proposal to locate in Śrem a big foundry of the H. Cegielski metalworks in Poznań. Originally, this establishment was to be erected in Poznań where the main factory operates. However, in view of the deglomeration policy pursued by the regional authorities a location outside the capital was sought. The town of Śrem offered locational advantages which turned the location decision in its favour: reserves of manpower in Śrem and vicinity (a sizeable number of workers commuted daily to Poznań), good water conditions (Warta river), favourable transport location and good site conditions.

The construction started in 1964 and production began in 1968. The construction of the foundry and its expansion continued further until 1974. In its final shape it is an establishment producing about 60,000 tons of cast annually and employing over 3,000 workers. The iron foundry substantially altered the development of Śrem. The new workers, together with the additional employment necessary in other sectors of the town's economy, have rapidly increased the population: from 10,400 in 1960 to 21,400 in 1979 and 25,200 in 1986. The increase between 1960 and 1980 was 142 per cent. A new settlement unit for 12,000 inhabitants, mainly for the foundry's workers, has been constructed and another of a similar size is planned. In addition, numerous single-family homes have been built by individuals in many parts of the town. In connection with the construction of the foundry the town obtained considerable investment funds from the regional budget for expansion of its urban infrastructure: the water supply system, a steam power plant, new streets, a bridge over the Warta river, a bus station, a new hospital, several schools and kindergartens, a sports hall, swimming pool, a boathouse and several other facilities. Without the new foundry the town would have had great difficulty in securing even one-third of the investment funds allocated up to 1986.

However, the overall development of Śrem's industry proceeded too fast for the town to keep pace in all aspects of urban life. Therefore, the housing shortages have not yet been eliminated, the need for school places exceeds the supply, and the supply of various services is insufficient. A difficult situation prevails on the local labour market. Labour reserves existing before the construction of the foundry were overestimated. In addition, wages paid at the foundry in view of the difficult working conditions there turned out to be less attractive than previously thought. Other industrial enterprises operating in Śrem suffered somewhat on behalf of the foundry. They had difficulties in recruiting labour but have reacted by modernization of their production processes and by raising wage and salary levels. This is particularly true of the two next-largest industrial firms: the Polimer chemical workers' cooperative (250 employees in 1986) and the Wlokniarz textile workers' co-operative (160 employees). The first produces bottles and other utensils out of plastics, the second textile furniture coverings, bedspreads and cotton battings and, since 1987, cosmetic cotton wool. Both have plans for further modernization and changes in their production profiles to meet market demand.

The foundry has a diversified production profile of cast for the chemical, automobile, shipbuilding and machine-tool industries and for railways. The bulk of its production is tailored to the individual demand of its customers. Innovations occur mainly in the form of many, often small-scale, improvements in the production process. The foundry, although one of the largest in the country, does not possess its own R&D unit. It relies on outside assistance in this matter, primarily from polytechnic colleges around the country. Despite satisfactory experience of cooperation it seems that the foundry cannot continue to work without its own R&D unit.

The construction of the foundry was initially resisted by the local community for fear of environmental pollution. However, the factory is equipped with modern protection devices which prevent most of the dust from reaching the atmosphere. The situation is worse with regard to air pollution by industrial gases. Measures have recently been initiated to reduce substantially the amount of gas released to the atmosphere. Also recently the amount of waste moulding has been reduced by half by recycling. The remaining waste mass is partly utilized by the local brickyard in the production of construction material.

Efforts to improve the local environment is one of the major contemporary aims of the century-old Society of Friends of Śrem. It is instrumental in preserving and expanding the town's parks, green areas and recreational facilities. It is also continuing its traditional activity, namely that of preserving the local heritage and promoting knowledge of local history. In the nineteenth century, when the region was occupied by Prussia, Śrem and its surroundings were the heartland of the Polish co-operative movement and economic resistance to Bismarck's fight against the Poles.

In conclusion, one can state that the centrally decided establishment of the iron foundry in Śrem decisively altered the development of the town. It boosted Śrem's overall development, helping it to invest in housing and many urban facilities. The foundry substantially changed the local labour market from a surplus one to a market with a sizeable deficit of manpower. The ties of the new establishment with the other local enterprises are, owing to their profile, sporadic. The foundry has not yet been fully integrated into the urban fabric. Śrem has paid a price for its rapid development in the considerable deterioration of its environment, which only now is being improved.

9.7 KOŁO: INDUSTRIAL DEVELOPMENT INDUCED FROM OUTSIDE

Koło is situated about 130 kilometres to the east of Poznań, and until 1975 was a county town. The administrative reform of that year incorporated Koło into the newly created *voivodeship* of Konin. In 1950 the number of inhabitants was 9,900. For some time the population had grown at a slow pace (11,400 in 1960 and 11,700 in 1965). Then it started to increase more rapidly (15,100 in 1975, 19,600 in 1980). In recent years Koło has attained a more stable growth rate and reached 22,200 inhabitants in 1986.

The growth rate of Koło's population is closely related to the town's industrial development. After initial postwar reconstruction and then a static period which lasted till the end of the 1950s, Koło entered a period of rapid industrialization. Not only did the existing plants (the pottery factory and dairy) expand, but new enterprises were established: a plant producing sanitary ceramics, an establishment producing outfits for construction enterprises (ZREMB) and an abrasive-material factory (Korund). These establishments were further expanded in the 1970s. In addition, in 1974 construction began on a large meat-processing factory, and a milk-powder plant was built for the existing dairy. In sum, from an insignificant number before the industrialization process began, total industrial employment in Koło increased considerably, to reach about 5,600 in 1984. It is only in the most recent years that a slowdown has been seen in the expansion of industrial employment and even a drop in the overall total.

The decisions to locate new industrial projects in Koło rested either entirely with the *voivodeship* authorities (ZREMB, milk powder) or with the central government, with the concurrence of the *voivodeship* administration. In both cases the local authorities were actively lobbying for the particular locational decisions. These decisions with respect to Koło were based on several general factors. Access to raw material was an important locational advantage to several plants. Historically, the availability of clay suitable for pottery production was the decisive reason why a plant was in operation as early as the nineteenth century. This production continued after the Second World War despite the exhaustion of local supplies. Skills available for pottery production caused a further expansion: construction of a sanitary ceramics factory and its separation from the parent factory. Nowadays the factories employ over 300 workers each, most of them women, and produce tableware which is well known both in Poland and abroad, and bathroom ceramics sought after in home construction.

Access to raw materials coupled with good access to surrounding agriculture were also locational factors for the dairy with the milk-powder plant and for the meat-processing plant. In 1986 the latter processed almost 60,000 tons of meat and employed about 1,800 people. It was believed that the surrounding region represents a high potential for agriculture. The processing plants, located in a region hitherto poor in processing facilities, should help to utilize this potential to the full extent.

Availability of surplus labour was high on the list of locational factors for a number of establishments, primarily those with the highest labour requirements, like the above-mentioned meat-processing plant or the abrasive-material factory, Korund, which employs much the same number as the former plant. The same factor played a role in the ZREMB plant with a staff of about 350. The labour reserves existed in rural areas around Koło, on numerous small farms which could not support all the people living on them.

Only a small number of the rural workers could find permament homes in Koło, because apartment construction was relatively small. This explains the fact that almost half of the total number of persons working in Koło's factories commute daily to work from the surrounding villages and small towns. Many of them belong to the peasant class as they still live on small-scale farms. This class is a typical by-product of rapid industrialization in overpopulated rural areas of the country.

The local labour market of Koło has already reached its saturation point. It cannot expand further because it is situated very close to another much stronger market some 30 kilometres to the west – namely the Konin industrial district consisting of three large electric power plants, brown-coal mines and an aluminium foundry. Demand for manpower there made inroads in the Koło labour market, the more so since the Konin mining and industry offered higher wages and apartments for all willing to work.

The industrial enterprises of Koło helped the local authorities in developing a number of communal facilities. They either contributed to the town's communal fund or assisted in construction of particular facilities. The factories also constructed apartments for a limited number of their employees. However, all this was not enough to overcome past neglect in communal infrastructure and to keep pace with the rapid overall development of the town. Hence, in 1984 there were still 27 per cent of inhabitants who were not connected to the communal water system and 39 per cent to the sewerage system. The construction of the city gas supply system has yet to begin. The future development of the town depends to a considerable extent on the closer cooperation of the industrial enterprises with the local government. For this purpose the directors of the six largest enterprises recently associated themselves with the mayor's office. The town will need all the help it can get from the industrial enterprises in order to improve the living conditions of its inhabitants.

9.8 KOŚCIAN: INDUSTRIAL DEVELOPMENT INDUCED FROM OUTSIDE AND INSIDE

Kościan is situated some 40 kilometres south-west of Poznań in a productive agricultural region. Its population has grown slowly but steadily since 1945. In 1960 it reached 15,500 inhabitants, in 1975 20,000, and in 1986 22,900. The increase over the period 1960–86 amounted to 48 per cent. Gainful employment in all sectors of Kościan's economy and services reached over 10,000 persons (1986), of whom about 5,000 are industrial employees. Initially, industry in Kościan was dominated by food processing (such as the meat-processing plant, sugar factory and dairy), and a few other establishments serving the local market (such as the furniture factory and tobacco plant). This industrial base was too weak to secure for the town firm ground for future growth. Apart from employment in this industry, work for the inhabitants of Kościan was available in relatively well-developed retail trade and handicrafts and also in various social sectors, of which secondary and professional education and health services were particularly strong for a county town. In the early 1960s the regional authorities, according to the deglomeration policy applied to industry in Poznań, did not allow a factory producing metal appliances for chemical industry (Metalchem) to be transferred to a new site within Poznań. Instead, Kościan was suggested as the new location. Despite protests from the staff of the old plant in Poznań, the new establishment was constructed in Kościan. At about the same time, Kościan was also selected as location of a new plant producing technical gases for customers in this part of the country. The two construction endeavours required relatively large sites not available in the traditional industrial grounds of the town. Therefore, urban planners designated a new area in the north-eastern suburb of Kościan as a new industrial park. Ground has been set aside there for the relocation of other plants from the congested inner areas of Kościan.

The Metalchem works in Kościan employ over 1,000. It evolved from relative simplicity to the production of a diversified profile of various appliances, some with a high degree of sophistication. Among its major products are appliances for the granulation of chemical materials, and cisterns and tanks for various chemicals. The enterprise is constantly innovating with regard both to products and to production processes (particularly in welding techniques). The factory employs a relatively high number of engineers and operates its own laboratory.

The construction of the plant producing technical gases, Polgaz, was started in 1965 alongside the previously mentioned Metalchem works. Its location in Kościan was also decided within the framework of deglomeration policies applied to Poznań. The plant produces oxygen and nitrogen and employs about 130 people. It has recently installed a closed water circuit in order to reduce pollution and save water. It has also increased the safety standard of production through innovations in its production process.

The new industrial park in Kościan attracted to its grounds another plant which previously operated for many years on another site within the town. This is Meprozet, an enterprise producing various utensils for agriculture. In particular its staff of about 280 persons produces sanitation carts, pumps and agricultural trailers. It has its own construction unit which helps the enterprise to respond quickly to changes in market demand. The enterprise also cooperates with R&D units located in Poznań and other parts of the country.

The location of new plants in an industrial park together with the construction there of two relocated ones (besides the above-mentioned Meprozet, a repair shop for agricultural machinery has also been transferred) helped to achieve considerable economies thanks to the juxtaposition of all factories (joint water and energy supply system, common industrial health service, transportation routes etc.). These economies, as well as the manpower rseserves of Kościan and its vicinity, were a major factor of location for the new plants.

The industrial expansion of Kościan was gradual, spread over a longer time and did not involve such a drastic increase of industrial employment as in Śrem. The regional authorities did not consider Kościan to be a town in need of especially increased investment assistance. Its budget has obtained the 'normal' proportion of investment grants based on past allocations, which meant a chronic shortage compared with the needs felt. The situation has been somewhat alleviated by the industrial enterprises themselves, who contributed to the development of some urban facilities. This pertains particularly to Metalchem, which helped to construct houses, streets, and a bridge, kindergartens, sports facilities etc. However, all this was insufficient to solve the major problems of the town, even though its industrial development has not been revolutionary.

One of the problems still to be solved in Kościan is the prevention of a further deterioration of the natural environment: the water and air pollution stemming from industrial activities, particularly those still located within the inner city (the sugarbeet factory, the meat-processing plant and the dairy), has continued to grow. The sanitary conditions of the town have deteriorated further: a sizeable portion of communal and industrial sewage flows to the Obra river canal without cleaning, the communal sewerage system serves only 61 per cent of the population and the air, particularly in the town centre, is being polluted by numerous heating installations burning coal.

The fight for a healthy environment in and around the town is currently one of the major aims of the local Society of Friends of Kościan, an independent citizens' organization which stands behind many local initiatives in the economic, social, but above all in the cultural field. This society sponsors noteworthy cultural events, publishes books and documentation concerning the heritage and development of Kościan, and backs local authorities in their efforts to gain more local autonomy. It successfully strives to make Kościan a better place to live.

9.9 OBORNIKI: INDUSTRIAL DEVELOPMENT INDUCED FROM OUTSIDE AND INSIDE

Oborniki is a town situated about 30 kilometres to the north of Poznań on the route to the ports and resorts of the middle section of the Baltic shore within the Polish border. It belonged to the smaller county towns of the pre-reform administrative structure. In 1960 it had a population of 8,100. In the 1960s and 1970s the population increased by 58 per cent to reach 12,800 at the end of 1979. Then the growth rate slowed down and in 1985 Oborniki's population was about 14,000.

As in other towns analysed here, the pace of population growth is closely related to the development of the town's industry. The latter was the outcome of two tendencies: the expansion (or lack of it) of existing industry and the location and development of new plant. Among existing enterprises the two largest are a meat-processing factory and a furniture plant. Both establishments are a continuation of traditional industrial activities existing in Oborniki long before the Second World War. The furniture enterprise, with a present-day workforce of about 1,000, has undergone many changes in production profile and organizational structure. Initially, it was a small-scale endeavour with diversified furniture production. Over the years it acquired other furniture plants located in nearby towns and started to specialize. In 1980 plants situated in the new *voivodeship* of Piła were separated from the parent Oborniki enterprise and the latter had to adjust to the new organizational situation which coincided with the economic crisis of the early 1980s. The enterprise found a new and very profitable direction of production: the supply of custom-made furniture for hotels, offices, restaurants etc. at home and abroad. This profile gives the more than 1,000 employees (1986) a sound base for profit and further development.

The meat-processing factory is also a traditional branch developed originally for the export of bacon to Great Britain and other Western countries. Since then the production profile has changed, to comprise mainly meat-processing for home consumption. In 1986 more than 400 employees processed approximately 12,000 tons of meat. The location of the plant is very unfavourable: the main production unit operates in the centre of the town. Plans have been prepared to move the plant to a new location in the suburb of Slonawy. However, for lack of sufficient funds (and concentration of the available funds on the construction of a new meat-processing plant in Koło) the project still awaits implementation.

A new development in Oborniki's industry was the location in 1971 of a new plant, Metalplast, producing light construction plates and walls. The availability of a labour force in Oborniki and its vicinity and the short distance to the Poznań construction market were the main locational factors which helped the regional authorities in making the locational decision in favour of Oborniki. The Metalplast factory opened up a new industrial horizon in the town. It started operation in 1974 and nowadays employs about 1,200 people. Its production, in high demand on the home market, has since the early 1980s encountered difficulties in expansion owing to shortages in material supply. This is why the capacities of the enterprise have not been fully utilized, and the plant is seeking new sources of supply at home and abroad.

In all, more than 3,000 people were employed in the industry of Oborniki in 1984. The manpower surplus which was an important locational factor in the 1970s, turned out to be overstated. Labour for the industrial enterprises and other sectors of the economy has to be drawn from relatively distant places. In 1983 more than 2,400 workers commuted daily to Oborniki and about 800 were commuting from Oborniki mainly to Poznań. The labour-market conditions in Oborniki reflected in the figures cited above should be analysed against the living conditions in Oborniki. In comparison to other cities Oborniki is noteworthy for its relatively intensive housing construction, both private, cooperative and government-sponsored. The average number of persons per room dropped from 1.68 in 1960 to 1.04 in 1984. On the other hand, its communal facilities have been relatively poorly developed. Only half of its population was served by the communal water-supply system and only 17 per cent by the sewerage system in 1984. Unfortunately, the existing industry in its present shape is not capable of helping to improve the situation to any great extent.

9.10 OSTRZESZÓW: INDUSTRIAL DEVELOPMENT INDUCED FROM INSIDE

Ostrzeszów is a former county seat of the old Poznań *voivodeship* located some 150 kilometres to the south-east of Poznań city. After the 1975 administrative reform it became part of the newly formed *voivodeship* of Kalisz. In 1960 it had about 7,000 inhabitants who earned their living mainly from servicing the agricultural surroundings. Manufacturing was represented by small-scale enterprises: brickyards, a sawmill, a dairy and several units producing various metal and chemical products for local consumption. All this was insufficient to guarantee growth and higher income. Therefore local enterprises started to look around for new products or production profiles. One of them, FUM, signed a contract with the Industrial Board of Machine Tools in Warsaw to produce coupling devices. Successful fulfilment of the first contract brought further orders and in 1965 a takeover of the factory by the board. The factory gained a monopoly in its field in Poland and started to export its products abroad, mainly to the Soviet Union. Export production helped the factory to expand and modernize. Today FUM employing about 1,400 people, is the largest factory in Ostrzeszów and one of the strongest in the region.

Another local industrial unit which in the 1950s produced press buttons and drawing-pins started to cooperate with the electro-machine industry. It took this factory a longer time and a series of profile changes in production until in the 1980s it became an independent enterprise under the name Ema-Elfa, producing hydro-electrical equipment for the metallurgical, electro-machine and shipbuilding industries.

The town of Ostrzeszów is also the headquarters of a dynamic construction enterprise, Techma, which specializes in constructing industrial units and petrol stations both within Poland and abroad. It also produces construction elements for its activities and for sale. Before 1971 Techma, under a different name, was a small-scale repair and construction firm.

Besides the enterprises already mentioned, Ostrzeszów is now the home of a chemical factory producing washing powder and soap, a unit producing spare parts for agricultural machines, a poultry-meat-processing plant, and several other enterprises. Altogether industry in Ostrzeszów employed over 4,200 workers in 1985, five times as many as thirty years previously. This growth has been achieved without the relocation of any major new plant. It was the existing enterprises which substantially changed their profile, expanded production of high market potential, modernized equipment and extended their facilities. Naturally, they were assisted in this by central industrial boards and branch ministries, but the initiative was local and so was the will to seize the opportunity.

Several local factors contributed to the successful industrial development in Ostrzeszów. According to a special sociological study conducted in 1984,[11] the most important seem to be the following:

1. Horizontally integrated local power structure comprising the heads of administration, party leadership and management of key enterprises. Most of these people were born here or have lived in Ostrzeszów for many years. They were not numerous, but were influential and effective because they worked for the same goals in concert.

2. A local pressure group in and around the Social Committee for Construction of Cultural Facilities. Members of that group were mostly former leading actors in various local organizations who, because of their great experience and good contacts, were very effective in organizing support for development of social and cultural infrastructure in Ostrzeszów. The town is known as the organizer of several cultural and sports events on a national scale.

3. Well-organized and well-trained labour. The workers' organizations and councils in many enterprises command great authority and have been effectively lobbying for better wages and working conditions.

4. Good cooperation among the enterprises in matters of construction, supply of materials and services. The formal organization for such cooperation took the shape of a directors' club, but many non-formal arrangements and gentlemen's agreements (as in matters of a non-competitive wage policy) were equally efficient.

The combined operation of the above-mentioned factors together with others helped to develop the town at a relatively fast pace and to attract immigrants. Development of numerous social and cultural facilities, many of them with the help of local industry, has made Ostrzeszów a good place to live. Efforts on the part of enterprises located in other towns (such as the Wroclaw conurbation) to attract engineers and skilled workers from Ostrzeszów were highly unsuccessful. The population of Ostrzeszów has a high record of self-help projects executed in many sectors.

NOTES

[1] The author analysed the scope of centralization and decentralization during different postwar periods, e.g. in 'Les dimensions spatiales et structurelles de développement polonais', *Economies et Sociétés, Cahiers de l'ISEA*, vol. 3, no. 1 (1969), p. 167–200; 'Regional development policies in Poland; ways of formulation and instruments of implementation', in A. Kuklinski and J. G. Lambooy (eds), *Dilemmas in Regional Policy*, Amsterdam: Mouton, 1983.

[2] Data cited here are taken from the official statistics published by the Polish Central Office of Statistics (Glowny Urzad Statystyczny). In particular they came from the *Statistical Annual* published each year by GUS. Other sources of data are: *Rozwoj wojewodztw 1960–1970*, Warsaw: GUS, 1971; *Statystyka Miast i Osiedli 1945–1965*, Warsaw: GUS, 1967; *Rocznik Statystyczny Przemyslu 1974*, Warsaw: GUS, 1974; *Rocznik Statystyczny Miast 1985*, Warsaw: GUS 1986; *Rocznik Statystyczny Wojewodztw 1986*, Warsaw: GUS 1987.

[3] Detailed analysis of the spatial structure of the region's industry and its origins is contained in the author's study: *Rozwoj przemyslu Wielkopolski w latach 1919–1960* (Industrial Development in the Poznań Region during the years 1919–1960), Poznań: Wydawnictwo Poznanskie, 1964.

[4] The process of industrial restructuring of the region is analysed in detail in: R. Andrzejewski, *Proces przebudowy struktury gospodarczej Wielkopolski 1945–1970* (The Process of Restructuring the Economy of the Poznań Region 1945–1970), Warsaw–Poznań: PWN, 1978.

[5] According to *Rocznik Statystyczny Miast 1985*, Warsaw: GUS Statystyka Polski, 1986, pp. 728 and 746.

[6] This opinion was overheard by the author in the discussion at a recent *voivodeship* conference devoted to the adoption of a provincial plan of innovations up to 1990.

[7] According to a special nationwide investigation performed in 1984. Source as in note 5, pp. 770 and 778.

[8] More details in a paper by Ewa Glugiewicz and Bohdan Gruchman, 'Le rôle des innovations pour la restructuration économique des régions dans les conditions spécifiques de l'Europe de l'Est', *Technologies Nouvelles et Développement Régional*,

Centre Economie-Espace-Environnement, Université de Paris I, Paris: GREMI, 1, 2, 3 Septembre 1986, vol. 2, pp. 23–40.

[9] One of the most fundamental changes in Poland has been the introduction in 1990 of territorial self-government at the lowest administrative level: in communes and townships. In many areas influencing local development and living conditions of the population, freely elected local authorities have become independent of the state administration. This fact, together with the introduction of free-market conditions throughout the economy, sets an entirely new frame for local development.

[10] The case studies are of towns with externally and locally induced industrial development. The examples were prepared on the basis of field reports prepared by P. Gruszka (Ostrzeszów, Śrem), A. Janc (Oborniki), T. Kowalski (Kościan) and K. Zawisny (Koło). The study has been performed within the national research programme CPBP 098, coordinated by the University of Warsaw.

[11] 'Zakres i charakter polityki spolecznej wladz terenowych oraz ekonomiczne podstawy jej realizacji' (Scope and character of the local authority's policy and the economic foundations of its implementation), *Raport z badan terenowych – Ostrzeszów*, vol. 9, 1984, Instytut Polityki Spolecznej Wydzialu Dziennikarstwa i Nauk Politycznych Uniwersytetu Warszawskiego, Problem wezlowy 11.10, temat VI-6, Warsaw (unpublished).

CHAPTER 10

Some local development experiences in Hungary: a socialist country in transition towards decentralization*

E. Baráth and P. Szaló

10.1 INTRODUCTION

The genesis of 'success stories' in Europe today is based in an almost concordant way on the detection of local/regional resources as well as on a successful testing of general theories – in spite of important historical, geographical, political and cultural differences. This coincidence provokes a certain uneasiness. The reason for this similarity, however, is the structural identity of changes taking place in the world system.

Historical differences in Europe range from dictatorial systems of extreme centralization at one extreme to the 'liberal'-democratic means of decentralization applied as another extreme. Both solutions – in their essence very simplified – represent in the last resort a subordination of the energy to be found in small communities to the interests of the so-called 'global' society. In spite of existing differences in ideology between Western and Eastern European countries, the different social systems are regressing, in the present moment of crisis, to very similar means because of the similarity of causes in the restructuring of the world economy occurring at present.

The events behind the phenomena have, in essence, a double structure: the one, vertically organized, is the functional economic dimension, linked to a central power. The other, horizontally organized, is the territorial-cultural dimension. The dialectical separation and relation of these two subsystems of given social and political systems, i.e. the *economic* subsystem and the *territorial* subsystem, offer a possibility of explaining – according to our view – the dilemma of identity and difference of local systems.

The vertical productive and economic subsystem has a sectoral character, oriented to efficiency, whereas the cultural-territorial subsystem is oriented towards broader values, being non-hierarchical, which guarantees the reproduction of the social and political system.

The different characters of the two subsystems cannot be better proved than by the fact that spatial economic processes require some basic condi-

* This chapter refers to the conditions just before the recently initiated drastic reforms towards a market economy and multi-party democracy in Hungary.

tions whose most essential components are infrastructure, qualified manpower and, last but not least, the natural and artificial environment. These three elements – social, natural and artificial environment, and infrastructure – are in an interrelationship that represents the structure of the territorial system. It ensures a much greater permanence for the territorial (local and regional) relations than can be observed in the case of the economic sectoral system. On the one hand, settlement and regional condition (natural and social values, historical and cultural experiences) determine the movements of economic factors (in better cases these are dynamic and flexible): on the other hand, economic development is a precondition, a forming element of the constant changes in territorial structures and cultural values. The two subsystems are carriers of dialectical relations between each other.

From the point of view of our study it is important to know that under conditions of external change, given internal constraints, the determining direction will change. In order for supplementary resources to be detected, so-called structural energies are required, which means that a given social and political organizational system (in this case the state) will be constrained to put into service any knowledge accumulated in the lower territorial layers. In our present situation, with external supplementary resources becoming exhausted, the economic system offers no increase in prosperity. The external crisis therefore requires us to address ourselves to the social-cultural subsystems and their historically matured elements, and to the local as opposed to regional communities. These are two extreme possibilities: a further centralization reinforcing past circumstances; and a further decentralization.

With regard to decentralization the question is whether the political subsystem has accumulated sufficient stability, and at what cost to the individual as opposed to the community, to satisfy this request? The Western democratic systems can quietly continue to increase their degree of 'freedom' as long-standing experiences support them. In such a case the so-called 'public interests', those of the whole society, can be put under protection, and the state can retire behind the 'local' interests. Our opinion is that the case studies on Hungary justify this possibility and, within this framework, simultaneously the European character, as well.

The perception has strengthened during recent decades, and has been transformed into planning and development aims, that the synthesis of local forces, their 'synergy' and, as a superior realization, 'local innovation complexes' are nothing other than the liberation of the peculiar structural energy hidden in a territorial system. This can also contribute to a change in the determining trend, as it offers a broader and freer scope to deploy an autonomy that seems to be less dangerous nowadays.

Building from 'inside' and 'below' is a natural reaction against any expansion of the economic system into a world system, a normal defence against all effects coming from the 'outside' and 'above'.

We have written our study as a 'view from below', for one of us has lived for a certain period in one of the settlements selected as venues of our analysis (in Zalaegerszeg as engineer-in-chief of the town); the other of us has been acquainted with this region (the county of Veszprém) as a manager of its regional plans. We have also, however, acquired general experience in one case as leader of the national territorial settlement plan, and in the other as a researcher into the territorial social and industrial structure of Hungary, in a 'view from above'.

10.2 MACRO-CONDITIONS FOR TERRITORIAL DEVELOPMENT

In Hungary development politics have been defined in recent decades by the system of redistribution, by withholding through the central budget, and redistributing the resources. Redistribution is carried out via several channels, but distribution realized via a vertical structure according to sectors is decisive as it determines horizontal territorial distribution. Consequently, the development of a given town is defined through the centrally managed development of sectors (industry, agriculture, public health etc.), as well as by development financed by the town's own means. These latter include both the resources obtained during global redistribution, and those arising from the town's own resources. During the 1950s, redistribution was completed by normative provisions including all fields; in the 1960s the normative regulation was successively reduced, although even so a considerable amount of utilization was still centrally defined (see Table 10.1).

Table 10.1 The share of investment (as a percentage)

	State investment				*Investments of enterprises*	*Total*
	Direct central	*Indirect centrally defined*	*Residue*	*Sub-total*		
1955	—	—		100.0	—	100.0
1968	18.8	21.9	8.4	49.1	50.9	100.0
1970	15.1	20.4	9.2	44.7	55.3	100.0
1975	13.5	20.0	10.4	43.9	56.1	100.0
1978	13.6	20.6	8.8	43.0	57.0	100.0
1980	13.8	22.2	9.7	45.7	54.3	100.0
1984	11.6	14.4	16.1	42.1	57.9	100.0
1986	6.9	11.1	20.5 (14.4)[1]	38.5	61.5	100.0

Note:
[1] The share of investment of local councils

Source: *Datastore of Investments 1950-1970*, Budapest: Central Statistical Office, 1979; *Statistical Yearbooks 1975, 1977, 1980, 1984, 1986*, Budapest: Central Statistical Office.

From the end of the 1970s the proportion of central subsidy diminished. The autonomy of local councils increased in respect of questions related to development. There are now trends to establish local, territorial taxation, and to create autonomous territorial management, this being one of the very explicit aims of the present reform aspirations.

In industrial development the first period after the Second World War was also extremely centralized, and industry's managing authorities executed entrepreneurial planning and operative management via administrative methods to the total exclusion of market factors. In those times enterprise autonomy was purely formal. During the 1960s, successive dismantling of the extreme planning system that had proved unsuccessful became possible, and in 1968, with the introduction of the economic management system, the autonomy of enterprises increased, and market factors, goods and financial, especially monetary, relations took on a more emphatic role.

At present, industrial investments are realized from two sources: the state's budget and investment made by industrial enterprises themselves. Since 1968 the share of central investment has shown a diminishing tendency: in the 1970s the investment launched by enterprises represented 55 per cent and in 1986 more than 60 per cent of total investment.

The second extensive period of industrial development of about 1968–78 was characterized by a territorial expansion on the part of enterprises and the opening of plants in less industrialized regions. Through the organizational relations existing between the different plants, less industrialized regions became subordinated to and dependent upon the more developed industrial regions. For instance, in 1980 approximately 1,500 plants, which represents 42 per cent of the total plants of this kind, had their management centre in Budapest and this dependence, which even nowadays is decisive, is the biggest obstacle hindering the deployment of local development strategies. The decentralization tendencies of the 1980s led to the liquidation of a few trusts, horizontal monopolies. A small number of business plants became autonomous, but they could not modify the general dependence structure present throughout industry.

As a result of this second stage a new type of territorial inequality emerged in Hungary at the beginning of the 1980s. One of the most essential features was that urbanization development and economic, industrial development were delinked from the territorial point of view.

In the 1950s and 1960s industrialization determined urban development: in the course of redistribution of development resources the industrially developed regions obtained a much higher proportion of resources in order to develop urban infrastructure, housing stock etc.

At the beginning of the 1980s the dynamic development of industrial towns slowed down. In these settlements a high standard of basic and medium-level supply was ensured, but higher-degree functions did not evolve. As their development was realized via central resources, they relied further on external resources in order to progress, or, in some cases, simply to maintain the level already achieved. Moreover an important part of industry depends also organizationally upon other industrial enterprises situated in the major centres. These so-called centre satellites possess, through their purely executive character, a low intellectual capital quotient that is not sufficient for industrial innovation, nor for exploring the possibilities for autonomy needed for the development of the settlement.

10.3 THE FORMER ROLE OF THE STATE

In Central and Eastern Europe the intervention of the state in social and economic processes goes back a long time. Although Hungary participated from the eighteenth century onwards in an increasingly intensive way in European trade, certain peculiar historical and structural circumstances of the region meant that neither a capitalistic system of production, nor the civil institutional system, was able to develop organically in the country. The difficulties of restructuring based on its own resources, the increasing lag between Central and East European development on the one hand and West European on the other, as well as the slowness to modernize, made it more or less imperative for the state to intervene. Thus, the role of the state increased in this 'historical region of Europe' (Szücs, 1983) and it must be judged by other criteria than is the case in the West. On the one hand, the state is the carrier of modernization; on the other, it builds up (and built up) bastions of *étatism* against liberal ideals.

In Western Europe the state played a considerable role in the development of mercantilism, since commodity production was initially under a protectionist shelter. The state intervened in an indirect way, by creating favourable conditions for entrepreneurial activities, only by assisting in the transformation. In the East it intervened directly, and the degree of intervention was considerably greater, and deeper. In the period from 1881 to 1899 three laws on industrial development were promulgated in Hungary with a whole system of subventions. It was the state that mobilized foreign credits by undertaking guarantees, participating in the construction of railways etc.

The rigidity of the historically established society, the unsolved questions related to feudal large-land property meant that broad masses of society demanded the intervention of the state, as there was no possibility of finding a solution to the land problem by individual efforts via an organic transformation. Thus, after the Second World War, the peasantry 'received' land, it did not acquire it itself. This difference is fundamental as regards the development carried out on an individual basis as proposed nowadays. Any society will develop a different mental structure when individual initiatives are more successful than is the case when it hopes to find a solution for its problems via outside interventions.

This is not to say that Hungarian society is entirely of the Oriental type, living in passive dependence; neither, however, is it purely of the autonomous-undertaking type. Both attitudes are simultaneously characteristic, however, in a degree that differs from region to region.

In the county of Bács-Kiskun – situated in the Great Plain – where no farmers' cooperatives were founded, where there was no collectivization, producers and farmers gathered in trade unions and only processing and marketing were done collectively. Here entrepreneurial traditions survived and gained strength. This is well illustrated by the fact that after 1982, when the central decrees made the foundation of small autonomous enterprises possible, the proportion of enterprises established in this county considerably surpassed the average.

After 1949, with the introduction of the one-party system and the model of centralized economic development, the state's interventions entirely determined social and economic life. An essential feature of the programme concerning socialist transformation was to modernize the country in a radical way, and the socialist model was legitimized by this necessity of transformation. (During the 1950s even in the West the expansion of state functions was accepted and certain elements of planned economy also made their appearance in Western countries.) Socialist modernization realized the industrialization of the country in a radical way via an unprecedented concentration of material and intellectual resources, and through the concentration of agriculture it broke up the historically established social structures.

However, if an acceleration of the modernization process was unavoidable, and was needed in order to remove structural obstacles, the fact of its drastic and sudden execution was correspondingly dangerous. Hungarian society suffered under external effects a transformation of its social structure and this was not the result of an internal organic process; on the contrary, the disruption of traditional society resulted in a weakening of social cohesion. The establishment of a monopolistic power and political system had as its consequence a reduction in social integrity and, thus, a certain 'double' society was created.

The large-scale industrialization solved the mobilization of resources, regrouping capital. It ensured manpower recruitment for industry, but it also developed one-sided local industrial structures dependent upon external central agencies which drove the economy for decades into a forced course. At the same time, the reorganization of agriculture created technical and organizational conditions for large-scale farming, but it also stopped the development of ownership and entrepreneurial attitudes.

In the long run, this compulsory modernization of society 'from above' resulted in prejudicial effects: a strong dependence on the centre, while self-determination and spontaneous actions were kept within minimal bounds. Today, when freedom is becoming a productive force, when information, knowledge and quick decision-making on the part of society are elementary requirements, it has become a disadvantage, causing slowness in restructuring and difficulties in adapting to the new circumstances.

With the exhaustion of the resources of extensive industrialization, from the beginning of the 1960s onwards, the central question of the renewed reform processes was to advance towards a multi-polar institutional model. There had in fact always been a hidden multi-polar decision system, functioning at the lower as well as at the upper levels of the decision-making hierarchy. The different territories and sectors articulated their interests, via their existing institutions, and coordinated and conciliated their interests.

For centuries local initiatives were pushed into the background in the small villages because of cultural disadvantages and underdevelopment, therefore associations to reinforce local initiatives were of great importance. At the end of the last century local cultural associations were founded in the villages. By the interwar period many kinds of associations were functioning: associations of farmers, reading clubs, church and secular associations and people's colleges. After the Second World War these associations fell victim to centralizing endeavours.

In 1980 a new law again authorized the foundation of cultural and other associations. Since then many cultural associations and some people's colleges have started to function again. The people's colleges aim to realize the reinforcement of local initiatives by the creation of cultural and social facilities. Their basic desires are to promote decentralization of culture, to enhance people's mental identity, and to enable them to represent their individual and public interests. Now the local cultural associations are organizing the Association of Rural Settlements from 'below' to try to reconcile their interests.

The division of power among distinct institutions and groups of interest is called 'latent pluralism' by Hungarian political scientists. The institutionalization of this pluralism is at a very low level; it does not offer any explicit representation for social layers and interest groups. However, one should not underestimate its indirect effects which take place mainly via informal relations that are exerted on the decision-making processes, *inter alia* on local development.

In our case studies we examine the effects on local development that were brought to bear by local institutions and leaders, the degree to which central decisions concerning local interests determined the development of settlements, and the way in which developments correspond to coherent, consistent local development policies. We present development policies that were successful and local policies that resulted in a broader effect enabling local communities to answer today's economic and territorial challenges.

10.4 TERRITORIAL DEVELOPMENT IN HUNGARY

After the Second World War the territorial development process in Hungary was a special result of several multi-directional influences. First, the given settlement structure was unbalanced, having its only centre in Budapest – as a result of the lost First World War. The territory of Hungary was drastically reduced in size by the Trianon Peace Pact and the largest towns that were the counterpoles of Budapest were annexed to neighbouring states. The settlement system of the country became a monocentric one including, first of all, agrarian towns with an undeveloped infrastructure. There were, furthermore, cultural development disparities inherited between east and west, having been dominant during centuries, as well as the clear differentiation between the northern and southern counties from the point of view of their industrial and agricultural character.

Second, the existing unbalanced structure was further distorted as a result of the political transformation pursued after the Second World War; ground was lost in the settlements carrying traditional civil values. On the one hand were the new 'socialist' towns having an industrial character in accordance with industrialization and the new social ideas; on the other were the new agrarian centres that arose as a result of the collectivization of agriculture.

Third, the considerable centralization of power increased further because of the role of settlements having administrative attributes. This phenomenon, accompanied by increased territorial and social mobility, regrouped the traditional settlement and social classes.

In spite of the fact that territorial planning had set itself the aim of creating a balance, diminishing the infrastructural differences and of approaching a more balanced urban system, practice proved that the overriding priority of sectoral economic objectives made the development of settlements a simple spatial projection of sectoral development policy. Territorial development was, on the one hand, a tool and on the other a result. Territorial organization became an objective in its own right only at the beginning of the 1980s (Baráth and Futó, 1984). It is, however, a fact that at the end of the 1960s sociologists drew attention to the dangers of an already established territorial development practice: infrastructure, the recognition of local social requirements and a complex conception of environment were incorporated into the national economic plans only as a certain expensive constraint. The increasing of local autonomy was a secondary product of a social and economic reform process; it was not a recognized tool allowing individual freedom and the establishment of local communities; and the territorial and settlement system, as a development objective, existed only as a second-order political possibility.

From the above we can arrive at the conclusion that the settlement and territorial characteristics in recent decades evolved under two extreme possibilities: on the one hand, entirely on the basis of central resources, based on the localities' political access to the central agencies and their related bargaining position, on the other hand under the effect of local resources and the possibility they owe to their peripheral situation, being 'beyond the horizon'. In our study we will examine the combinations of this double possibility, and their variants.

According to our studies, industrial development exerted a broad and long-lasting positive effect on the settlements where local society was more developed, better articulated in the traditional intellectual centres, where a strong territorial and political power status was ensured and, consequently, a strong human capital concentration could be achieved (Szaló, 1985).

We examined by factor analysis the indices of 143 urban districts of Hungary in 1980. It was found that 25.8 per cent of the phenomena described by all variants were explained by the first factor, that of human capital. This means that the concentration of human capital, and its spatial distribution, determine best the process of territorial differentiation (the factor of human capital is defined by a distribution of highly qualified population having a high social status, by good social infrastructure (public health, education, trade) and by a high consumption level) (Baráth and Futó, 1984). The majority of quality indices indicating an urbanization process show a repartition identical with this factor. Whereas during the 1950s and 1960s industrial development induced strong regional migration, in the 1970s the attraction of intellectual capital became the main determinant.

Table 10.2 The coefficients of human capital factor with the original variables ($R > 0.5$)

0.891	Proportion of white-collar employees
0.884	Proportion of higher-grade qualified persons
0.870	Proportion of employees of service sectors
0.843	Proportion of physicians (per 1000 inhabitants)
−0.788	Proportion of skilled and semi-skilled workers
0.787	Daily trade turnover (in thousands of forints)
0.775	Immigration ratio from 1970 to 1980 (per 1000 persons)
0.747	Food shops' turnover per household (in thousands of forints)
0.731	Number of kindergarten places per 1000 inhabitants (places/thousand persons)
0.666	Secondary education supply
0.5999	Places in infants' nurseries per 1000 inhabitants (places/thousand persons)
0.549	Ratio of divorced persons

Source: E. Baráth and P. Szaló (eds), *Spatial Structure in Hungary and the Structure of Its Complex Energy System*, Budapest: Vati, 1983.

Table 10.3 The coefficients of the industrial main component with the original variables ($R > 0.25$)

0.8814	The ratio of fixed assets per employee (in thousand forints per capita)
0.8783	The ratio of machines and equipment per employee (in thousand forints per capita)
0.7566	Electrical energy consumption per employee (in kW per capita)
0.7428	Exports in convertible (hard) currencies per employee (US$ per capita)
0.6758	Value added results per employee (in forints per capita)
−0.4927	The ratio of female employees (%)
−0.4532	Ratio of persons employed in the textile, garment and shoe industries (%)
0.4066	Proportion of persons employed in the extractive industry (%)

Source: P. Szaló *et al. Research on the Territorial Industrial Structure of Hungary*. Budapest: Vati, 1985.

The leading variable of the human capital factor (the ratio of white-collar employees to total population) and the main component values of industry (Tables 10.2 and 10.3) were used in order to present a breakdown of the territorial centre–periphery system. In Figure 10.1, a hypothetical 45° line describes the axis along which industrial and human-capital development are mutually proportional. On the right of this axis are situated the centre satellites where industrial production is better developed than the human potential. On the left are the centres where the inverse is the case. The figure shows that human capital in part developed independently of industrialization in Hungary. The scissor between the two axes shows that industrialization aimed at modernization has frequently not resulted in a comprehensive development of settlements even if they achieved economic advantages in the short run.

The *centres* are characterized by a high rate of highly qualified people, by high levels of supply and services, and richness of cultural life. The rate of industrial employment, however, has been decreasing for ten years in the centre regions. The extent of industrial production is around the national average, and their products are suitable for export. The assortment of jobs is large and the social and technical infrastructures are good but their very concentrated form creates pressure to settle in centre regions and causes permanent bottlenecks. Budapest and some county towns belong to the centres.

In the *semi-peripheral regions* the incidence of highly qualified people is

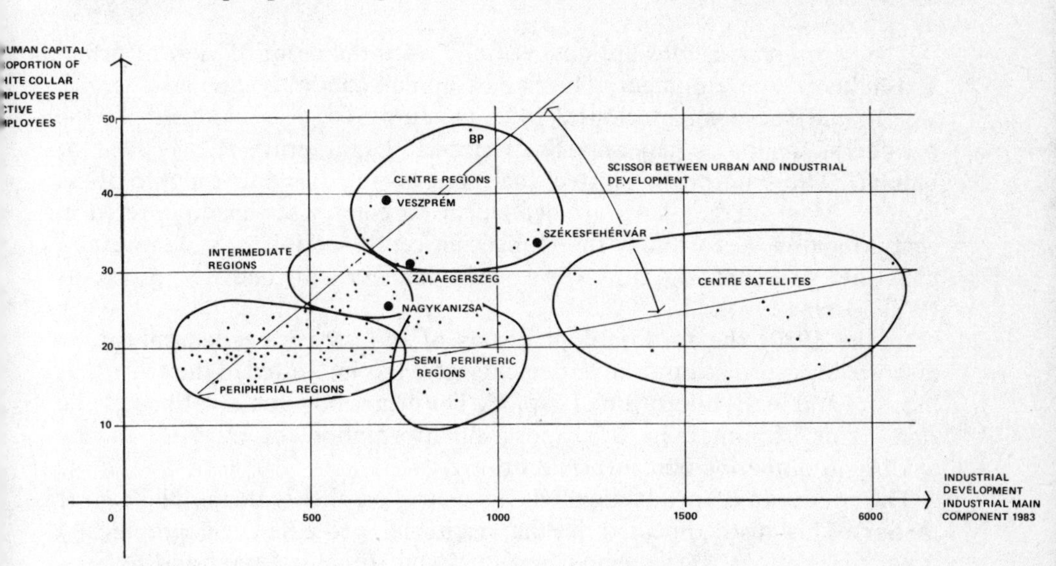

Figure 10.1 Distribution of 143 regions of Hungary by indicators of human capital and industrial development. Source: research on the industrial structure of Hungary carried out by P. Szaló *et al.* (1985), Budapest: Vati.

above and around the national average, which indicates the high level of human capital. Local industry is diversified and consists of small-scale plants. The value of gross industrial production and investment per capita is around the average. Both social and technical infrastructure are developed to high levels. Other county towns and traditional towns are in this group of settlements.

The *centre satellites* were mainly developed during the great industrialization wave in the 1950s and 1960s. In these new towns production is high, but the development of social functions reached a deadlock. These settlements are satellites in the meaning of outside regulation, subordinated to the centre regions. They are dependent on management situated in centre regions in a very centralized form. The management - generally the human capital - is segregated from production and the human conditions for innovation are missing in this group of towns. In Hungary the potential of innovation is restricted to a few complex urban and industrial centres. In these regions the value of gross industrial production per capita is extremely high, but the structure of industry is one-sided and specialized. The basic supply and the medium-level functions are developed in quantitative respects. The lack of qualitative supply, that lack of high-level functions, the lack of the very values of urban life can be observed in centre satellites. The great extent of immigration and the one-sided demands of industry on labour homogenized the structure of local societies. These centre satellites have the least flexibility in the case of economic crises linked to the determinant structure of industry.

The *peripheral regions* are undeveloped from the point of view of urban and industrial development. The rate of human capital is low, as is service supply and social and technical infrastructure. The local industry in the peripherial region is generally undeveloped; its structure is one-sided or bipolar. The enterprises shifted their out-of-date production into these regions. Most of the plants are dependent on enterprises headquartered in centre regions. Accordingly the Hungarian centre–periphery system resembles the structure of the world economic system and its problems (Wallerstein, 1974).

In the 1980s the territorial processes of industry were determined by disposal of information. The settlements were less interested in the growth of industry and in the allocation of capital. The centre position is defined by the right of decision-making, by disposal of information, of expertise and by priority in obtaining new information.

The processes of production fell into parts parallel with the division of labour. This also appeared in the territorial processes. Geographically isolated functions were created by functional, organizational and technological integration. The organizational integration appeared in Hungary as the colonization of the factories in the countryside and the establishment of horizontal trusts and monopolies. The technological integration appeared as the transfer of one part of the production process with no market of its own;

therefore the transfers are determined by the institute controlling the technological function.

The structure of activity became differentiated: the highly qualified activities reached the high levels of the territorial centre–periphery system, while low-level activities and out-of-date branches were relegated to the peripheries. Industry was differentiated by sectors in this national centre–periphery system. The preferred, so-called 'drawing', sectors (electronics, vehicles, precision instruments, telecommunications) were located in centre regions, the backward branches in the peripheries.

In our study we examined two towns that are among the centres: Veszprém has a more balanced position while Székesfehérvár is near to the centre satellites. The second pair, Zalaegerszeg and Nagykanizsa, is close to and within the intermediate regions (see Figure 10.1).

10.5 CRITERIA FOR SELECTION OF CASE-STUDY AREAS

In an earlier study we investigated the territorial structure of Hungarian society (Baráth, 1987). The urban level of settlements was analysed along three dimensions: the structure of production and employment; social infrastructure and the qualification and cultural level of the population; and the natural and artificial environment of settlement and its demographic conditions and circumstances.

About eighty indices were used per settlement, having recourse to factor and cluster analysis. The territorial–social stratification was approached with the aid of eight status indices (Kolosi, 1985) elaborated by means of statistical data: cultural level, consumption level, division of labour, material conditions, residential settlemental condition, interests prevailing (access to institutions, decision-making), housing, and magnitude of parallel economy.

The final result of the enquiry was a model-like description of the system of territorial–social inequality in Hungary. This research offered a good basis for the purposes of this study in order to select settlement pairs that had approximately similar rank. We selected two pairs of settlements where, in spite of a more or less identical starting position, development paths have been different and – according to our hypothesis – chances are different with respect to the response to be given to the challenge of the recent past and where it is possible to prove – on the basis of experiences accumulated during the last decade – that the integrative forces are rather different.

On our ten-degree scale describing the level of urbanization – where only Budapest, the capital, obtained a place in class 10 – the chosen neighbouring towns Székesfehérvár and Veszprém figured in classes 8 and 9 respectively, while Nagykanizsa and Zalaegerszeg were also in classes 8 and 9 respectively (see Map 10.1).

This is an important element, as our point of departure is that only settlements having reached a well-determined balanced and multifaceted

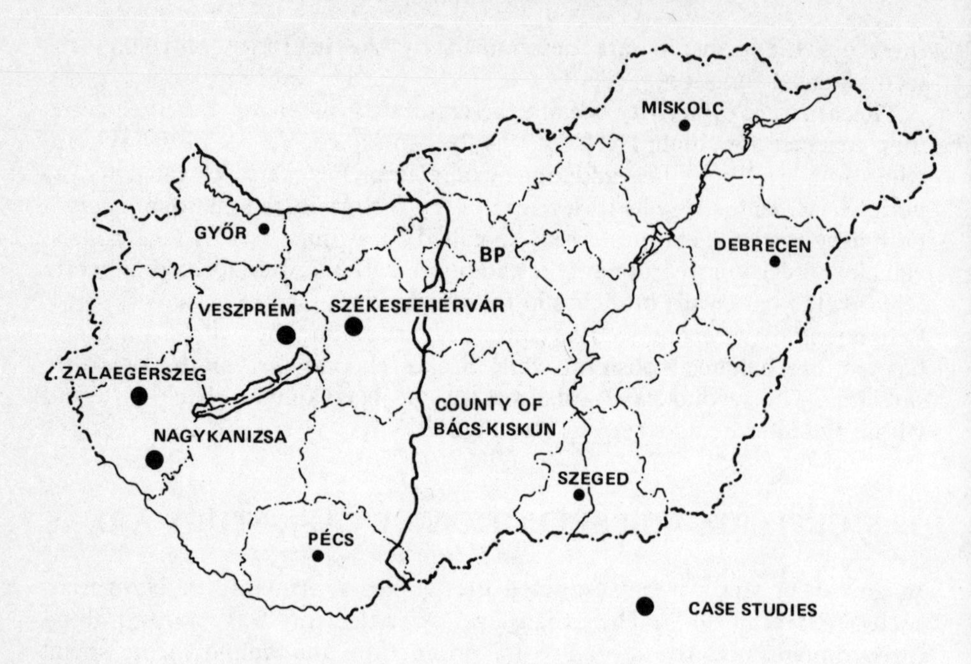

Map 10.1 Case study areas.

development can respond to the broad exploitation of all their resources – as envisaged in our study – and to the criteria of autonomy.

An important criterion of selection was that the settlements represent a typical life-path with regard to the redistribution mechanism that was characteristic during the four last decades. We compared pairs of towns with similar starting conditions but one town of each pair was influenced mainly by central resource allocation, the other more by local resources. The primary reason is that, as has been proved by several studies, under socialist circumstances any central distribution system is in a definite way differentiating. We know from prior research that in a socialist society redistribution determines not only the considerable class and social differences, but also territorial ones. Moreover not only is redistribution of investments guided by such a differentiation, but redistribution itself also determines their reproduction.

Therefore during the past decades, in the majority of cases the spatial allocation of production was the spatial determinant of local social structure and the principal former of structure in a settlement system.

The chance of obtaining or lobbying for central resources to be distributed, the existence or non-existence of local key persons or institutions ensuring access to the central position, played an important part in the modalities of development in a given settlement. An alternative mechanism is offered by the relations existing within specific localities between the elements making up these local actors (persons, institutions) and the indigenous resources they can mobilize.

The first pair of towns to be compared is Székesfehérvár and Veszprém (Map 10.1). These are two towns having a central position (Figure 10.1) with many similarities. The difference is that Veszprém's development has been based much more on local forces, whereas Székesfehérvár's was more centrally determined, and is more dependent on external sources.

Székesfehérvár is close to the centre satellites (Figure 10.1), which have high indicators on the industrial development axis but lower ones on the human capital axis. It developed mainly on account of central government decisions taken without regard for local forces and interests: industry was developed for its own sake, whereas the remaining settlement scarcely profited from it. This quick industrial growth did not touch the essential features of broader local modernization, its close intertwining with culture. Normative modernization attributed the development only to quantitative economic terms. The hypothesis is that regions which have grown on the basis of such external forces are unable to protect themselves against economic fluctuations; if restructuring is indispensable they must turn once more to external sources.

Zalaegerszeg and Nagykanizsa are towns that have been competitors for a hundred years. Zalaegerszeg preceded Nagykanizsa in its development owing to its more integrated development politics giving preference to economy and culture, and because of its high concentration of intellectual resources. With the aid of its entrepreneurs and its redistribution position it was able to lay the foundation for a more self-sustained development.

10.6 CASE-STUDY PAIRS OF CENTRALLY DETERMINED v. MORE INDIGENOUSLY LED LOCAL DEVELOPMENT

The important characteristics of economic, social and political structure are as follows.

10.6.1 SZÉKESFEHÉRVÁR AND VESZPRÉM

The past of these towns goes back to similar periods – the time of establishment of the kingdom of Hungary around AD 1000. Veszprém, the queen's town, was already an episcopal see (AD 1009), whereas Székesfehérvár was declaredly the seat of the king from the eleventh to the fourteenth century.

At the turn of the century, Székesfehérvár was a small town of 33,000 and Veszprém of 15,000 inhabitants. Both members of the first pair became county seats. In 1985 Székesfehérvár had 111,500 and Veszprém 64,000 inhabitants. Both towns had by then doubled in population as compared with 1960.

Any comparison of the two towns at first sight seems to be made difficult by the fact that Székesfehérvár has twice as many inhabitants as Veszprém. However, considering the data, reinforced by more subjective experts' evaluations, Veszprém had more urban characteristics so that its smaller size does not entail a disadvantage in the development competition. Further on, Veszprém, taking into account its attraction area (including Füzfö, Balatonalmádi), is diversified and possesses a broader, expanded attraction sphere of very heterogeneous structure, beginning with an important settlement having a chemical-industry character (Balatonfüzfö) to the centre of tourism at Lake Balaton, Balatonalmádi. Accordingly on the 10° scale applied for examining the social stratification of settlements (Baráth, 1987), Veszprém figured among the settlements of class 9, while Székesfehérvár only belonged to class 8. As to their speed of development, their access to administrative and political functions of industrialization and relations to the central development politics concerning these towns, the differences observed can be attributed to deviating local development strategies.

Székesfehérvár - Predominantly Externally Determined

In the case of Székesfehérvár, external interventions played an important role even before the Second World War. It was a weakly industrialized country town. Its industry included a dyeworks, a tannery, an electrical power station, gasworks, a flour mill and railway repair works. In 1935 a total of 495 employees worked in nineteen industrial units. Simultaneously with the war boom, rapid development of industry took place. Aeroplane repair works, an aluminium plant and an armament factory were founded. These three plants were built up on the basis of external central goverment decisions and with the help of external capital investments.

The large-scale industrial works that developed on the basis of the three factories correspond to approximately 70 per cent of Székesfehérvár's industrial employment. Just before the war, the legal predecessor of Videoton, the biggest plant of the town, was about to be established in Veszprém by the shareholders. However, because of 'difficulties encountered during the administration of the affairs by the competent authorities' (Boha, 1970) a decision was taken in favour of Székesfehérvár.

Later on, the factory became unpopular with the inhabitants of Székesfehérvár, as the plant was blamed for heavy bombings during the war. After the war the range of products was changed, but the determinants of development remained external. It was under the effect of special central investments, central development programmes and external developments that the works became really huge. The research and development department of the firm, however, was located in Budapest, the capital.

It is again as a result of external decisions that the former aeroplane repair works became one of the biggest bus factories in Europe. Around the end of the 1950s it merged with the Ikarus works in Budapest and today it is a branch plant of this firm. It is also via external efforts that the light-metal works were rebuilt here after the war, with Soviet assistance.

The Székesfehérvár plant of the garment factory was established as a company headquarters, the same being true for the light-industrial machine-tool works. Besides these rather specialized and concentrated industries mammoth organizations were created in the building industry related to the state's housing construction programmes. As a result of the central development programmes, the per capita values of fixed assets, technical equipment and gross production greatly surpass the national average. The same applies to the central government's budgetary subventions.

Veszprém – More Indigenously Led

Veszprém's economic structure is more diversified and the proportion of industry is rather modest (Table 10.4). In the past, local initiatives strongly predominated; this resulted, because of the shortage in local capital, in slower but more sustained development. The local traditions of small-scale industry were very strong (in the last century there were twenty-four guilds). Development of manufacturing industry began during the last century, and among nine factories four produced goods for export (Hornig, 1912). In 1860 twenty-six milk cooperatives jointly founded a milk-processing plant, again producing for export. Savings banks, foundations, and local monetary and financial institutes were able to mobilize local capital.

The town's industrial structure is comparatively multifaceted and diversified. A certain shift could be observed in favour of manufacturing of electrical machines at the expense of the building material industry (in 1983 40 per cent of all industrial manpower worked in one of these two sectors). The production value of industry, its supply in fixed assets and technical equipment vary around the national average. On the other hand, central governmental subventions are well below the average.

Veszprém's industry is not outstanding in terms of quantity of production, but its settlement environment and social infrastructure are more developed than those of Székesfehérvár. In Veszprém one can find autonomous industrial research. The Heavy Industries Research Institute and the Mineral Oil and Natural Gas Experimental Institute are located here, and the Bauxite Research Enterprise is nearby. The Chemical Industrial University of Veszprém opened in 1948. The Hungarian Academy of Sciences has a Committee in Veszprém. In Székesfehérvár – after the 1960s – three technical secondary schools, and later a high school, were established.

Székesfehérvár's characteristic dependence on externally oriented industry appears, too, when considering the proportion of white-collar employees (in industry, Székesfehérvár has 35.1 per cent and Veszprém 36.5 per cent in 1980, and in all sectors Székesfehérvár has 35.1 per cent and Veszprém 40.3 per cent). Moreover, its proportion of technical employees is higher, whereas in Veszprém the proportion of public health workers, and employees in the cultural sector is particularly high. The infrastructure is more developed in Veszprém; 88.4 per cent of the apartments are connected to the drainage system, the figure for Székesfehérvár being 75.2 per cent. In Veszprém,

Table 10.4 Sectoral structure of industry: the ratio of employees in different industries in the attraction spheres of the case-study towns in 1983

Sectors	*Székesfehérvár*	*Veszprém*	*Nagykanizsa*	*Zalaegerszeg*
Industrial research	—	6.43	—	0.07
Mining	0.23	3.10	5.85	6.45
Metallurgy	1.33	2.93	1.32	1.32
Electric energy industry	14.7	0.22	0.12	0.0
Mechanical engineering	6.65	7.84	18.04	8.26
Manufacturing of transport vehicles	14.21	0.18	0.22	5.97
Electrical engineering	0.07	25.51	0.58	2.73
Telecommunication, vacuum technique	39.38	8.56	29.56	6.29
Instruments, measuring set	1.16	0.92	0.26	4.42
Iron and metal mass production	0.02	0.0	0.0	0.0
Building material industry	0.56	15.32	8.77	6.3
Chemical industry	1.28	4.91	6.43	8.2
Wood processing industry	1.77	5.13	9.19	10.65
Paper industry	0.0	0.0	0.0	0.0
Printing trade	0.53	2.15	0.24	1.31
Textile industry	2.85	0.0	0.0	2.24
Leather, fur and shoe industry	1.37	0.0	2.03	1.35
Garment	2.29	0.01	1.94	11.58
Handicraft	0.29	0.42	0.33	0.0
Diverse	2.14	4.70	2.47	1.84
Food industry	9.79	11.60	12.67	20.52
Total	100.0	100.0	100.0	100.0

Source: Central Office for Statistics: Industry statistics data base 1983.

kindergarten, public infants' nursery, secondary school, library and cultural facilities, as well as commercial supplies are better.

While both towns had a theatre before the war, the aspirations of the politicians in Veszprém are well shown by the fact that the county council in 1961 founded an independent theatre, organizing its own company, whereas in Székesfehérvár the reconstruction of the theatre was not finished until 1962 and has no resident company even now.

Székesfehérvár chose the way of 'normative' modernization, which created a temporary boom: the productive sphere was placed first and

Table 10.5 Breakdown of active employees according to the type of activities and the school qualifications (per cent) 1980

	Székesfehérvár	*Veszprém*	*Nagykanizsa*	*Zalaegerszeg*
Among the blue-collar workers				
Finished at best 8 classes	31.9	28.3	39.7	31.9
Qualified in a secondary vocational training school	14.5	15.8	14.7	16.8
Finished a secondary high school	9.6	7.2	7.1	6.6
Qualified at a higher education institution	0.2	0.2	0.1	0.1
Total blue collar workers	56.2	51.5	61.6	55.3
Among the white-collar employees				
Finished at most 8 classes	7.2	8.5	6.8	7.2
Qualified in secondary vocational training schools	1.6	2.0	1.4	1.5
Finished a secondary high school	21.9	23.6	18.7	22.4
Qualified at a higher education institution	12.0	13.4	8.8	11.6
Total white-collar employees	42.7	47.5	35.7	42.7
Independent assisting family members	1.1	1.0	2.3	2.0
TOTAL	100.0	100.0	100.0	100.0

Source: *Population Census 1980*, County volumes. Central Office of Statistics.

broader interests of the settlement were regarded as secondary. Given these priorities, capital flows into the region concerned, but it does not penetrate into all spheres. In general terms a quick development of 'manpower maintenance' is likely. In the meantime there is development of 'not directly productive' spheres which represent social-cultural development, and mean real modernization. Labour culture is also an integral part thereof, however, and a real induction of development can be brought about when people live in an autonomous culture, unfolding their efforts in all dimensions and directions.

As already mentioned, there was little likelihood that an autonomous local development strategy would be realized under central management; there was only a narrow margin of local manoeuvre by management.

Nevertheless their exploitation could lead to essential changes that may have effects even in the longer run.

This is well illustrated by the fact that after 1982, when small private enterprises were authorized to function in accordance with a new decree, local management in Veszprém aspired towards greater autonomy, a stronger role for social-cultural traditions and value-models, whereas Székesfehérvár showed a conscious appearance of external dependence.

Possibilities of exploiting new resources exist in both settlements; however, in view of their divergent characteristics, they prevail in very different ways.

Székesfehérvár is a town possessing a typically national strategic industry, with big production units to which the settlement administration is 'subordinated'. As to the urban development projects, the role of industrial plants is considerable in decision-making. But precisely because of the strategic importance of the productive sector, the major determinants of production development are concentrated in Budapest. This proximity also provokes a shortage in the quarternary sector.

Infrastructure and relations with the capital are good – there is a motorway, driving time is just one hour by car, the local telephone exchange belongs to Budapest – but all this does not result in the settlement having good management and educational institutions. Székesfehérvár is a Hungarian town of very special industrial conditions where all productive bases are given, but where there is a lack of a creative–administrative–intellectual capacity producing the required synergy (Stöhr, 1984). Among urban local public personalities, the production managers dominate, and in the social and national hierarchy the politicians.

Veszprém, as we have seen, is more 'urban'. All criteria of territoriality are given, and it is multicentred. Science, politics and the church play a considerable role in the town. Moreover, there is a chemical industry, tourism and agriculture and these play a 'direct' strategic role (not in a transmitted way via Budapest as in the case of Székesfehérvár). The party secretary of the county

Table 10.6 General statistical data, 1985

	Székesfehérvár	*Veszprém*	*Nagykanizsa*	*Zalaegerszeg*
Population on 1 January 1985	111,478	64,071	55,326	61,456
Population aged 0 to 14 years	25.2	27.1	24.1	24.6
Percentage of persons older than 60	12.4%	10.3%	14.8%	11.9%
Socialist industry				
Number of employees (per 1000 population)	283	146	231	248
Number of blue collar workers per plant (persons)	298	122	239	222

Table 10.6 *Cont'd*

	Székesfehérvár	*Veszprém*	*Nagykanizsa*	*Zalaegerszeg*
Gross value of fixed assets (1000 forints per capita)	608	526	477	464
Net value of mechanical equipment	59.1%	61.3%	54.6%	57.1%
Average salary of blue-collar workers (in forints)	5,694	5,278	5,383	5,278
Basic shopping space per 1000 inhabitants (m^2)	560	657	621	627
Restaurant space	325	322	254	427
Retail trading turnover per capita (in forints)	64,088	72,780	60,810	78,610
Number of dwellers per 100 apartments	298	304	293	309
Number of apartments built per 1000 inhabitants	7.3	13.4	8.0	10.0
Proportion of privately built apartments	89.3	71.0	94.5	76.6
Ratio of three-room and larger apartments	58.1	52.2	48.9	62.0
Average apartment area (in m^2)	73.5	64.1	68.0	73.0
Percentage of apartments linked to sewers	74.9	63.7	65.4	62.3
Ratio of physicians (per 10 000 people)	39	47	33	56
Hours of consultations by specialists (physicians)	1,004	1,273	1,083	1,393
Number of secondary school students (per thousand people)	49	45	44	53
Library units (per thousand people)	3,794	3,934	6,140	4,646
Readers (per thousand people)	164	176	170	225
Cinemas (per thousand people)	27	44	92	69
Number of private enterprises per 100 000 persons on 1 January 1984	8.25	8.43	5.45	8.13

Source: County Statistical annuals, 1985. Central Office for Statistics.

is traditionally the leader of a highly respected commission in Parliament. The town is characterized by transferred central functions.

The relatively great distance from the capital increases its own political economic and scientific weight. The region is, in its prospects, enjoying preferences; considerable economic losses are endured only by its large enterprises, but there is considerable local building activity and the role of the local centre is also ensured by continuous reproduction of resources. The statistical data and analyses presented here do not give sufficient idea of the political vigour expressed by the unity of the local organs - mainly during the recent period - in the course of restructuring industry.

Zalaegerszeg and Nagykanizsa

Zalaegerszeg and Nagykanizsa are the two most important towns - of 61,000 and 55,000 inhabitants respectively - of the county of Zala situated close to Hungary's western border. There is a degree of rivalry between them that manifests itself in the acquisition of administrative and power attributions, as well as in the development of decision-making functions, making it possible to study numerous variants of the local strategies.

Until the turn of this century Zalaegerszeg was a town of 15,000 inhabitants, whereas Nagykanizsa had a population of 25,000. The development of Zalaegerszeg continued remorselessly whereas Nagykanizsa was stagnant from 1930 to 1950; in spite of the increase launched in 1970, the initially smaller Zalaegerszeg remained the winner. Both settlements have above-average urbanization, but the evaluation showed a qualitative difference in favour of Zalaegerszeg: on the 10-point scale of social stratification we used in our research, Zalaegerszeg figured in settlement class 9 and Nagykanizsa belonged to class 8.

According to the National Settlement Network Development Concept, which was approved in 1971, both settlements are co-centres of the superior degree, which means that they satisfy together the superior-degree functions of their respective regions. Thus, their reproduction and redistribution positions ought to coincide. However, Zalaegerszeg has been the county town since 1919, which helped it to acquire a more advantageous position and gave it better access to development resources allocated to the county. The presence of the county administrative machinery increased development chances owing to the presence of a stronger group able to obtain certain favours for Zalaegerszeg. The local administration had good relations with the central government via the person of an important Deputy Secretary General of the Hungarian Socialist Workers' Party who was at the time of writing the President of Hungary's Presidential Council.

The development of Nagykanizsa was launched by the oil discoveries of the 1930s. The high-quality oil attracted considerable external, particularly foreign, capital to this region. The structure of industry in both settlements is rather diversified. During the large-scale industrial development of the 1960s and 1970s almost all branches established enterprises or plants in both towns.

In Zalaegerszeg the development of the engineering industry and of telecommunications and vacuum technology was particularly important, whereas in Nagykanizsa food, textile and wood-processing industrial branches underwent a large-scale increase. Even in the field of industrial fixed assets, in technical equipment and in the level of technology there was little distinction between the two towns.

Nor was there much difference in the qualification of the work executed and the external dependence of the industry. Although the bigger industrial companies in both settlements have external management, the number of enterprises having local management in Zalaegerszeg is double that for Nagykanizsa.

Within industry, in Zalaegerszeg 58.6 per cent of the workers are skilled and 39.0 per cent are semi-skilled; in Nagykanizsa these proportions are 50.3 per cent for skilled workers and 46.9 per cent for semi-skilled ones. The proportions of intellectuals is 28.6 per cent in Zalaegerszeg, whereas in Nagykanizsa it corresponds to 26.7 per cent.

There are similar differences in other sectors as well. Trade, individual and collective services, public hygiene and social and cultural facilities are much better developed in Zalaegerszeg. Not only are there more people working in these sectors, but the services are of a higher quality, which is also indicated by the qualification level.

For instance, in Zalaegerszeg in the field of individual and economic services, the proportion of manual workers is 48.7 per cent, 38.6 per cent of whom have completed eight years at primary school, while in Nagykanizsa 61.4 per cent are manual workers and the proportion of similarly low-qualified persons was 50.4 per cent in 1980. A total of 40 per cent of the economically active employees are white-collar in Zalaegerszeg, whereas this proportion is 32.7 per cent in Nagykanizsa. Similar differences in their school qualifications can be demonstrated (Table 10.7).

Because of the similarity in economic and territorial conditions, the differences in qualification levels of employees indicate (beyond the redistribution position) the conscious and consequent enforcement of local urban development policies, although they are also strongly influenced by certain prestige factors.

Table 10.7 Breakdown of active employees according to school qualification in 1980

Highest school level reached	*Zalaegerszeg*	*Nagykanizsa*
Eight years of primary school	40.5	48.4
Qualified at a secondary professional training school	18.6	16.6
Qualified at a secondary high school	29.2	26.1
College or university education	11.7	8.9
Total	100.0	100.0

Source: Population census of the Central Office for Statistics; County of Zala, KSH, 1980.

Zalaegerszeg – Local Cooperation for Innovation

The ambition of Zalaegerszeg's leaders was to establish and develop a computer centre. Such institutes have been set up in all county towns, but here a special emphasis was put on the establishment of local horizontal relations, particularly through the foundation of local higher-education institutions. The construction of a factory manufacturing computing hardware offered employment to qualified specialists, as well as 'joint' employment of the managers between the computer centre and the high school.

In connection with the considerable industrial food production, research and development activities in this field were developed. The leaders of the local meat and cold-storage industry are among important local social entrepreneurs. The local organizations of the technical and natural-science associations play a special role in organizing the local cultural élite and its integration into urban life.

Zalaegerszeg was the first town in Hungary after the Second World War to invite Western technologists to reorganize the technology of the garment trade. This took place at the end of the 1960s and the beginning of the 1970s.

The more global, culture-oriented character of urban planning is stressed by the fact that Zalaegerszeg built a theatre and organized a resident company. Although Nagykanizsa's culture centre is an important cultural factor, too, particularly for the different interest groups organized around it, it seems that Nagykanizsa did not exploit the periodic possibilities quickly enough. Even now the most important local enterprise is the oil industry. During the 1950s and 1960s the oil company's central management allowed only a small margin for prevailing local interests. This margin belonged to the organizations assisting local interests, but the boom generated by oil exploitation came to an end with the diminishing of local oil reserves. During the boom period living conditions were improved – infrastructure was much more developed than in Zalaegerszeg, but no vertical elements and institutions (education, culture) were built that could have reinforced the town's capacities to bring about a restructuring.

However, as we mentioned earlier, if this margin had been exploited consciously, and in a more constructive way, it could today, in a period of structural crisis, have made a fundamental difference from the point of view of further development. So far we have investigated the differences in the characteristics of industry, labour and the structure of their qualification. However, there are also differences in the organization of the service and settlement infrastructure. In Zalaegerszeg the supply of trading and restaurant units, and the structure of dwellings and public hygiene are all better, the proportion of children studying in secondary schools is higher and so is the number of library-users.

Only the level of sewage systems is higher in Nagykanizsa. However, these data are insufficient to make an evaluation between the two towns because there are important differences in their central redistribution positions, in the sense that Zalaegerszeg had access to more central resources than did Nagykanizsa. Although this shows the possibilities of also letting local interests prevail, the results are not so much the consequence of local forces, rather they indicate the influence of the central authority (for example the rate of housing construction is higher in Zalaegerszeg, entirely because of the state's budget, while the proportions of private apartment construction are the same in both towns).

The orientation towards modernization, considered in the broad sense of the word, is a further development of the town as a social and cultural entity. The stronger structure-transforming force of society in Zalaegerszeg, which can be characterized by a rather culture-oriented development, is shown by the fact that in 1984, two years after the authorization of private enterprises, there were one and a half times as many autonomous small private enterprises in Zalaegerszeg (8.13 per 1,000 persons) as in Nagykanizsa (5.45 per 1,000 inhabitants).

To sum up the situation, the dynamics of Zalaegerszeg are not a result of the 'directly' placed central power functions (as in the case of Veszprém). In view of the geographical distance from the capital, the role of the local centre assumes greater importance, and its aim is to obtain an equilibrium, based on tolerance. All this makes it possible for the local (individually insignificant) organs, on the basis of the local availability of 'entrepreneur' functions, to combine their forces and efforts.

Among the local party organizations, the council and factory managers, the leaders of other organs of less importance, such as the church, the bank, the savings bank etc., maintain consultative (that is to say non-hierarchical) relations. Considering that there are no dominant factories and workshops, settlement development as such becomes a general task. It increases the sense of solidarity among the local forces: development of community consciousness is assisted by an extremely good information situation – the town can receive television channels from three other countries (Yugoslavia, Czechoslovakia and Austria) – and by tourism.

An illustration of local solidarity occurred when Zalaegerszeg's football team entered the first division of the national league; a stadium with 20,000 seats was built during a single summer in the 50,000-inhabitant town and the cup final was watched by 25,000 spectators.

Nagykanizsa – Handicap of Industrial Monostructure

The lagging behind of Nagykanizsa – in spite of its favourable infrastructure – is caused by its industrial monostructure: there is a single, very large factory with initially 9,000, later 6,000 employees, and relatively lower qualification requirements. The inherited higher cultural level, the traditionally more developed urban citizenry and local trade could not counterbalance the deforming structural effects of a one-sided and excessively dynamic development, as well as the lack of internal cohesion between the organizations caused by it. In a period of industrial restructuring, lower flexibility and the dependence on the big organizations in Budapest represent a downgrading of resources to be allocated in favour of broad local development. It offers less chance to follow the lead of events in the field of politics and economy.

10.7 CONCLUSIONS

The analyses, investigations and interviews made on the spot seem to reinforce our hypotheses. The 'double structure' mentioned in the introductory part could be perceived. The territorial-cultural subsystem is the value preserver of dynamic sectoral-economic processes; in critical periods it will be this subsystem that determines the future of the local development processes. At the same time, it is this subsystem that offers, on the basis of historical experience and knowledge embedded in social structures, a force opening towards the future.

This is what ensures – even in the case of European societies with different political systems – a similar reaction capacity, and comparability. This is the force that mobilizes the hitherto latent resources in the territorial organization, into the economy, if synergetic relations exist between the local economic and cultural actors, be they organizations or persons.

If the large-scale economic system plays a more dominant part over a long period in a given region or settlement, this dominance may have negative effects on the local social-cultural organization, as well as the relations of the two spheres, particularly if local identity is eroded. The boom period may be followed by a structural crisis and there will be a possibility of reversing this crisis only by introducing external resources.

The lesson to retain is that within the Hungarian homogeneous political-economic system, the territorial-cultural subsystem maintains a very broad variability as to its acting capacities. This testifies in an uncontested way – if making a comparison with the territorial-cultural analogy existing in different political-economic systems – that dialectic interrelations characterize the functioning of our double structure.

That even under these conditions there is room for local manoeuvring, however, is well illustrated by the intentions currently observed in Székesfehérvár. With already-reduced dynamics of the production sphere, municipal leaders and managers of plants urge in common the development of research activities; in doubtful fields they even take the initiative of 'implanting' creative intellectuals. The very first steps are made in order to establish a 'technopolis' organization in Hungary, sustained by local forces.

The reintegration of the internal local structure in such a monostructured locality can be successful only over a long period. Until then it has to rely on external sources. Székesfehérvár is becoming connected with state central development programmes such as development programmes of vehicles and electronics. Thus the Videoton factory developed the production of computers, integrated robotized production of colour TV sets in common with Thompson, and the bus factory has been buying expertise and licences from different units.

At the same time an overcentralized enterprise has been liquidated in Veszprém. The liquidation of a firm which was not indigenous to the structure of Veszprém was inevitable: the large state construction firm was incapable of surviving while the small firms developed successfully. This apparent negative process showed the possibility of reorganization from inside. The more diversified structure, the broader cultural-social basis, and the efforts towards independence make possible the continuous restructuring process.

The central regions dispose of local - intellectual and economic - sources of power for restructuring. The centre satellites have insufficient local intellectual forces, and peripheral regions have neither human nor economic resources. Even so, the success of these regions' development will depend on mobilizing the potential indigenous resources that they do have. The success of reorganization will depend on the perception of their own factors of development and the creation of social-cultural conditions reinforcing local initiatives. Certainly the central state will indirectly have to promote and facilitate this.

REFERENCES

Baráth, E. (1987), 'La structure sociale du système des agglomérations hongroises', *Bulletin de la Société Languedocienne de Géographie*, vol. 21, pp. 105–17.

Baráth, E., and Futó, I. (1984), 'Regional planning system based on artificial intelligence concept'. *Papers of the Regional Science Association*, vol. 54.

Baráth, E., and Szaló, P. (1982), *Planning and Research of Interaction between the Complex Energic System and the Territorial Structure of Hungary (Methodological Conclusions)*, Budapest: Institute for Town and Regional Planning.

Boha, J. (ed.) (1970), *The History of the Videoton Factory, 1938 to 1970*, Budapest: Atheneum.

The county of Veszprém from 1963 to 1966, Veszprém: Municipal Committee of the Hungarian Socialist Workers' Party 1966.

Gerschenkron, A. (1984), *Backwardness of Economy from a Historical View*, Budapest.

Hornig, B. K. (ed.) (1912), *Past and Present in Veszprém*, Veszprém: Egyházmegyei Nyomda.

Kolosi, T. (ed.) (1984), *Economy and Stratification. Stratification Model Testing IV*, Budapest: Central Committee of the Hungarian Socialist Workers' Party, Institute of Social Sciences.

Kolosi, T. (ed.) (1984), *Means and Methods. Stratification Model Testing II*, Budapest: Central Committee of the Hungarian Socialist Workers' Party, Institute of Social Sciences.

Kolosi, T. (1985), *Status and Stratification. Stratification Model Testing III*, Budapest: Central Committee of the Hungarian Socialist Workers' Party, Institute of Social Sciences.

Lébényi, P. (ed.) (1970), *Székesfehérvár 1945–1970*, Székesfehérvár: Székesfehérvár Városi Tanács (local council).

Muegge, H., and Stöhr, W. (1987), *International Economic Restructuring and Regional Community*, Aldershot: Gower.

Ránki, Gy (1986), *State and Society between the Two World Wars in Central-Eastern Europe*, Budapest: Hungarian Academy of Sciences.

Stöhr, W. B. (1987), 'Regional innovation complexes', *Papers of the Regional Science Association*, vol. 59, pp. 29–44.

Szaló, P. *et al.* (1985), *Research on the Territorial Industrial Structure of the Country*, Budapest: Institute for Urban Planning.

Szücs, J., and Hanák, P. (1986), *The Regions of Europe in History*, Budapest: Hungarian Academy of Sciences.

Szücs, J. (1983), *A Sketch on Three Historical Regions of Europe*, Budapest: Magvetô Editing House.

Wallerstein, I. (1974), *The Modern World-System*, New York: Academic Press.

PART IV

LOCAL RESTRUCTURING IN CENTRAL AND SOUTHERN EUROPE

CHAPTER 11

Externally induced regional development on the western side of the 'Iron Curtain': attempts at indigenous regional development in Austria's rural areas

Günter Scheer and Anneliese Zobl

11.1 THE ECONOMIC AND POLITICAL FRAMEWORK IN AUSTRIA

The development of the Austrian economy since the Second World War can be divided into five periods.

11.1.1 PERIOD I: RESTORATION AND POSTWAR BOOM (1953–62)

During this period the GNP in Austria grew by an annual average rate of 6.3 per cent. Of all the OECD countries, during this period only Germany and Japan were able to outdo Austria in this respect. The restoration after the Second World War, a rapid growth in the demand for exports, and the slowly growing buying power were significant driving forces during this extraordinary period of growth, the extensive nature of which is underlined by the rapid decrease in the unemployment rate in Austria from almost 9 per cent in 1953 to less than 3 per cent in 1962.

The founding of the Commission for Wage and Price Parity (1957) also took place during this first phase. This is the core element of the Austrian 'social partnership'. All of the larger professional associations, the Austrian Chambers of Trade and Industry, of Agriculture, of Labour, the Austrian Trade Union Federation (ÖGB) and the government all work together on this commission. For the last thirty years it has assured a social climate largely free of conflict and has made possible a union wage policy oriented to the conditions of the respective economic and political situation.

11.1.2 PERIOD II: STRUCTURAL CRISIS (1962–67)

In this period the Austrian economy showed a clearly lower and, compared to the other OECD countries, below-average rate of growth.

The causes of this slow-down were diverse. The fact that the industrial investment rate greatly declined played an important role. After achieving full employment, a conversion to a capital-intensive, labour-saving form of economic growth had to be made. This conversion was not completed quickly enough by Austrian industry. Moreover, the nationalized iron and steel industry neglected to push up production of final products at the critical time.

The negative effects of this structural crisis on the labour market were compensated for by an expansion of the tourist industry. Of primary importance was the increase in the number of visitors from abroad to tourist areas in western Austria (Salzburg, Tyrol, Vorarlberg). Thus, the surplus of foreign currency from the tourist industry grew from 1.5 per cent of the GNP in 1955 to 4.4 per cent of the GNP in 1965, despite the simultaneously declining number of natives seeking overnight accommodation. During this second phase of the development of the Austrian economy an economic imbalance between east and west in favour of the latter began to develop which today has become a major and ever-growing problem for regional policy.

11.1.3 PERIOD III: THE MODERNIZATION OF AUSTRIA. ECONOMIC BOOM (1968–74)

During this phase the Austrian economy again showed above-average rate of growth compared with the other OECD nations. Manufacturing industry, whose net product increased by an annual average of 7 per cent, gave the most important impetus to and support of this boom period. From 1971 on there was a change in economic policy due to the takeover by the SPÖ (Socialist Party of Austria): a policy of hard currency and high government investment and intervention were the cornerstones of this new course of action.

Table 11.1 Changes in real GNP in selected OECD countries between 1967 and 1974

Country	*Average change (%)*	*Country*	*Average change (%)*
Austria	+5.2	Sweden	+3.7
Belgium	+5.2	Switzerland	+3.9
FRG	+4.4	Turkey	+6.7
Denmark	+3.3	Japan	+7.9
Finland	+5.5	Canada	+5.3
France	+5.3	USA	+2.8
Great Britain	+2.7		
Italy	+4.8	OECD-Europe	+4.6
Netherlands	+5.2		
Norway	+4.0	OECD-total	+4.3

Source: Butschek, 1985, p. 144.

11.1.4 PERIOD IV: AUSTRIAN KEYNESIANISM (1975–80)

The oil crises of 1973 and 1979 brought the international boom to a halt. Austria also suffered relatively severe collapses in 1975 and 1978. During this period real economic growth was at an annual average of 2.5 per cent and was thus more or less at the same level as the OECD average in Europe. Unlike the majority of the OECD countries, however, Austria succeeded in maintaining full employment, even though the potential labour force grew significantly during just these years. This Austrian 'labour-market miracle' was made possible by a special 'mixed policy' consisting of a hard-currency policy, growing deficit spending, relatively low wage settlements which were made politically possible by the social partnership, an active labour-market policy, and holding back on the reduction of labour in the nationalized iron and steel industry, in which, after all, 17 per cent of all industrial labour was employed. As a result of this Austro-Keynesian course the unemployment rate in Austria remained at 2 per cent, while the average in the other OECD countries was 5 per cent. The other side of the coin, however, was that the budget deficit rose from less than 2 per cent of the GNP at the beginning of the 1970s to 6 per cent by the end of that decade.

11.1.5 PERIOD V: STAGNATION AND POLITICAL CHANGE (1981–)

Since the end of the boom period in 1981 the Austrian economy has been unable to recover fully. The real growth rate for this period until the end of 1987 stands at an annual average of 1.3 per cent. The continuation of an expansive fiscal policy has been blocked by the absence of the necessary budgetary and political conditions. The rapidly rising national debt and the very change-resistant and inherent dynamic of public spending (tertiary-cost pressure) have led to a growing burden of public spending on the tax-payer. In 1976 public spending made up 38 per cent of GNP, and by 1986 it was already 44 per cent. Along with the change in government at the beginning of 1987, a return was made to the 'Grand Coalition' between the SPÖ (Austrian Socialist Party) and the ÖVP (Austrian People's Party). The primary goal of this new administration is to balance the budget. This will most likely mean significant cutbacks in assistance for regional development programmes and jobs which have been created over the past ten years. At the same time, the administration places strong emphasis on restructuring nationalized industry, which will mean a reduction of jobs by one-third. Owing to the traditional cooperation within the social partnership, little if any significant resistance to this policy is to be expected from the unions.

11.2 REGIONAL PROBLEMS AND REGIONAL POLICY IN AUSTRIA

The following features mark the regional distribution of economic activities in Austria:

1. Austria lies on the very edge of OECD territory. Its long border with relatively closed East European COMECON countries has greatly hindered eastern Austria's economic activity.

2. Tendencies towards strong regional concentration were the result of structural changes in the agricultural sector, which over the past four decades has lost more than two-thirds of its labour force.

3. In western Austria with its predominantly mountainous and rural geographic situation this did not lead to the otherwise normally resulting loss of income or population because the tourist industry was able to compensate for the above-mentioned losses there. In eastern Austria, however, this tourist development, other than for Vienna, had little success and the typical disadvantaged rural regions in the periphery with high population drain, lower incomes and high unemployment rates were the result.

4. Until the mid 1970s it was possible to attract manufacturing industries to the agricultural, problem areas of eastern Austria with the aid of assistance programmes and subsidies. However, with the end of the boom period in 1975, and then the crisis since 1981, this wave of industrial settlement has come to a standstill. Many of the firms which settled in these areas in the 1960s and early 1970s were unable to survive the crisis.

5. The older industrial regions in eastern Austria, however, have proved to be the major problem for regional policy. Among them are the iron-producing regions in the provinces of Lower Austria and Styria. An imbalanced, large-scale and static iron and steel industry has left these regions with difficult structural problems. The location in narrow valleys and the lack of an appropriate socio-economic climate impede the initiation and establishment of a diversified industrial structure.

6. These five factors together have resulted in a decline in economic growth rates from west to east. In recent years the economic imbalance has become even worse, and the gap will grow wider still, as the crisis continues. Between 1974 and 1986, eastern Austria, comprising the provinces of Vienna, Lower Austria, Burgenland, Styria and Carinthia, lost one quarter (!) of its industrial jobs, while western Austria, comprising the provinces of Upper Austria, Salzburg, Tyrol and Vorarlberg, lost only 7 per cent.

In view of the unequal economic development of these two larger regions, the failure of the traditional regional policy, and the new administration's placement of a consolidated budget at the top of their priority list, the central government's regional policy is getting further and further away from a programme of assistance and compensation for smaller, individual areas, which were the targets of the previous regional policy (cf. Map 11.1). For the last few years this withdrawal has been accompanied by a gradual pattern change. Today, regional policy pursues, above all, structural and innovative

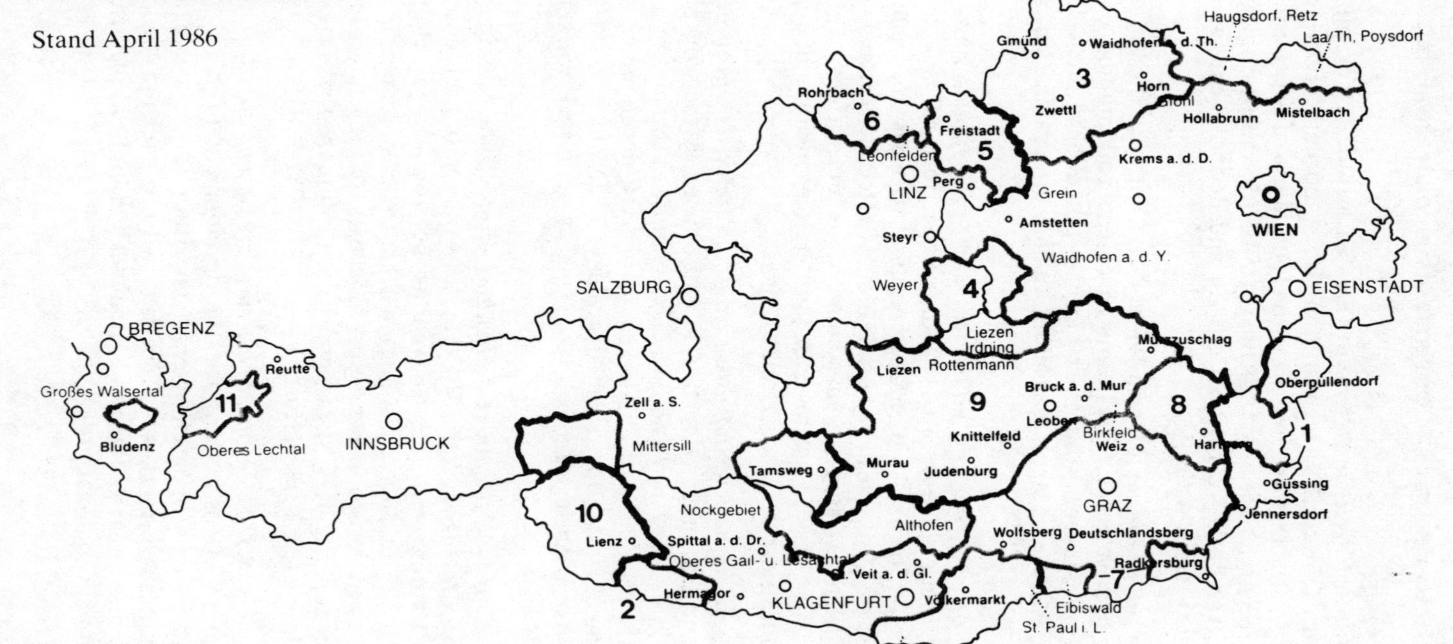

Map 11.1 Assisted areas. Regions: 1, Southern Burgenland; 2, Upper Gail and Lesach Valley; 3, Waldviertel; 4, Eisenwurzen; 5, Lower Mühlviertel; 6, Upper Mühlviertel; 7, Southern Styria; 8, Eastern Styria; 9, Upper Styria; 10, Eastern Tyrol; 11, Lech Valley. Reproduced by courtesy of the ÖAR.

economic goals. It is oriented to the principle of 'internal revival' and has the following features:

1. *Mobilization of regional potential*: this can include for example the promotion of innovative measures in firms and enterprises already established in the region stimulating the foundation of new businesses, and also the organization of unconventional job creation initiatives. All in all, the basis for regional growth and development should be strengthened 'from within'.

2. *Structural policy above job creation policy*: Central to the beginnings of new innovative processes is the improvement in adaptability of firms and enterprises by boosting their business potential in areas such as research and development, marketing, quality of organization, ability to cooperate etc.

3. *Transfer of information above transfer of capital*: Favourable conditions should be created by organizing regional networks of industry-wide information and counselling services for small and medium-sized businesses. The promotion of effective regional centres having sufficiently advantageous connections and, offering the right services, gains new importance in light of the creation of regional innovation, networks, and organizations.

4. *Promotion of human capital above material capital*: The promotion of training and continued education in and outside of the workplace is certainly not new; however, it gains significantly in importance through the pursuance of an indigenous regional development as well as through an innovation-oriented strategy. The transfer of ideas and innovations is most effective when highly qualified workers are employed.

5. *Promotion of co-ops above individual firms*: More co-operatives should be organized vertically (e.g. raw material processing and finishing/refining services) as well as horizontally (development, marketing, education and training) in order to achieve a more favourable competitive position in a region. Concepts for indigenous regional development particularly stress cooperation within the firm in the form of self-administered enterprises.

6. *The complementing of central structures by decentralized ones*: The establishment of institutions for the development of their specific region and a 'policy of development from within' should create favourable conditions for those concerned with organizing their own initiatives and integrate them to a greater extent in the implementation of government regional programmes.

11.3 LOCAL AND SELF-RELIANT DEVELOPMENT IN AUSTRIA AS A MOVEMENT FROM BELOW

11.3.1 HISTORY

In addition to the major economic and political changes which have led to a change in regional policy and the increased utilization of local initiatives, there is also a history of local development from below.

In the mid 1970s a regionalistic movement arose in Austria which spread rapidly until the early 1980s and included a wide spectrum of groups and trends. New-left-wingers, 'Greens' and conservatives all belonged to this movement. Regional planners and architects, scientists and educators, politicians and officials as well as social workers, students of various subjects, sociologists and biologists could also be found among its members. Yet there were only a very few of the people who were actually affected by the conditions in the disdvantaged regions to be found among them. Those involved in the movement, however, were all the more active and enthusiastic. Primarily they were farmers of the Mühlviertel.

This regionalist movement had diverse, initially non-economic origins:

- Those new social movements which share the opinion that smaller associations are more easily managed and better able to take notice of the social interests than anonymous, large-scale organizations or governmentally established corporations.
- The strong conviction that policy and the state consider the needs of the common man only when he exerts pressure on them through citizens' initiatives and/or other forms of non-parliamentary opposition.
- The growing awareness of responsibility for the environment and the growing scepticism toward the industry's policy of exploitation, along with the resistance to the destructive effects of industry, government and military, has intensified the search for examples of some form of economy which can operate in harmony with nature. It is not merely a coincidence that in the 1970s 'biological farming' became the first and most popular symbol of a method of production in harmony with the ecology.
- The growing belief, especially in times of crisis, that self-help is a better way to solve problems than waiting for the state to take action.

In the mid to late 1970s countless initiative groups with various aims emerged from the above-mentioned trends and convictions primarily in the rural problem areas of Salzburg and Upper Austria. They shared the principle of cooperative self-help above and beyond the existing limited professional and political methods.

In the course of the 1980s the economic situation has changed significantly – to the disadvantage of the rural problem areas:

- Agricultural surpluses have grown and the cost of utilizing them has risen. While the farmers' income in the divisions of grain, milk and wheat production is sinking, the costs of governmental subsidies from public funds are rising to exorbitant amounts.
- The prices in the forestry sector have stagnated at an extremely low level because the demand continues to be low owing to the poor economic situation of the building and construction industry and the increased accumulation of damaged lumber due to acid rain.
- The pressure on industrial production to adapt itself to the international market continues to increase, and at the same time 'Austro-Keynesianism', a tradition which has been practised for years, has ended quite suddenly.
- Even the tourist industry no longer functions as a primary source of growth. More and more those areas just beginning to develop tourist trade are falling behind (e.g. Lower Austria, Styria and Burgenland).

As a result of these developments farmers have had to suffer huge losses of income. In 1987 unemployment rose to its highest level since the postwar years. The disadvantaged and poorly structured regions have been most affected by this.

The economic aspects of the concept of indigenous regional development have become increasingly important as a result of these changes. And at the same time, the number of people living in those disadvantaged regions who have become active in regional development initiatives has grown.

However, growing pressure due to economic difficulties does not necessarily lead to increased participation in local and regional development projects. Internal regional reconstruction is impeded by several obstacles in the way of motivation, information, education and culture-related problems. In many rural areas and iron and steel regions in eastern Austria there exists a long tradition of dependency. Individual initiative, the willingness to take risks, the ability to work together, and the courage to try new things, are, to a great extent, seriously lacking there. Austria's political history, from its feudal system of inherited subservience to the present day corporatism, has taught people to wait for assistance from above rather than to work together to help themselves. The socialization process in family, school, church and professional organizations promotes obedience and toleration far more than criticism and initiative. And lastly, the erosive, decade-long process of migration has also played a damaging role as it affected the most mobile and most committed people in particular.

The concept of indigenous regional development therefore views the disadvantaged regions as the products of a multidimensional process.

The social structures and behaviour of the disadvantaged regions, as well as their economy, policy and culture, are closely connected and cannot be seen independent of one another. That means there are no simple, clear-cut remedies to the problem. Integrated changes in the economy, politics and culture are called for here. And the major role in effecting these changes must be played by the residents of these disadvantaged regions themselves. Others can help, but cannot act for them.

Successes and noticeable improvements are essential for indigenous development really to get off the ground. Without success, every movement stagnates, particularly as failures tend to confirm prejudices and increase fear and mistrust.

All of these problems make regional development a slow and long-term process. However, early successes are greatly needed, particularly in the economic and social sectors.

The aims of indigenous regional development are:

- to develop new possibilities of economic development;
- to found innovative projects that can raise the regional net product;
- to build on new possibilities for intra-firm cooperation;
- to develop new possibilities for industry-wide cooperation;
- to develop new, ecologically and socially useful possibilities of employment;
- to solve public and social problems by means of cooperative self-help;
- to create new and integrated training programmes (e.g. for environmental and recycling advisers);
- to set up regional centres for the exchange of information and experience between local development projects;
- to coordinate and establish communications between the various initiatives in regional centres.

Despite all the emphasis placed on the principle of self-help and local development, it is necessary to stress that indigenous regional development is neither a regional self-administration of decline nor a strategy of the central administration and regional governments to free themselves of the fiscal burden. On the contrary, intensive and long-term outside assistance is essential for the promotion and support of potential regional development. This includes the mobilization of the regional population (by regional advisers), the exchange of information and know-how, and the promotion of various organizations and cooperatives, specific to individual regions. However, these measures require the establishment of regional organizations to advise and supervise. It also includes the maintaining of initial assistance programmes, such as grants, for the development, production and marketing of new products, and the continued training of employees/workers.

11.3.2 STRUCTURE OF THE ÖAR

Often there are gaps in communication between the subsidizing institutions of the government and local and regional development programmes. The governmental institutions generally behave in an inflexible, bureaucratic and sovereign manner. In the rare instance that feedback from the local initiatives actually reaches the bureaucracy, it is usually viewed as disruptive and improper interference. Furthermore, Austria's political tradition impedes cooperation between central government and institutions more closely involved with the local initiatives.

Thus it was necessary to establish an organization at the intermediate level, between the innovative basis and the governmental institutions. This is the ÖAR's task in Austria. The ÖAR (Austrian Association for Self-reliant Regional Development) was founded in 1983 and today it serves as the mouthpiece for indigenous regional development in Austria. At the same time it works together with offices of the administration's regional policy for the promotion of local development projects.

There are four elements of indigenous regional development in the ÖAR:

- local and regional development potential represented by regional associations;
- a decentralized staff of regional advisers;
- a central unit for the rapid exchange of information and know-how and access to governmental offices responsible for planning and promotion; and
- the running of a promotional institution which conforms to the particular situation of disadvantaged regions.

Table 11.2 illustrates the organizational scheme of the ÖAR.

Table 11.2 The organizational scheme of the ÖAR

State administration		Financing Control	State
ÖAR		Coordination Information	
Regional counselling	FER	Training and continued education programmes Promotion	Intermediate level
Regional associations		Potential for innovative regional base development	

In the following sections a short survey will be given of the specific components of the ÖAR's organizational scheme starting from below.

Regional associations

Regional associations are the basis of the ÖAR and they play an important role in the mobilization of the potential for self-reliant development in the diadvantaged regions. With the support of an ÖAR adviser they carry out a significant amount of the communication and public-relations work for indigenous regional development and they organize educational and cultural programmes for the regions. An extensive knowledge of the regions on the part of the association's members is necessary to ensure that the measures taken are effective and appropriate for the individual regions.

People from all walks of life come together in the regional associations and together they look for solutions to the regional problems. They discuss their own plans, exchange experiences, look for advice and support, plan presentations, working groups and seminars, visit other projects, produce a paper, or simply stop by to meet interesting people. Regional associations provide a basis for people who contribute either in a social, cultural or economic aspect to the benefit of the regional development and who, in doing so, want to work together with others from the region.

Regional associations are non-partisan organizations in which problems of differing political confessions are put aside in order to be able effectively to initiate incentives for development in the region. They activate and unite committed people of the disadvantaged rural areas. Their activities are an important contribution to improving the social and cultural climate of resignation in the regions which impedes the implementation of innovative measures. They stir the potential for innovative development of disadvantaged regions. Furthermore they are an important support mechanism for new and unconventional development projects in these regions.

Regional counselling

Regional counselling differs from the usual form of counselling in many respects. It is activating, comprehensive, long-range and directly related to the region and its projects.

Projects are supervised from the initial idea through to actual operation. The form of counselling changes in content and method in the course of this process. In the initial conceptual stage the emphasis is put on motivation and incentive, whereas during the actual project planning phase organizational counselling and assistance providing the necessary qualifications come to the fore. Expert advice concerning business management and financial and promotional assistance are the primary points of concern in the investment stage, and finally, in the consolidating phase.

Advisers are committed to and actively involved in the projects in a kind of partnership, although they do not, in any of the stages of development, actually carry out the tasks of those responsible for the project. Regional counselling, however, entails more than just extensive project assistance and advice. Regional development is also an important feature. Regional counsellors find themselves confronted with the following tasks:

1. Search and informational strategies: regional advisers actively search for people with ideas, in addition to developing their own concepts which can be realized on the basis of local and regional self-help (e.g. concepts for tourism). They also help to draw up plans for regional development. They do not see themselves as planning experts, but rather as catalysts for developmental impulses and democratic planning processes.
2. Creation and support of a regional base: self-reliant regional development requires many actively involved people who can think, learn and work together. Individual project supporters alone are not sufficient for the creation of this kind of innovative atmosphere. The economic projects in particular usually are completely occupied with trying to realize their specific goals. Therefore one of the adviser's responsibilities centres on the creation of regional platforms and networks where people and initiatives can meet and discuss, examine and consider new efforts in all areas of life. As is to be expected, the cohesion of these networks is loose, owing to the diverse goals of the different initiatives. They do represent, however, an important meeting place for the exclusively local and/or occupational or group-oriented types of organizations.
3. Public relations: in this case the object is to win the acceptance of a wider public for the concept of indigenous regional development. Aside from the projects and regional centres, a permanent information network on a provincial and nationwide level should make supporting the regional initiatives easier.

The training and employment of regional advisers was initiated in 1981. In 1986 twenty regional advisers supervised 132 projects and initiatives, forty-seven of which are in the final stages. The projects are diverse in nature. There are thirty farm co-ops, approximately thirty farmers' co-operative power plants fuelled with chopped wood collected in the process of thinning out forests, around fifty initiatives active in the health, educational and culture-oriented aspects of tourist industry, as well as some twenty entrepreneurial initiatives in which people who have started their own businesses want to put their products on the market.

The ÖAR Central Office

In the ÖAR's central organizational unit, experts in the fields of agriculture, energy, tourism and trade are working together. Their tasks include:

- assessing the possibility of sponsorship for proposed projects by the ÖAR;

- establishing and introducing favourable connections for the regional organization;
- documenting experiences made in the implementation of projects;
- preparing and organizing training and continued education programmes for regional advisers and other interested parties;
- gaining national and international experience (keeping up to date on progress);
- editing ÖAR publications;
- carrying out the administrative tasks of the ÖAR;
- advising those in charge of the organization's regional policy;
- influencing the government's economic policy with regard to the goals of indigenous regional development.

The FER (Governmental Committee for the Support of Self-Reliant Regional Development)

The FER was established by the Federal Chancellery in 1979. Its goal is the support of innovative economic development projects which will promote business and have a positive effect on the region's economy. Subsidies are provided for the initial costs. Only those cooperative projects which have a democratic infrastructure (one member–one vote system) receive assistance from the FER.

Social and educational projects, purely infrastructural organizations and job-creation initiatives which will not establish new businesses are excluded from assistance, as are innovative projects implemented by conventional, individual firms.

In order for a proposal to be important for the region it must have the following features:

- use and refinement of the region's raw materials and the skills of its residents;
- production of high-quality products and services specific to the individual region;
- inter-regional communication and cooperation;
- improvement of the regional economy's market position;
- filling gaps in the market;
- environmentally safe products and production methods;
- positive effects on the regional unemployment rate and income level.

Investments in intangibles can also be sponsored. That means that in addition to tangible assets and buildings, expenditures for planning and consultation, development of prototypes, market introduction of products are also eligible. Assistance is given in the form of grants for the initial costs only, and may cover up to 50 per cent of actual cost, but may not exceed one million Austrian schillings. At least a quarter of the investment costs must be financed by the applicant through human resources, work, bank loans etc. That means that a grant from the FER can be combined with grants from other organizations or the government.

11.3.3 A GENERAL OVERVIEW OF THE ÖAR PROJECTS

Between 1980 and 1986 grants were given to a total of 127 projects. Of those, eleven were double grants given twice for the same project, and four projects failed, so that 112 projects remain for evaluation. In 1980 eleven projects were completed. Since then the number has grown continuously, reaching twenty-three in 1986. Table 11.3 shows the structure of the projects which have been completed thus far. From other sources approximately AS 230 million was invested in the 112 projects. Another AS 60 million in grants and subsidies was added by FER. That amounts to a grant quota of 26 per cent.

Eighty per cent of the projects are newly founded. In nearly every case the existing personal resources were insufficient and the investment risks too great for the realization of the project to have been possible without a grant.

The projects create an additional yearly net product (based on the figures from 1986) of approximately AS 140 million. Assuming that an additional AS 100,000 in additional net product is necessary to secure one job in farming

Table 11.3 Special grants for rural problem areas: the structure of sponsored projects

Sector/branch	*Number of sponsored projects*	
	absolute	*in %*
Agriculture (total)	43	38.5
Refining	16	14.2
Marketing	15	13.4
Cooperative production	12	10.9
Bio-energy projects	11	9.8
Tourism projects	37	32.9
Educational	12	10.7
Health (spas, resorts)	25	22.3
Skilled handicrafts	5	4.5
Business and industry	16	14.3
Cooperation	2	1.9
Founding business/takeovers	11	9.8
Job creation (initiatives)	3	2.6
Total	112	100.0

Table 11.4 Regional distribution of special grants

Area[1]		*No. of projects*		*No. of regional advisers*
		absolute	*in %*	
1	Mühlviertel	29	25.9	6
2	Waldviertel	25	22.4	5
5	Styria east and borderland	14	12.5	4
7	Lower Carinthia	14	12.5	0
4	Upper Styria	7	6.2	2
3	Eisenwurzen	6	5.3	2
9	Salzburg	6	5.3	0
	Upper Carinthia	3	2.7	1
	Eastern Tyrol	3	2.7	1
10	Northern Tyrol	3	2.7	1
11	Vorarlberg	1	0.9	0
6	Southern Burgenland	1	0.9	0
Total		112	100.0	22

[1] See Figure 11.1.

or tourism, and AS 200,000 for a job in industry, the ÖAR/FER have created or secured some one thousand jobs.

Table 11.4 shows the regional distribution of the sponsored projects. In reading this table two facts become apparent: most projects were sponsored in those regions where there is an active regional adviser (Mühlviertel, Waldviertel and Styria); the willingness to implement new and unconventional economic initiatives is greater in regions with an unfavourable economic situation than, for example, in western Austria.

11.4 CASE STUDY OF THE MÜHLVIERTEL

11.4.1 RELEVANT CHARACTERISTICS: NATURAL, SOCIAL AND ECONOMIC CONDITIONS

The Mühlviertel, forming the part of the province of Upper Austria which lies between the river Danube and the Czechoslovakian border, covers an area of 3,083 square kilometres, which is about a quarter of the total area of Upper Austria.

North–south valleys typify the landscape of this region, where roads were built early and business and trade developed. Most towns are located on hills more than 500 metres above sea level.

Naturally the Mühlviertel is woodland. About one-third of its area is forest. And although the ground is often suitable only for forestry, 62 per cent of the total area is used agriculturally.

The climate in general is rough and cold. The growth period of vegetation lasts between 190 and 230 days. Small and medium-sized farms are dominant in these typical highland areas, and the population has always depended on additional income. The oldest and most important industry is textile production. Another important source of additional income developed in the field of forestry. The third sector which has always played an important role is the transportation system. Other industries in the Mühlviertel include granite industry, paper mills, flour mills and breweries.

Since the middle of the nineteenth century there has been a steady change in two ways. The reduction of non-agricultural employment opportunities led to more intensive farming (to survive without additional income) on the one hand, and on the other to rural exodus and commuting. The limited scope of other industries, the unfavourable traffic network, and the fact that the Mühlviertel was occupied by the Soviets between 1945 and 1955 together have led to a substantial economic decline (Aistleitner, 1986; Leimlehner, 1974).

Administratively the Mühlviertel is divided into four districts which comprise 122 local authorities. However, the geographic conditions suggest a division of the Mühlviertel into a western and an eastern half. Traditionally we speak of the Upper Mühlviertel (west) and the Lower Mühlviertel (east).

Although the Mühlviertel covers about a quarter of Upper Austria, only 18 per cent of the population live in this region. The last census in 1981 recorded 229,425 inhabitants in the Mühlviertel (Upper Austria 1,269,540, all of Austria 7,555,338 inhabitants). There were no particular concentrations of population within the Mühlviertel. In 1981 there were only three local authorities with more than 5,000 inhabitants; another eleven had between 3,000 and 5,000.

Directly adjoining to the south, however, is the industrial city of Linz with about 200,000 inhabitants, a dominating factor for most of the Mühlviertel.

The population density ranges between 58 and 94 inhabitants per square km. The density increases with decreasing distance to the Greater Linz area. The highest population growth range in Upper Austria (1971/81), 16.4 per cent, was recorded in the district of Urfahr-Umgebung, the area surrounding Linz to the north. The districts of Freistadt and Rohrbach are below average. Notice that the growth rate in Freistadt and Rohrbach is a result of the relatively high birthrate, for both districts shows negative migration balances.

The working population of the Mühlviertel comprises 101,387 persons or 44.2 per cent, which is somewhat below average (Austria 45.7 per cent). A differentiation by economic sectors shows the important role of agriculture and forestry as compared with the Austrian average. The service sector on the other hand is substantially below average. The percentage of farmers in the Mühlviertel is nearly 18 per cent, in Austria as a whole only 8 per cent.

A similar picture is drawn by the social status of the working population: the percentage of blue-collar workers in the Mühlviertel is much higher than the average, while the percentage of white-collar workers is essentially below average. The share of self-employed and working family members is above average as a result of the great number of farms and small enterprises characteristic of the region.

Since the number of job-seeking people has increased more than the number of jobs, there has been a drastic increase in the number of commuters who go out of the district for their work. Nearly 70 per cent of the working population in the Mühlviertel commute, which means that 70 per cent of the working population have their jobs in a community other than their area of residence. Almost 12 per cent commute overnight, which is more than 4 per cent above the Austrian average. For 5.5 per cent of the commuters in the Mühlviertel it takes more than an hour to get to their job. The most important commuting centre is Linz, the capital of Upper Austria. Half of the Mühlviertel workers commute to Linz.

Wage earners in the Mühlviertel have the lowest average income compared to other districts of Upper Austria, with the districts Freistadt and Rohrbach at the bottom of the scale. The income in the district of Rohrbach is about one-fifth lower than the average income in Upper Austria.

There are about 17,600 farms in the Mühlviertel; more than 76 per cent are mountain farms. The situation of farmers is due to the great number of small farms. More than half of the farmers cultivate an area of less than 10 ha. Only about a quarter cultivate more than 20 ha. The dominant form of agriculture is determined by the existing conditions for production; pasture (43 per cent), followed by a combination of arable land and pasture (32.9 per cent) and forestries (12.6 per cent), while the pure crop-farms are very rare (11.5 per cent). The percentage of part-time farms increased from 30 to 60 per cent between 1970 and 1979 while those who earned their income from farming only decreased from 46 to 34 per cent.

There is a high percentage of commuters (84 per cent) among the part-time farmers. Only one in three employees has a job close to home, 45 per cent commute to the Linz metropolitan area. More than a quarter of the part-time farmers travel over 60 km to work every day. The overall situation for farmers is so bad that we can confidently predict a continuous shrinking process in the number of farms. In January 1987 there were 3,784 enterprises in the Mühlviertel, 13 per cent of all Upper Austrian enterprises. But the percentage of employed is only 7.2 per cent. There are far more blue-collar workers (66.5 per cent) than the Upper Austria average (55 per cent). In regard to firm size, small enterprises predominate: two-thirds of the enterprises employ fewer than five persons, and 80 per cent of the enterprises have between one and nineteen employees.

The average productivity in the Mühlviertel is distinctly below the Upper Austrian and Austrian average. At the same time, the costs for wages and salaries are also well below Upper Austrian and Austrian average standards.

Tourism is one of the prospective sectors for economic development in the peripheral regions. But its importance should not be overestimated. The

Mühlviertel belongs to the one-seasonal summer tourism regions in which tourism is declining. In 1986 lodgings declined by 6 per cent (Upper Austria 3.5 per cent). The Mühlviertel has attractive natural landscapes, health resorts, a nature preserve and the like to offer. The region is ideal for cross-country skiing and, to a lesser extent, for downhill skiing. It is ideal for outings.

The political structure of the Mühlviertel

The political picture in Austria is dominated by three political parties: SPÖ (the Social Democratic Party - Austria 43 per cent, Mühlviertel 34 per cent), ÖVP (the Christian Social Conservative Party - Austria 41 per cent, Mühlviertel 54 per cent), FPÖ (the National Liberal Party - Austria 10 per cent, Mühlviertel 7 per cent).

While the SPÖ is the strongest party in the Federal government, the Mühlviertel is dominated by the conservative ÖVP. The FPÖ plays a minor role. Ninety-five of 122 local authorities show an absolute majority for the ÖVP (OÖN, 7 October 1985). The influence of the SPÖ is higher in the suburban area of Linz. It is not unusual for commuters to be ÖVP members at their place of residence and SPÖ union members at work in Linz.

Associations and voluntary organizations often have a party political character as well. There is a chamber system with compulsory membership of a quasi-corporate type in Austria, which in fact has high political authority. Both the chamber of commerce and the chamber of agriculture are ÖVP-related. The dominance of agriculture in the Mülviertel and the fact that the chamber of agriculture is ÖVP-related indicate the ÖVP dominance in the region. There is a distinct block-thinking between the members of the two major parties, ÖVP and SPÖ.

11.4.2 SELF-RELIANCE MOVEMENT IN THE MÜHLVIERTEL

Historical overview

The initiation of self-reliance in the early 1970s differed between the Upper (Western) and Lower (Eastern) Mühlviertel.

1969: The infrastructure-oriented AKOM (Action Association of Upper Mühlviertel) was founded. Its success includes the building of a new hospital in Rohrbach and of a new bridge over the Danube in Rannaried. Its most important success, however, was within the conservative ÖVP: critique taken seriously by members of the same political party mobilized primarily mayors and other politicians in different local authorities. Members of AKOM were all-round engaged people. Therefore it is sometimes difficult to draw a line between what was AKOM and what not.

1975: The ÖBV (Austrian Association of Mountain Farmers) was founded. It originates largely in the Upper Mühlviertel and can be defined as agrarian opposition of small farmers and mountain farmers to the traditional agrarian policy and the big agrarian co-operatives (such as Raiffeisen or the

Dairy-Fonds). Pioneer projects were established by its members.

1978: Foundation of the Bergland-Aktionsfonds (BAF) by mountain farmers and urban people. Among this organization's major tasks are information work, scientific studies and practical project work.

Early 1980s: There was an increasing number of regional activities in the Mühlviertel (e.g. a peace festival and an exhibition called 'demonstration of new ways'). Supporters in the Upper Mühlviertel were mainly ÖBV farmers and the Catholic working youth (KAJ), whereas in the Lower Mühlviertel they were critical members of the Socialist Party (SPÖ), students and, to a smaller extent, ÖBV farmers. Committed individuals united to form the Mühlviertel-Aktiv group in Freistadt.

1980/1: The BAF sent the first regional advisers to the Mühlviertel. The first activities in the Upper Mühlviertel were supported by the ÖBV farmers, who searched for attractive marketing alternatives. Together with farmers and critical urban people they founded a producer–consumer co-operative, called MÜLI (*MÜ*hlviertel–*LI*nz, the first element being the name of the region and the latter the name of the capital of Upper Austria).

The first efforts by regional advisers in the Lower Mühlviertel were made in connection with the community project in Schönau, which was carried out by critical farmers (partly ÖBV) in search of agricultural alternatives. The approach of the regional adviser succeeded with the support of ÖBV farmers under the protection of local politicians (e.g. a mayor) and officials. The concrete result of their efforts was and is the direct sale of meat products.

1982/3: Foundation of the regional associations in the Upper and Lower Mühlviertel, VEROM (*V*erein zur *E*igenständigen *R*egionalentwicklung im *O*beren *M*ühlviertel) and FREI (Verein zur *F*örderung *R*egionaler *EI*gen-initiativen im Unteren Mühlviertel).

1983: Foundation of the ÖAR (*Ö*sterreichische *A*rbeitsgemeinschaft für Eigenständige *R*egionalentwicklung). Since a growing number of women and men from various disadvantaged regions have been working to realize the concept of a self-reliant regional development, they joined together in nine regional associations and founded the ÖAR.

Present state

There are two regional associations in the Mühlviertel:

VEROM

The association in the Upper Mühlviertel comprises about sixty members and thirty sympathizers who are actively informed. VEROM can be described as an open platform, a wide civic movement for activities in the region. Its supporters are to a great extent intellectuals who have moved to the region in search of a natural life. Catholic working youth (KAJ), ÖBV farmers and union officials act as promoters as well.

There is a variety of activities: VEROM publishes a periodical (four times/year), the *Saurüssel* (800 copies), which is the critical organ of public opinion in the region. VEROM also organizes seminars (e.g. 'Erdsegen' in 1987), discussions, informational events and excursions.

There is a rather loose connection between regional advisers and members of VEROM. The members of VEROM work relatively independently and emphasize ecology, culture and education, while the regional advisers counsel new economic projects. The objective of VEROM is to strengthen the critical potential in the region. VEROM strives for an increased socio-political consciousness in general.

FREI

The association in the Lower Mühlviertel comprises about forty members and a group of active sympathizers. Supporters of this open platform are students, social democrats, farmers (ÖBV) and employees of different occupations, all of whom are native to the region. Activities include discussion about special current topics (especially agrarian ecology) and cultural events (e.g. 'Mühlviertler Frühling' in 1986). There is an emphasis on expert-oriented work.

There is a close connection between regional advisers and members of FREI. The main objective and strategy of FREI is to activate the region via professional support and continued education.

11.4.3 REGIONAL COUNSELLING IN THE MÜHLVIERTEL

Between 1980 and 1986 twenty-five projects were subsidized in the Mühlviertel within the framework of the FER. This was 22 per cent of all FER subsidized projects in Austria. However, not all projects are subsidized by FER; some receive no subsidies whatsoever. The projects subsidized by the FER are just a small part of what happens in regional self-reliance. Regional counselling concentrates on four fields: agriculture, energy, tourism and trade and industry. In reality the contrasts are not so harsh: farmers produce heat from their waste wood and sell it to the community. Nevertheless, the traditional sectoral terms will be used in the following sections depending on where the focus of a project lies.

Agricultural projects

Principles and objectives

There is a great surplus production of milk, grain and meat. As a result, producers' prices remain constant while the investment and operational costs continue to rise. Most farmers try to increase their production, but an alternative is to convert to production of goods for which there is a greater demand. Farmers can process these either alone or together with others and market them directly. Three points are to be emphasized:

1. processing agricultural goods: a return of industrial functions, i.e. the production of bread and pastries, dairy products, wool, sausages, fruit-juices and the like back to agrarian communities;

2. direct marketing: there is a market niche for Farmer's Markets where farmers can sell their produced and processed products directly;
3. special production: there is a wide variety of possible alternatives in production such as the plantation of herbs, hops, special sorts of cereal (e.g. spelt), lambs, goats milk and honey. The objective is to produce high-quality products.

FER-subsidized products in the Mühlviertel

Between 1980 and 1986 fifteen projects were advised and subsidized by the FER, six of which belong to category 1, four to category 2 and five projects belong to category 3. The amount of FER subsidies totals AS 5.65 million. Together with other support and private capital stock, AS 12.84 million has been invested. The added value totals about AS 15 million (see Table 11.5).

Table 11.5 Subsidized projects, investment and added value (in thousands of AS)

	Investment				*Added value*
	FER	*Other subsidies*	*Capital stock*	*Sum*	
Category 1	1,574	73	1,691	3,338	2,683
Category 2	1,427	710	1,091	3,228	9,830
Category 3	2,650	384	3,241	6,275	2,473
Total	5,651	1,167	6,023	12,841	14,986

The efficiency varies between the different projects. The hops projects for instance require large investments at the beginning and a relatively long set-up period until the plants run at full capacity. The cereal-seed-dressers projects, on the other hand, have comparably low investment costs and a high net value. Direct marketing initiatives also show a higher range of efficiency.

The area of job creation must be seen in terms of income protection and improvement. New jobs can be created only in exceptional cases. But the crucial point in agriculture is to maintain the number of existing farms and ensure agricultural production in depressed rural regions. With those projects a number of farms can run as full-time farms again whereas before the farmers were part-time.

Promoters

Promoters of the initiatives are primarily ÖBV farmers, farmers who produce on a biological basis, and seldom officials of traditional agricultural interest groups (e.g. chamber of agriculture). Regional advisers gain access to farmers through the ÖBV.

So far there has been no or only inadequate counselling by traditional institutions for finding new ways for farmers. But there is an innovative potential among the farmers themselves who were willing to search for alternatives to increase their income. There is a great demand for innovation since small farms, low income and poor production conditions threaten the survival of many farmers.

There is also an increasing demand for products from organic farming by consumers and the catering trade.

Role and tasks of the regional adviser

- Regional advisers often have an advantage in expertise as compared with traditional institutes, because the latter are not specialized and trained in cooperative innovation development processes.
- They can refer to well functioning pilot projects in other regions.
- Counselling is oriented to needs and is always given wherever initiatives are taking place.
- The FER is one of the few special subsidies in agriculture for peripheral regions.

Conclusions and outlook

- The initiation of small community projects provides experience and a good basis for bigger projects later on.
- Producer–consumer co-operatives of farmers and urban people failed owing to structural problems. Direct marketing in the form of farmer's markets, on the other hand, is easier to organize and carry out.
- The target group of ÖAR and regional associations is not only the agrarian opposition leaders but has been expanded to reach traditional farmers as well.
- Biases of the traditional agricultural officials are slowly diminishing.
- The willingness of farmers to try alternative methods of production is increasing as they realize the limits of the present economic system.

Energy projects

Principles and objectives

Austria depends heavily on the use of fossil fuels for energy, causing alarming damage to health and environment. Two oil crises demonstrated its dependence on supply monopolies. Therefore a decentralized, renewable, ecological and economical power supply for the region seems advisable. There are three types of projects which have proved successful:

1. Chopped waste wood. In a jointly run heating plant, chopped waste wood from the farmers' forests is burned, and a supply line then distributes the heat to private and public buildings. Thus, the money spent on heating remains within the region and the farmers have the opportunity to earn some additional income.
2. Solar energy. Some villages have started building solar panels themselves. The technical know-how is provided in jointly organized lectures and presentations. The finished panels are then installed by experts.
3. Biogas from manure. Liquid manure is transformed into biogas which in turn is used to produce electricity and heat. This system has not quite paid for itself yet. In future, however, biogas will be an environmentally sound source of energy contributing to the independence of rural areas.

FER-subsidized projects in the Mühlviertel

In 1985/6 four chopped-wood projects were supervised and subsidized by the FER. The amount of FER subsidies totals about AS 2 million, the total investment nearly AS 10 million. The added value is more than AS 3 million. The result has been a reduction in costs and a greater amount of self-help. For the participating farmers these projects provide a new source of income (see Table 11.6).

Table 11.6 Wood projects, investment and added value (in thousands of AS)

	Investment				*Added value*
	FER	*Other subsidies*	*Capital stock*	*Sum*	
Five chopped-wood projects	2,250	2,497	5,046	9,783	3,265

There is also a solar-energy project which has been supervised but not subsidized. In 1986 more than 125 collectors were organized and installed. The collector area comprises about 1,400 square metres. A biogas project is still in the initial phase.

Promoters

Farmers and officials joined together to form co-operatives for chopped-waste-wood projects. People of different background who shared an interest in energy technology supported solar-energy projects since it is a good investment for houseowners. There is overall public interest in natural energy among a great part of the population.

Role and tasks of the regional adviser
They have an advantage in expertise and can offer efficient counselling wherever needed. The organization of solar-energy projects – from the joint purchase to the final instalment – is carried out by the regional adviser.

Conclusions and outlook
Chopped-wood energy as innovative pilot projects as well as solar energy projects have proved successful. Meanwhile regional advisers are acknowledged for their expertise and competence.

For decades we have been exploiting our limited energy sources such as oil, coal and natural gas. These fuels are burnt in huge plants from which unfiltered emissions are often released, thus polluting the air. Furthermore, the transport of energy over long distances results in high energy losses. Air pollution and dying forests are just two of the results of our carelessness. If our future is to be worth living, we will have to learn from our mistakes.

This means: energy consumption is to be reduced to a minimum; alternative and environment-oriented energy production; small power plants instead of large, central ones; avoiding waste of energy.

Are coal and oil really the only energy sources for heating in the rural areas? Specifically in these regions renewable energy sources are in sufficient supply. Examples show that there are new, responsible means of decentralized energy production.

Tourism-related projects

Principles and objectives
There is a structural change in tourism within Austria: a concentration of tourism in the two-seasonal regions in the western Alps and at the same time a decline in most other regions. Tourism should be an integral element of the regional economy. There are also alternative forms of tourist development that can be put into practice at reasonable cost, although it is wrong to see tourism as a substitute for regional development. The following alternatives, however, do show particular promise: soft tourism (landscape oriented); special offers (e.g. sports, action, health, recreation); offers for certain target groups.

FER-subsidized projects in the Mühlviertel
So far two projects have been realized in the area of educational and cultural tourism and one recreation project is now in its test phase. AS 4.8 million has been invested with an added value of nearly AS 2.2 million (Table 11.7). The FER subsidies amount to almost AS 2 million. There has been just a small effect on direct job creation but there are indirect results in the regional economy (e.g. gastronomy) which must not be forgotten. In the long run there will be an increasing demand for recreation in environmentally intact regions.

Table 11.7 Tourism projects, investment and added value (in thousands of AS)

	Investment			Added value
FER	*Other subsidies*	*Capital stock*	*Sum*	
1,993	906	1,896	4,795	2,185

Other related projects

The tourism association for the Mühlviertel is mainly concerned with marketing and public relations, but there have been no efforts to create new offers for alternative forms of vacation. In the future it will be important to coordinate the activities of different local authorities more effectively.

Promoters

Promoters of the projects are, on the one hand, traditional tourist organizations, local politicians and officials, as well as other sectors of the economy. On the other hand, there are new initiatives which have little in common with the traditional view of tourism. There exists a potential for conflict as it seems the traditional groups are not yet ready for innovative ideas. This can be seen as particularly characteristic of peripheral regions.

Role and tasks of the regional adviser

Counselling is innovation-oriented. There is pressure on tourism officials to offer activity programmes according to new demands. Therefore officials are open to and welcome new ideas. Counselling therefore concentrates on subsidizing paths for such projects. Since there is not sufficient counselling from the traditional establishments it is a great opportunity and the responsibility of the regional adviser to work in this direction.

Conclusions and outlook

Regional advisers find new alternatives and thereby enrich the variety of possibilities a region has to offer tourists. Good projects earn acknowledgement for loosening rigid attitudes toward tourism. There is good potential for teamwork with officials in tourism associations. The best perspectives lie in interregional cooperation, the development of new region-oriented programmes (offers) and well planned marketing. A difficult but very important task lies in the creation of the right mentality of the population towards tourism.

The tourists have high expectations regarding the quality of the offer. The projects have to provide this quality, but at the same time they have to preserve their consciences and their independence as well as demand respect towards their traditions. Apart from counselling projects the regional advisers have the task of supervising this process of learning and in the first place making it possible.

Trade and industry projects

Principles and objectives

In general there is very limited development of trade and industry in the Mühlviertel. In some areas there is even a decline. Unemployment is a growing threat for many people, especially in the peripheral regions. But there is a possibility of strengthening and improving the existing economic structure. Of course the establishment of new enterprises has caused considerable problems. Yet it is possible with the cooperation of advisory and promotional institutions such as the ÖAR, ÖSB (Österreichische Studien- und Beratungsgesellschaft) and the Labour Market Promotion Organization (Arbeitsmarktförderung AMF). However, in the special case of the Mühlviertel, where everything is dominated by the adjoining industrial centre of Linz, the prospects for the development of these sectors are rather poor. There are only a few exceptions.

The main objectives of promotion therefore are: the cooperation of different enterprises in the region; setting up of co-operatives; and directed transfer of information and expertise to the region.

FER-subsidized projects in the Mühlviertel

Three projects have been implemented with a total investment of AS 3.3 million and an added value of AS 3.8 million (Table 11.8). FER subsidies amount to AS 967,000. The Chico Hamocks co-operative has been in existence since 1985 and has a turnover of about AS 3 million. The project provides seven jobs, some of them part-time. The Freiwald workshop, which produces furniture and is also a leather manufactory, employs eight, some of them also part-time. The third project is a self-managed and self-owned company of seven persons who produce 100 per cent wood furniture with a special design.

Table 11.8 Entrepreneurial projects, investment and added value (in thousands of AS)

	Investment				*Added value*
	FER	*Other subsidies*	*Capital stock*	*Sum*	
Three projects	967	851	1,465	3,283	3,791

Promoters

Promoters are critical Catholics, social workers and unemployed. So far traditional entrepreneurs have not become involved in this new type of enterprise. Overall there are few new activities in the area of trade and industry in the Mühlviertel.

Role and task of the regional adviser

The access of the regional adviser is difficult because he does not have the reputation of belonging to a competent counselling organization on trade and industry. There are no specific strategies for producing project promoters who are qualified leaders and have technical skills as well. So the realization of new ideas finds little response. But there is an effort on product development.

Conclusions and outlook

The transformation of enterprises on a self-management basis by the labour force is problematic. It requires a long and careful period of preparation of the projects. A concrete concept of the product and project is essential for success as is a homogeneous group of promoters. In the future specific consideration must be given to job creation because we expect many workers to be laid off from the adjoining nationalized industries in Linz within the next few years. Many previous commuters will then be looking for work in their own residential areas.

11.4.4 OTHER NEW COUNSELLING ARRANGEMENTS

Besides the regional advisers who concentrate primarily on the effects of economic projects on the regional added value, there are other kinds of adviser as well.

Labour-market advisers

Since the labour-market administration is confronted with high unemployment it was necessary to find new instruments for employment policies. Public intervention should help mainly those who are affected most by the labour-market crisis. The long-term unemployed and youths without any job experience do not meet the expectations of employers. Therefore experimental labour-market policies have been developed. And although there are still some precarious factors it has been proved that an experimental labour-market policy is more innovative and shows better results in some areas than the more conventional forms (Lechner and Reiter, 1987). Because of the change of strategy of the governmental policy, whose highest goal is to diminish the budgetary deficit, those experimental initiatives of labour-market policy look as though they may disappear.

One instrument of experimental labour-market policy is the appointment of regional advisers. The Mühlviertel belongs to those regions which have the advantage of a labour-market attendant. This is a trained person who carries out developmental work in social matters and looks for labour-market promotion. A labour-market attendant works on behalf of the Ministry of Social Affairs as part of the labour-market service. The advantage of this arrangement is the mobility, so work can be done where needed, either in the region concerned or in the capital city. Right now there are some projects in their developing phase, for this is a very difficult area (BMfSV, 1986).

Experimental farm projects

This project is an initiative of FREI and is financed by the Ministry of Social Affairs (carried out by the Provincial Labour Office, LAA). The objective is to create long-term jobs for the long-term unemployed. The objective is also to find out under which circumstances this is possible in the area of agriculture. For this reason the project is experimental.

Unemployed youths are being employed and trained on selected farms. The employers are the farmers, but the assistance and coordination is taken care of by a project adviser. The project plan is set up as follows: a probation period of two months (paid for by the LAA); sixteen months' employment (costs are split between the farmer and the LAA); a break for two months in the winter for an educational course. On average ten unemployed youths are working in the context of this project. There are no final results yet since the project has not ended; but after one year of experience some problem areas can be identified: it is difficult to define the occupation of an agricultural labourer, for this is a dead occupation. Traditionally labourers were stablemen and did not have high status. Also, the Regional Labour Office primarily sends socially inadequate persons (according to para. 16 of the Labour Market Promotion Law) and it is very hard to create long-term jobs for them. The basic expectation is that the farm contributes substantially to the preparation of the unemployed for integration to a labour market which is not subsidized.

Association for Regional Culture (ARGE Region Kultur)

ARGE Region Kultur is an association of initiatives and models of base culture. Its objective is to promote innovative folk-culture as well as regional and international interrelations of ethnic diversity. Activities cover areas all over Austria in the fields of cultural work and adult education (national education) with an emphasis on depressed economic regions. The working team organizes seminars, carries out scientific projects (e.g. village analyses) and cooperates with regional counselling (ÖAR) in the economic domain.

The ARGE Region Kultur and ÖAR work together, with the ÖAR stressing the economic dimension and the ARGE Region Kultur emphasizing the cultural dimension. The latter has not yet developed to the same extent as the former, but it is crucially important for a comprehensive regional development.

Farmers' wives work for their autonomy (a BAF project)

This project also represents the socio-cultural dimension of regional self-reliance. Two (female) counsellors are employed (part-time) to encourage the women to strive for self-confidence and more independence. The members of the group meet once a month, where they have the opportunity to get together with other farmers' wives, talk about their experiences and about their life on the farm and at work with various topics (e.g. health, family, agriculture, environment). They organize workshops and discuss various possibilities of changing their position as women in the family and in society.

Problems arise with traditional (farmers') wives organizations such as those of the Chamber of Agriculture or of the Catholic women's movement.

The social pressure upon this new group is quite strong since it is not yet accepted that farmers' wives (or even women in the countryside in general) should be able also to live their own lives.

11.5 CONCLUSIONS

It is very difficult and frequently also unsatisfactory to measure the success of local development initiatives in terms of net product figures. Furthermore, the net product produced by the FER-sponsored projects only hints at the effect the much larger number of other initiatives have on the economy. If one considers the effect on the economy created by projects which are not sponsored but which cooperate with the local initiatives and the ÖAR, then one can figure on 400–500 jobs in this relatively small region. Compared to other regions this is a positive trend. Since a majority of the projects in the Mühlviertel have to do with agriculture and they produce marketable goods rather than surpluses, one can assume that all of the governmental funding for these projects was financed with the savings they allow in the budget for the sale of surpluses. Thus, the jobs created in agriculture by local development programmes cost the public next to nothing.

Aside from the economic and fiscal effects, it is important to consider the qualitative changes:

- All of these elements of internal restoration have broad effect.
- The wide range of activities of the regional platforms/centres for development is resulting in a growing sense of regional identity and concern in which more and more people see themselves as the inaugurators of their cooperative development.
- These projects create many new and highly qualified jobs in the areas of consultation, organizational development and marketing. This opens the door for young, well-trained people originally from the Mühlviertel to return to their home communities.
- The many new local and regional initiatives help create new cultural life in these regions helping to make them more attractive to young, qualified people as a place to live.
- Many of the initiatives promoted by the regional associations VEROM and FREI act as an effective impetus to reorganize the traditional professional and political organizations. In the vicinity of the Chamber of Agriculture and agriculture cooperatives it is possible to see a definite, yet conflict-filled and slow process of change.
- The network of projects and initiatives in the regional associations and their public relations and information efforts serve as an example of a creative and open climate of innovative competition.

The prerequisites for this indigenous development process are:

- committed regional catalyst groups that are open to members of all professions and political parties;
- a certain degree of institutionalization of the regional catalyst groups so that they can develop a minimum of organizational authority (i.e. for cooperative programmes, the production of a newspaper, public relations work, projects etc.). However, at the same time, the group must be loosely knit to prevent them from appearing as some kind of parent organization for the many diverse initiatives;
- the fundamental principle of abstaining from exercising authority over the members of the group or other initiatives;
- the success of such a regional group depends on its ability to include the widest possible range of potential actors for regional development;
- access to advisory expertise and financial assistance for the regional initiatives. The catalyst groups should not, however, have these assets in their own possession. This could very quickly lead to distribution problems and the emergence of an exclusive or elite bureaucracy;
- the ÖAR's organizational structure which respects the above-mentioned principles. Its regional advisers are at the disposal of initiatives for consultation without being employees of the regional organizations;
- a regionwide network and access to knowledge and experience at home and abroad;
- access to governmental planning and subsidizing institutions.

The experience gained from the activities in favour of indigenous development in the Mühlviertel shows a variety of unresolved problems:

- The regional exponents of an indigenous development cannot be seen as a homogeneous group. There is a potential of innovative/socially oriented entrepreneurs, ambitious officials and politicians as well as a wide range of new social, cultural and environmental movements. Very often there are fundamental differences of opinions and views. There is a danger that this might lead to polarization instead of joint activities.
- This danger is increased by the lack of tradition of a regional policy of alliances in the Austrian system, which is dominated by the chambers and the major political parties in general and the rigid political and cultural situation in rural areas in particular.
- In Austria the term 'region' is not a real criterion for economic and political activities. People are either oriented towards their company or project, towards their village or community, or towards interregional

organizations (party, church, clubs). The question is whether the *region* can be a decisive and identity-creating principle of organization (as is the policy pursued by the *self-reliant development* movement).

- An important characteristic of indigenous development is the fact that regional proponents - instead of exterior forces (e.g. central political authorities or experts) - are responsible.
- The nomination of professional regional advisers who are paid by the state as well as publicly sponsored projects may paralyse the force of initiative of the population.

Regional (extragovernmental) development organizations like the ÖAR in their function of intermediate (transmissive) organs have to satisfy the demands of the public sponsors and the regional platforms of indigenous development, which are in many regards incompatible. They run the risk of becoming regionalistic lobby-organizations; of becoming semi-official bureaucracies with external offices in the regions; and of becoming publicly or privately financed consulting agencies.

An important condition for successful cooperation between local and regional development initiatives, experts, advisers and public financiers is that indigenous development in the disadvantaged regions is jointly financed by the federation, the regions and the communities. This way dependency on a sponsor as well as political discussions between the parties can be reduced.

Nevertheless, in spite of the installation of the Grand Coalition at the beginning of 1987, the situation for such inter-regional cooperation is not very favourable.

REFERENCES

Aistleitner, J. (1986), *Formen und Auswirkungen des bäuerlichen Nebenerwerbs*, Innsbruck: Das Mühlviertel als Beispiel, Selbstverlag des Instituts für Geographie der Universität Innsbruck.

Bernfeld, A., Butschek, F., Kutzenberger, E., and Neidl, R. (1979), *Arbeitsmarkt-entlastende Massnahmen am Beispiel Oberösterreichs*, Linz: Trauner Verlag.

Bratl, H., and Scheer, G. (1987), *Regionalbetreuung und Förderung innovativer Wirtschaftsprojekte für eine eigenständige Regionalentwicklung in benachteiligten Gebieten Österreichs*, Vienna: Österreichische Arbeitsgemeinschaft für Eigenständige Regionalentwicklung.

Bundesministerium für Soziale Verwaltung (ed.) (1986), *Arbeitsplätze selbst schaffen*, Vienna.

Bundesministerium für Soziale Verwaltung (ed.) (1984), *Forschungsberichte aus Sozial- und Arbeitsmarktpolitik*, no. 6, Vienna.

Bundesministerium für Soziale Verwaltung (ed.) (1987), *Arbeitsmarkt*, no. 1, Vienna.

Butschek, F. (1985), *Die Österreichische Wirtschaft im 20. Jahrhundert*, Stuttgart: Fischer Verlag.

Gerdes, D. (1981), 'Dimensionen des neuen Regionalismus in Westeuropa', in *Österreichische Zeitschrift für Politikwissenschaft*, Vienna.

Glatz, H., and Scheer, G. (1987), 'Regionalpolitische Strategien gegen die Krise. Neue Wege oder Kosmetik am traditionellen Instrumentarium?', in *Geographische Rundschau*, Munich.

Herzog, H., and Tödtling, F. (1982), *Versuch einer Einschätzung der österreichischen Regionalpolitik für periphere entwicklungsschwache Regionen* (Diskussionspapier), Vienna.

Höpflinger, H., and Wagner, M. (1984), 'Experimentelle Arbeitsmarktpolitik', in *Österreichische Zeitschrift für Politikwissenschaft*, Vienna.

Institut für Höhere Studien (ed.) (1985), *Analyse der Zielerreichung und Effektivität der Berggebietssonderaktion*, Vienna.

Kammer für Arbeiter und Angestellte für Oberösterreich (ed.) (1987), *Arbeitsmarkt-information*, 31 March.

Kammer der Gewerblichen Wirtschaft für Oberösterreich (ed.) (1986), *Oberösterreich in Zahlen*, Linz.

Lackinger, O., and Kutzenberger, E. (1984–5), *Die Bevölkerung Oberösterreichs – Volkszählung 1981; Beiträge zur oberösterreichischen Statistik*, vols. 1 and 2.

Landesarbeitsamt Oberösterreich (ed.) (1986), *Jahresbericht 1985*, Linz.

Lechner, F., and Reiter, W. (1987), *Aktion 8000 – Eine quantitative Evaluation aktiver Arbeitsmarktpolitik*, Vienna: Institut für Wirtschafts- und Sozialforschung.

Leimlehner, E. (1974), *Das Kriegsende und die Folgen der Sowjetischen Besetzung im Mühlviertel, 1945–1955*, Zürich: Juris Verlag.

Oberösterreichische Gebietskrankenkasse (ed.) (1987), *Gemeindestatisktik*, 1/1987.

Oberösterreichische Nachrichten, 7 October 1985.

Österreichisches Institut für Raumplanung (ed.) (1985), *Gebietsprofile der politischen Bezirke Freistadt, Perg, Rohrbach und Urfahr-Umgebung*, Vienna.

Österreichische Raumordnungskonferenz (ed.) (1984), 4. *Raumordnungsbericht*, Vienna.

Österreichisches Statistisches Zentralamt (ed.) (1987), *Statistische Nachrichten*, 42. Jg., 3/1987.

Reith, J. (1984), 'Kooperative Regionalpolitik und Hilfe zur Selbsthilfe – Ansätze in Österreich' (lecture to the Landesarbeitsgemeinschaft Bayern der Akademie für Raumforschung und Landesplanung in Coburg, Bavaria).

Scheer, G. (1986), 'Eigenständige Regionalentwicklung: Ursachen für das Entstehen eines neuen Konzeptes', in *Eigenständige Regionalentwicklung Verein zur Förderung der Eigenständigen Regionalentwicklung in Hessen* e.v., Hessen.

Schmidt, E. (1986), *Agrarpolitik 1983/1986*, Vienna: Bundesministerium für Land- und Forstwirtschaft.

Szopo, P. (1986), *Subventionen in Österreich – Im Auftrag des Bundesministerium für Finanzen*, Vienna: Österreichisches Institut für Wirtschaftsforschung.

Zobl, A. (1987), 'Project on local employment. Development in a rural framework', manuscript by order of the OECD, Linz.

CHAPTER 12

Towards a definition of the manoeuvring space of local development initiatives: Italian success stories of local development – theoretical conditions and practical experiences

Roberto Camagni and Roberta Capello[1]

12.1 THE 'DEVELOPMENT-FROM-BELOW' APPROACH

Before we examine the specific situation of Italian success stories in local development, it is worth putting forward a short overview of the theoretical approaches which underline the most important causes of local development in the new industrial regions.

Many theories have dealt with the problem of regional development from different points of view: the major discrepancy is between the 'endogenous' and 'exogenous' approach, that is to say between the idea that propulsive causes of development are inter-regional movements of capital and labour, and the idea that endogenous resources are the main driving forces of development. This latter approach, recently evolved into what is known as the 'development-from-below' model, seems to be the right one during years of reduced spatial mobility of both capital and labour (Stöhr and Tödtling, 1977).

This second approach focuses on the internal dynamism of an area and its ability to use its internal resources. With such a theory it is possible to interpret growth-rate differentials on the basis of two different sets of elements: 'subjective elements', such as local entrepreneurial skill, and 'objective' ones, such as different kinds of locational advantages. The presence of both elements is a prerequisite for the establishment of local firms and for the sustaining of regional growth.

The existence of just one of them will tend to cause firms to go elsewhere; in fact, the presence of locational advantages in an area without local entrepreneurial capabilities will pull external firms into the area, and the contrary will be true if we just find entrepreneurial capability in an area (Cappellin, 1983).

It is very interesting to study which variables affect these two elements influencing the development of an area. 'Objective' elements are determined by availability and cost of skilled factors, local demand and supply of raw material: in more general terms by local productivity levels. 'Subjective elements', on the other hand, are linked to the presence of local entrepreneurship and to the local capability of shifting resources from traditional to innovative uses. This capability in its turn comes from a host of structural characteristics of the local socio-economic milieu: structural diversification, the presence of small autonomous firms in agriculture or trade, the presence of a longstanding general mentality of 'self-help' and possibly of some local variations of Weber's Protestant ethic.

These 'subjective' features are deeply rooted in local societies and concern their structural characteristics. The problem with employing them alone in the interpretation of local development is that of a certain level of circular reasoning ('there is industrial development because there is entrepreneurship') and that it leaves aside the problem of 'why now and not before?'. Matching together subjective and objective elements may, however, lead to a better understanding and a sounder explanation of time-specific local success stories.

12.2 CHARACTERISTICS OF RECENT ITALIAN REGIONAL DEVELOPMENT

During the 1950s, the Italian economy was characterized by a clearcut dualism between industrially developed regions in the north-west of the country and backward, mainly rural areas in the south, centre and north-east. The former were characterized by the presence of capital-intensive production and large firms, while the latter were devoted mainly to handicraft production.

During the 1960s, and particularly during the 1970s, the spatial situation changed and nowadays the strict dichotomy between rural and industrialized areas is becoming blurred. During the economic crisis of the late 1960s and 1970s two new elements emerged, namely the crisis of large firms and the extraordinary dynamism of medium-sized and small ones, which became the most important factor in the accumulation process and in regional development.

During the 1970s, north-eastern and central regions became the propulsive elements of Italian economic development, thanks to the primary role played by small and medium-sized industries. At the beginning of the 1980s, the Italian economy was structured according to a threefold pattern, that is to say the north-western old industrial regions, the north-eastern and central fast-developing 'newly industrialized' areas and the southern ones.

The model of production organization in newly industrialized north-eastern areas of the so-called 'Third Italy' is completely different from that of central areas. In fact production is organized in 'system areas', characterized by sectoral specialization, physical proximity of firms and a non-metropolitan or mainly rural environment.

The homogeneity of such a productive structure in restricted geographical areas guarantees the achieving of rapid technological innovation and high degrees of labour skill, allowing at the same time benefits from scale economies at the district level and productive flexibility (Del Monte, 1982). For the sake of an acceptable simplification, Italian regions can be aggregated in four macro-areas, which have similar economic, social and industrial characteristics:

1. the north-western area (NW), which is constituted by regions such as Lombardy, Piedmont, Liguria, Valle d'Aosta and represents the old industrial area of the country, where large firms have been locating for many decades, and industrial activity is characterized by high productivity levels;

Map 12.1 Italian regions.

2. the north-eastern area (NE), encompassing regions such as Trentino, Friuli, Veneto and Emilia-Romagna, where the industrial fabric is characterized by the presence of new small and medium-sized firms and where the production is based on labour-intensive sectors and on spatial specialization.
3. the central area (C), encompassing Tuscany, Umbria, Marche and parts of Lazio, which are regions similar to the north-eastern ones, but more recently developed (with the exception of Tuscany);
4. finally, the southern area (S), geographically extended to all other regions (Abruzzi, Molise, Campania, Puglia, Basilicata, Calabria, Sicily and Sardinia), which contains the poorest areas of the country, with a low degree of productivity and a dichotomized industrial base: mainly big-branch plants of external firms and handicraft production.

12.2.1 SUCCESS INDICES OF PERIPHERAL REGIONS

The most important success stories of local development in the Italian economy took place in the 1960s and 1970s, mainly in the north-east and central regions. In this period a new pattern of industrial organization was starting to show and at the same time the old industrial regions were becoming characterized by the emerging weakness of concentrated, mainly metropolitan development.

The early sections of this chapter give an outline of the structural change taking place in the Italian regions during the 1960s and 1970s and underline how locational advantage shifted during these decades in favour of newly industrializing regions.

A first index of structural change is the share of agricultural employment (Table 12.1), which shows the rapid change in NECs (north-east–central regions) after 1960, with some two decades' delay with respect to north-western regions. To measure the superior performance of these areas, three different elements are shown: the relative growth rate of regional value added and employment, productivity growth, the number of employees in small and

Table 12.1 Share of agricultural employment in selected regions (%)

Regions	*Macro-regions*	*1961*	*1971*	*1981*
Veneto	(NE)	29.3	16.0	10.7
Emilia Romagna	(NE)	33.2	21.2	13.7
Tuscany	(C)	25.5	13.1	9.4
Marche	(C)	49.6	30.5	18.3
Abruzzi	(S)	48.9	31.6	19.7
Lombardy	(NW)	11.9	5.8	3.8
Piedmont	(NW)	24.1	13.1	8.9
Italy		29.7	18.4	12.8

medium-sized firms and a migration index. The results are presented in Tables 12.2–12.5. All these tables show that the reduction in regional disparities took place through the last growth of NEC regions, and in particular through their vital milieu of small firms.

So far as the balance of migration with other regions and abroad is concerned, the interesting element which emerges is that in the second half of the 1960s the sign of the movements in the NECs becomes positive (where the balance is calculated as the difference between the number of immigrants and the number of emigrants), some years after the economic takeoff.

The final success index of peripheral regions is the growth rate in industrial productivity of those years; it shows once again the better performance of parts of the Third Italy (NE) during the 1960s and the early 1970s as compared with the behaviour of old industrial regions such as Lombardy.

Table 12.2 Growth rate of value added in selected regions (constant prices) (%)

Regions	*Macro-regions*	*1961-6*	*1966-71*	*1971-6*	*1976-81*
Veneto	(NE)	16.0	28.5	21.4	15.8
Emilia-Romagna	(NE)	13.1	23.7	24.8	17.5
Tuscany	(C)	11.5	26.6	16.5	15.6
Marche	(C)	11.6	26.9	24.6	17.8
Abruzzi	(S)	9.9	33.6	21.9	16.5
Lombardy	(NW)	11.7	27.9	14.5	16.9
Piedmont	(NW)	9.7	28.5	19.8	8.0
Italy		12.7	27.9	16.8	14.8

Source: *Annuario di Contabilità nazionale*, 1976, 1981; *I conti economici regionali*, 1963, 1966, 1971.

Table 12.3 Growth rate of non-agricultural employment (%)

Region	*Macro-region*	*1961-71*		*1971-81*	
		Tot.	*SMF*	*Tot.*	*SMF*
Veneto	(NE)	13.7	42.0	10.3	27.0
Emilia-Romagna	(NE)	11.8	35.0	12.4	65.0
Tuscany	(C)	7.3	22.6	10.9	20.0
Marche	(C)	15.7	42.0	16.5	40.0
Abruzzi	(S)	20.0	37.8	20.9	18.0
Lombardy	(NW)	10.2	26.9	7.2	7.5
Piedmont	(NW)	11.4	12.8	8.8	10.7
Italy		10.3	22.5	13.3	10.0

SMF = small and medium-sized firms of fewer than 100 employees.

Table 12.4 Balance of migration with other Italian regions and abroad

Region	*Macro-region*	*1961*	*1966*	*1973*	*1980*
Veneto	(NE)	−34,243	−11,360	8,282	7,232
Emilia-Romagna	(NE)	1,850	−8,698	15,052	14,004
Tuscany	(C)	8,842	2,327	13,987	13,814
Marche	(C)	−16,617	−9,936	214	4,548
Abruzzi	(S)	−24,543	−11,360	906	2,512
Lombardy	(NW)	110,353	18,437	39,832	6,529
Piedmont	(NW)	91,855	25,508	24,383	2,250
Italy		13,986	−115,365	52,915	38,096

Table 12.5 Annual growth rate in industrial productivity (%)

Region	*Macro-region*	*1963-6*	*1966-71*	*1971-6*	*1976-81*
Veneto	(NE)	4.0	4.9	4.2	3.5
Emilia-Romagna	(NE)	3.4	5.7	6.1	2.8
Tuscany	(C)	6.3	4.8	4.0	2.9
Marche	(C)	9.7	5.7	3.9	3.5
Abruzzi	(S)	1.9	7.9	4.0	2.3
Lombardy	(NW)	5.6	4.0	3.1	3.9
Piedmont	(NW)	4.8	4.1	4.2	4.5
Italy		5.7	4.7	3.5	2.7

12.3 AN INTERPRETATIVE MODEL OF PERIPHERY DEVELOPMENT

One of the main features of the longstanding debate on the Third Italy miracle may be seen, in our view, in its mainly descriptive and *ex-post* nature. Interpretations of local success stories have so far been mainly based on micro-economic and micro-spatial elements (transaction costs, local synergies etc., as we shall see later), whereas we think that a more general theoretical framework is needed, capable also of considering some macro-economic preconditions for development. In particular, according to our view:

(a) An interpretative model should be a *general spatial equilibrium model* which takes into consideration the general interdependence of economic phenomena in the territory. In other words, it is of real interest to analyse the periphery's development within the general framework of inter-regional development and in comparison with the relative performance of advanced regions (Benedetti and Camagni, 1983).

(b) This spatial interdependence comes from both micro and macro elements: the path of the spatial diffusion of innovation and the exchange rate and balance-of-payments constraints. In this last respect, we have to remember that the situation of national trade balance derives from the sum of the single *regional* trade balances, and these represent in synthesis the competitive situation of the different regions and their role in the inter-regional division of labour.

During the 1970s, the general weakness of the Italian trade balance depended mainly on the particular situation of rigidity and conflict found in the large national firms located in the north-west. The consequent devaluation of the 1970s had a positive effect on all regions, but for two reasons peripheral areas received a greater relative advantage: first, advanced regions showed diffused constraints in expanding physical production, and secondly, the production of peripheral areas is mainly labour-intensive and in the medium run the effect of a devaluation is to reduce the relative cost of internal labour.

In fact, variations in the exchange rate heavily influence the economic structure of a country. As the 'Scandinavian' model of exchange-rate variation suggests (Onida, 1980), a devaluation in the real exchange rate helps the development of 'tradeable' sectors in comparison to what happens to the 'non-tradeable' ones, and particularly of those characterized by the highest price elasticity of demand.

(c) Another necessary characteristic of our interpretative model should be the dynamic nature of the approach, in order to grasp all the relevant feed-back effects. Such an approach in fact could supply us with a rationale for the so-called 'regional life-cycle' model: structural change brought about by true development processes may in the short run strengthen them through synergy and scale effects, but in the medium and long run it is likely to destroy previous preconditions of success, obliging regions to rely upon new and different success factors (Camagni, 1984; Camagni and Cappellin, 1985).

Industrial advantages in factor availability and costs rapidly vanish: local products get older, spatial agglomeration economies may turn to diseconomies. Hopefully, but in a less predictable way, crisis conditions may also stimulate revitalization processes, through a reverse path: less labour conflict, reduced urban activity density, lower rents and factor costs, etc. (Camagni and Gibelli, 1986; Gibelli, 1986).

12.3.1 THE MODEL

A first, simplified attempt to build such a model, taking into consideration both spatial interdependence and feedbacks that take place over time, may be undertaken through the concept of 'regional comparative locational advantage'. It is in fact a 'relative' concept, which may account in synthesis for all elements that constitute the competitiveness of each local economy, and which may be measured empirically by comparing two aggregate indices: an index of *local productivity*, encompassing elements like the performance of private enterprises, effectiveness of local infrastructure and social overhead capital, quality of the labour force and presence of entrepreneurship, and an index of the general level of labour costs, or of what Marxists used to call the *reproduction cost of the labour force*. We are in fact interested not just in the relative level of official contractual wages, but in a more general measure of purchasing power and real wellbeing of the labour force, coming from black and part-time job availability, industrial-agricultural job sharing and the like.

In a formal way, the regional locational advantage may be derived as follows. Let us suppose that product prices are the same throughout the country and that they are fixed through a mark-up (1 + p) on direct costs which differ among regions; if n means the nation and r the regions, we have

$$P_n = (1 + p)_r (W_r / X_r)$$

where:
P_n is the general national price level;
W_r is regional wage level, defined in the broad sense as stated before;
X_r is regional productivity level, also defined in a broad sense;
W/X is unit labour cost; and
p is the rate of gross profit.
Rearranging the elements, we find that a regional locational advantage, or a positive profit rate is found only where $X_r > W_r/P_n$; in relative terms a higher than average profit rate is found in those regions where

$$(X_r/W_r) > (X_n/W_n)$$

In purely abstract terms, these relationships may be stretched across the Italian regions in the last thirty years as in Figures 12.1 and 12.2. Putting regions on the X axes in increasing order of 'peripherality' and our two aggregate indices on the Y axis, we may hypothesize the inter-regional situation of comparative advantage in the first post-war period of concentrated

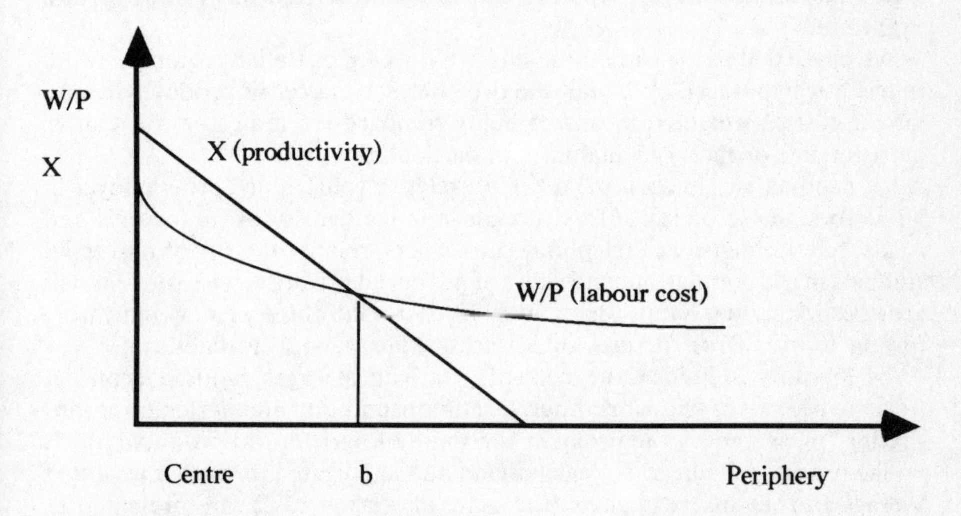

Figure 12.1 Abstract regional locational advantages of Italian regions, 1950-64.

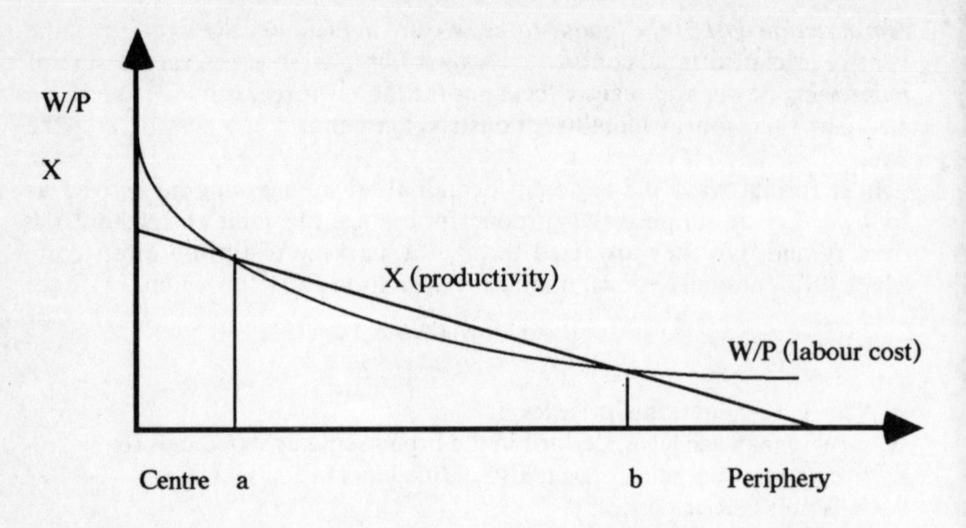

Figure 12.2 Abstract regional locational advantages of Italian regions, 1965–80.

development showing a locational advantage for central regions (0–b in Figure 12.1); in the second period of catching-up by Third Italy regions the situation could have evolved in the sense shown by Figure 12.2, with lower inter-regional disparities in productivity levels and a growing disadvantage on the part of the central regions in terms of labour costs.

Point b, which we may call the 'periphery frontier', to the right of which lie the underdeveloped areas, may have been moving outward from the first to the second period; on the contrary, in the second period an area of locational disadvantage may have shown up in the central regions (0–a in Fig. 12.2).

We have tried to measure these shifts in the case of Italian regions in order to test our hypothesis. We built the two complex indices of productivity and labour cost as weighted means of many standardized indicators of relative performance or factor availability; in particular:

(a) productivity index is the mean of: relative value added per employee in all sectors, share of rail infrastructure, relative density of motorways and roads, relative density of telephone subscribers, relative density of university students in the population, and share of advanced private services on regional gross product; the same weight was given to the three sets of indicators (productivity, infrastructure and services/skilled labour availability);

(b) labour-cost index is the mean of: relative real wages, industrial conflict (relative density of lost work-hours), self-consumption and agricultural job-sharing (measured by the proxy of the share of agricultural production).

The results, calculated for each region and aggregated through a weighted average in four macro-regions (the same of section 12.2), are presented in Figures 12.3–12.5 for the years 1961, 1971 and 1981. The curves behave in an

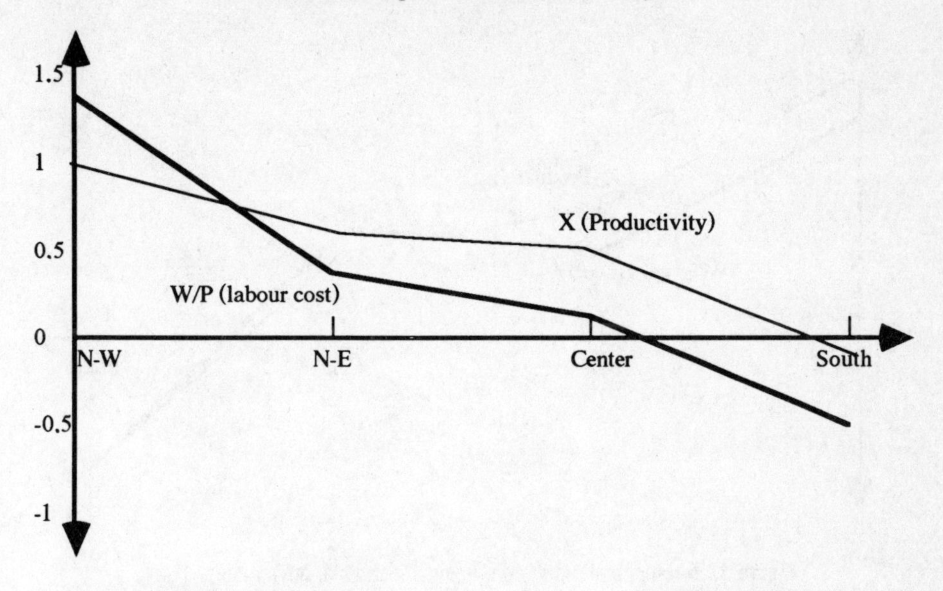

Figure 12.3 Empirical results on regional locational advantages, 1961.

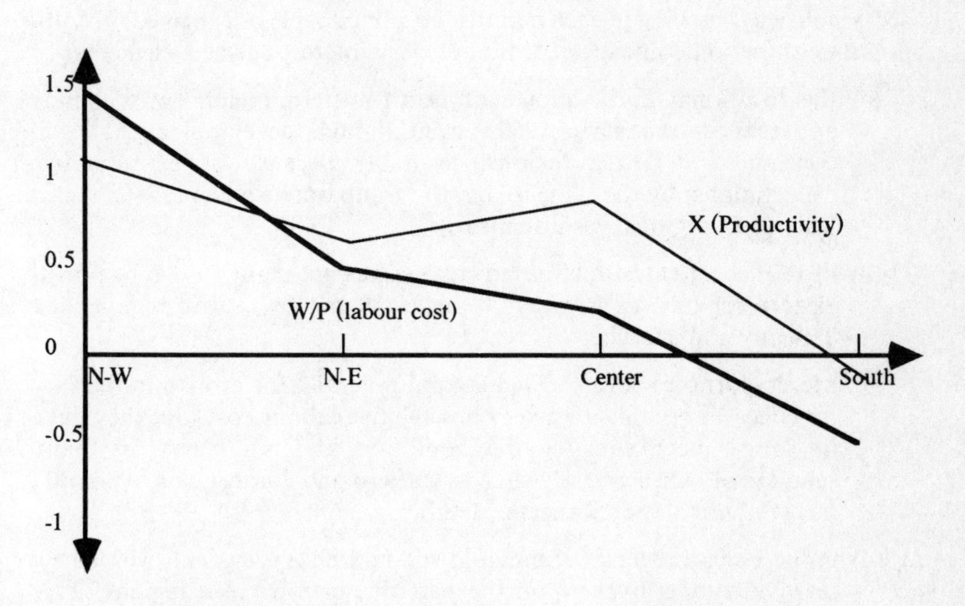

Figure 12.4 Empirical results on regional locational advantages, 1971.

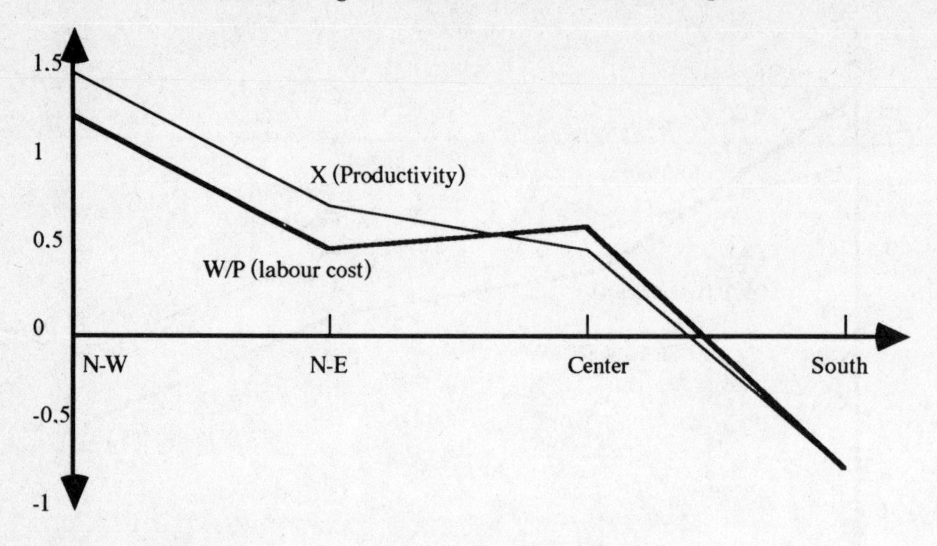

Figure 12.5 Empirical results on regional locational advantages, 1981.

acceptable way showing in each period a negative shape, as expected; in addition they show some unexpected, though very interesting, new elements:

(a) the locational disadvantage of north-western regions was already apparent in the early 1960s, even though development was still continuing in these regions, probably because of extrapolative expectations; by the same token, the comparative advantage of Third Italy (NE) regions was also clear;

(b) in 1971 the greatest relative advantage was not really shown by north-eastern regions, as is generally believed, but by central regions like Tuscany and Marche;

(c) Mezzogiorno regions showed a good potential for growth in the first two decades, thanks to a very low relative labour cost, but they have no longer been showing this condition in recent years, as many scholars of southern Italy such as Cafiero and Saraceno have pointed out (Cafiero, 1986; Saraceno, 1986);

(d) in the 1980s a reversal of previous spatial trends is evident, with a new relative competitiveness on the part of north-western regions. This situation was predicted by us some time ago, looking at the upswing of new-technology adoption in north-western regions (Camagni, 1986b, 1987);

(e) in the 1980s some problems of lack of relative competitiveness appear in the regions of central Italy, while north-eastern regions look like sharing with north-western ones a condition of strong and modern economies.

In conclusion, we may say that our model may be considered a good interpretative tool of regional behaviour, sometimes also showing in advance the economic roots of subsequent development processes.

12.4 TAXONOMY OF LOCAL SUCCESS FACTORS FOR NON-METROPOLITAN AREAS

In this section our intent is to present a taxonomy of local success factors, that is to say to analyse and classify the economic, social and political elements which determined the positive situation of the 1960s and 1970s in the Third Italy. The taxonomy takes into consideration two different sets of factors: the nature of the specific *spatial* conditions of the local context and the nature of the relationships involved (political, social and economic) (Table 12.6).

The economic and spatial conditions that may generate success factors are seen in the following: spatial proximity, sectoral specialization, small urban dimension and rural/urban integration. Among all the success factors listed in our double-entry table (both objective and subjective elements mentioned in section 12.1), a strategic importance is attributed to those which imply *synergy* and *integration*; local synergies are in fact seen as the most typical new general factor explaining the innovativeness and creativity of many local milieux (Aydalot, 1986; Camagni, 1988b; Federwish and Zoller, 1986; Gremi 1987, 1988; Perrin, 1988; Stöhr, 1986).

These elements are presented in bold characters, as opposed to more traditional success factors, coming from structural characteristics of the areas.

Among those which derive from the condition of spatial proximity we find reduction in transport costs, innovative information, networks of interpersonal relationships, market externalities (as economic elements), political cohesion (as political elements), and entrepreneurship and inter-personal communication (as social elements) (Innocenti, 1985; Piore and Sabel, 1984).

So far as sectoral specialization is concerned, the economic success elements which are generated are rapid circulation of information, common services for firms, production flexibility, labour specialization and adaptive elasticity to changes; political elements are found in the political cohesion and social relations.

The spatial condition of small urban dimension generates lower labour cost, lower land costs, local finance/industry integration and traditional entrepreneurial expertise; as political elements we find cooperation capability and as the social ones social cohesion and traditional cultural background. Finally, from urban/rural integration, agricultural–industrial part time, lower labour reproduction costs, labour availability are the economic success

Table 12.6 Taxonomy of local success factors in non-metropolitan areas

Spatial conditions	*Spatial proximity*	*Specialization*	*Small urban dimension*	*Urban/rural integration*
Economic relations	1a Transport costs reduction	2a **Information mobility**	3a Labour costs	4a Agricultural industrial part time
	1b **Innovative information**	2b **Services in common to firms**	3b Land costs	4b **Labour reproduction costs**
	1c **Networks interpersonal relationships**	2c **Production flexibility**	3c **Local finance industry integration**	4c **Labour availability**
	1d **Market externalities**	2d Labour specialization	3d Traditional entrepreneurial expertise	
		2e Adaptive elasticity to change		
		2f Ability to organize		
Political relations	1e Political cohesion	2g Political cohesion	3e **Cooperation capability**	4d Conservatism
				4e Political stability
Social relations	1f Entrepreneurship	2h Entrepreneurship	3f Social cohesion	4f Traditional values
	1g **Interpersonal communication**		3g Traditional cultural background	4g Traditional mentality

elements, while conservatism, political stability, traditional values and traditional mentality are the political and social ones.

Another important distinction is between objective and subjective factors. The former are represented by lower labour costs, labour availability, labour specialization and market externalities, all determining, in general terms, local productivity levels. The latter are linked to the presence of local entrepreneurship and to local capacity for employing internal resources in innovative uses; this ability is due to structural characteristics such as traditional mentality, traditional values, cooperation capability, ability to organize, traditional entrepreneurial expertise and social cohesion.

The horizontal elements on which local growth is based are in most cases economic: lower labour costs, major elasticity to change, local finance/industry integration, and traditional entrepreneurial expertise which guarantees the entrepreneur a relative advantage in comparison with other areas. All these elements are structural factors, while the synergic factors are mostly typical of single areas. As we will see, these factors have a different importance in different local realities in the explanation of local economic development during the period of general economic instability and recession of the 1970s.

We also wish to examine where in the Italian literature these success factors were pointed out. Tables 12.7 and 12.8 list references to these success elements.

Let us now consider in particular three case studies of local development: Prato (Tuscany, in the second macro-area), Bassano (Veneto, in the north-east) and Marche (in the centre).

12.4.1 PRATO: INNOVATION BY REORGANIZATION

The present economic structure of Prato is the result of a period of crisis which reached its apex during the 1950s: the solution to the economic problems of those years was found in the fragmentation of the big local textile firms into numerous small and medium-size firms specializing in a specific phase of the industrial process. This allowed the level of investment required to be lower than in the previous vertical structure.

We suggest, after reading most of the literature on the Prato area,[2] that the mechanisms which allowed the local fragmented structure to grow during the last twenty years are typically horizontal success factors: production flexibility, elasticity to change, finance/industry integration, an effective network of inter-personal links for contract sharing and sub-contracting are at the base of the present local success. But the literature also underlines the importance of the integrated social background as fundamental to the acceptance and fast assimilation of economic change: absence of class division and industrial conflict, presence of typical firm structure based on autonomous family and handicraft work. By contrast with other Italian local success areas, political elements in Prato do not seem important and the literature on

Table 12.7 Bibliography of economic local success factors, following the codes given in Table 12.6

Spatial conditions	*Spatial proximity*	*Specialization*	*Small urban dimension*	*Urban/rural integration*
Economic relations	**1a** Cori and Cortesi, 1977 **1b** Niccoli, 1982 Bagnasco and Trigilia, 1984 Cori and Cortesi, 1977 Berardi and Romagnoli, 1984 **1c** Cori and Cortesi, 1977 Niccoli, 1982 Niccoli, 1984 **1d** Balloni and Vicarelli, 1979 Cori and Cortesi, 1977	**2a** Bagnasco and Trigilia, 1984 Cori and Cortesi, 1977 Giovannelli, 1983 Niccoli, 1984 **2b** Giovannelli, 1983 **2c** Bagnasco and Trigilia, 1984 Cori and Cortesi, 1977 Lorenzoni, 1979 Lorenzoni, 1980 Niccoli, 1984 Bernardi and Romagnoli, 1984 Giovannelli, 1983 **2d** Fuà and Zacchia, 1983 Giovannelli, 1983 Lorenzoni, 1980 **2e** Giovannelli, 1983 Lorenzoni, 1980 Bernardi and Romagnoli, 1984 Niccoli, 1982 **2f** Giovannelli, 1983	**3a** Bagnasco and Trigilia, 1984 Brunetta and Segre, 1976 Anastasia and Rullani, 1982 Bagnasco, 1977 Maccelli and Romagnoli, 1979 Mazzoni, 1981 Niccoli, 1982 Niccoli, 1984 Lorenzoni, 1980 **3b** Anastasia and Rullani, 1982 **3c** Bagnasco and Trigilia, 1984 Anastasia and Rullani, 1982 Cori and Cortesi, 1977 Balloni and Vicarelli, 1979 Maccelli and Romagnoli, 1979 Jacobucci, 1986 Lorenzoni and Sandri, 1982 Niccoli, 1984 Berardi and Romagnoli, 1984 **3d** Bagnasco and Trigilia, 1984 Cori and Cortesi, 1977 Brunetta and Segre, 1976 Bagnasco, 1977 Niccoli, 1982	**4a** Anastasia and Rullani, 1982 Chiozzi and Gabbi, 1985 Agostinelli, Russi and Salmoni, 1983 Brunetta and Segre, 1976 Lorenzoni, 1980 **4b** Zaninotto, 1978 Anastasia and Rullani, 1982 **4c** Becattini, 1987 Mazzoni, 1981

this topic does not mention them as one of the key elements for the local success situation.

The two key success factors which the vast literature on the development of the area underlines are the flexibility generated by the high division of labour among small firms and the integrated social tissue capable of absorbing change and creatively adapting new technologies and new forms of commercial organization to the needs of a 'traditional' industry.

As the various researchers underline, the barriers to entry in these activities were very low and did not discourage the host of individual trials to respond to the economic crisis. There was a change from an integrated to a dis-integrated structure, with short internal cycles and a wide decentralization of activities. Prato maintained this economic situation and had positive results thanks to the fundamental equilibrium among social and economic forces; from the social point of view the traditional values of loyalty and friendship produced cooperation among economic operators and the production division in different phases could be sustained. At the same time the presence of strong family ties promoted the reproduction of a specialized labour force because traditional expertise was handed down from father to son.

In recent years new reasons for local success have appeared in the literature: the high degree of integration between industry and finance, and the role of local 'collective agents' (the chamber of commerce, the industrialists' associations, the local banks themselves) in taking care of general interests and in promoting general services like the local telematic network, SPRINT. In addition, thanks to easy information flows throughout this 'systems area', rapid diffusion of both organizational and process innovations allows the prompt and widespread adoption of leading-edge technologies.

What the researchers suggest in the case of Prato is that it is the *organizational model* in itself which determines a particularly innovative model. In fact, the fragmented but spatially integrated production structure allows the reproduction of both internal organization cost and of external transaction costs. This element is true not just in the traditional sense (*à la* Williamson) of reduction of operating costs, but also in the dynamic sense of reproduction of barriers to innovation diffusion and of resistance to change. The dialectic between vertical dis-integration and spatial integration of all production phases, from marketing and finance, to shipment and manufacturing, and even to advanced capital goods production, allowed by a fairly homogeneous social context has built the historical success of this area. The social integration element is particularly important, and makes this experience comparable with if not similar to the Japanese one, not so much with the recent technological phenomenon, but rather with the postwar origin of what are now big firms as cooperatives between management and labour.

Table 12.8 Bibliography of political and social local success factors, following the codes given in Table 12.6

Spatial conditions	*Spatial proximity*	*Specialization*	*Small urban dimension*	*Urban/rural integration*
Political relations	**1e** Trigilia, 1985 Bagnasco and Trigilia, 1984	**2g** Trigilia, 1985 Bagnasco and Trigilia, 1984	**3e** Bagnasco and Trigilia, 1984 Niccoli, 1982 Cori and Cortesi, 1977	**4d** Bagnasco and Trigilia, 1984 **4e** Bagnasco and Trigilia, 1984
Social relations	**1f** Bagnasco and Trigilia, 1984 Niccoli, 1982 Niccoli, 1984 Balloni and Vicarelli, 1979 Fuà and Zacchia, 1983 **1g** Giovannelli, 1983 Bagnasco, 1988	**2h** Bagnasco and Trigilia, 1984 Niccoli, 1982 Niccoli, 1984 Balloni and Vicarelli, 1979 Fuà and Zacchia, 1983	**3f** Bagnasco and Trigilia, 1984 Berardi, 1984 Niccoli, 1984 Berardi and Romagnoli, 1984 **3g** Bagnasco, 1983 Lorenzoni, 1980 Bagnasco, 1988	**4f** Fioravanti, 1982 Bagnasco and Trigilia, 1984 Bagnasco, 1983 Balloni and Vicarelli, 1979 Chiozzi and Gabbi, 1985 **4g** Bagnasco and Trigilia, 1984

12.4.2 BASSANO (VENETO, NORTH-EASTERN ITALY): LOW LABOUR COSTS, HIGH SPECIALIZATION AND SOCIAL COHESION

In all research concerning the economic and social structure of the Bassano area,[3] the major local success factor mentioned is the low cost of labour, allowed by the flat urban hierarchy of the Veneto region and the absence of big cities. This seems to be the crucial element: in fact it gave local firms an economic advantage in comparison with old industrial regions.

Beyond this, the economic success factors most often mentioned are strong labour specialization, mobility of information and the possibility of part-time agricultural–industrial employment. According to the specific literature concerning the area, political cohesion and political stability are fundamental success factors which, combined with the economic and social ones, led to the development of the industrial tissue. The high degree of social cohesion is generated by reciprocal cooperation among all industrial operators and between industrialists and other social actors such as local bankers, politicians and members of the Catholic Church hierarchy.

Moreover, the wide network of internal contacts allowed by the small-firm dimension and the relative importance of non-routine transactions may strengthen the relations of loyalty, esteem and in some cases friendship among the economic operators.

The last success element mentioned in the literature is the traditional cultural background, a mixture of entrepreneurial expertise, labour specialization and skill, traditional religious and rural values and a strong sense of regional identity.

12.4.3 MARCHE, CENTRAL ITALY: ABUNDANT LABOUR FORCE AND STRONG ENTREPRENEURIAL EXPERTISE

The third case study is the Marche region, where the economic structure is characterized by the presence of a host of medium-sized and small firms, the importance of social principles in industrial organization and the region's political stability.

In the literature concerning this area,[4] of the many success factors behind regional growth the most relevant are considered to be low labour costs, labour availability, the fact that combined industrial–agricultural part-time work is allowed even in big industrial firms, and the information mobility and traditional entrepreneurial expertise distinguishing the economic structure of this area and guaranteeing that the production system's transaction costs will be low. Other factors behind the region's success are social and political elements, such as political cohesion, willingness to cooperate, the presence of public *agents collectifs* that widen the spectrum of externalities available to the firms (common exhibitions and marketing initiatives), common guarantee funds for exports, and so on.

The family guarantees the transmission of traditional expertise and social values and the local associations guarantee the control of the cooperation mechanism, not just in typical specialization sectors (shoes, furniture) but also in new advanced ones (musical instruments, electric appliances).

The presence and maintenance of all these economic, social and political elements are, according to the main literature, the key to perpetuating the success of this area.

Table 12.9 represents the factors, of those mentioned in the vast bibliography on this topic, that are most important for the three local case studies we are analysing directly.

Table 12.9 Success factors of three specific non-metropolitan areas

	Prato	*Bassano*	*Marche*
Economic relations	Production flexibility Adaptive elasticity to changes Finance/industry integration Information mobility	Lower labour cost High labour specialization Information mobility Agricultural/industrial-part-time	Lower labour cost Adaptive elasticity to changes Finance/industry integration Entrepreneurial expertise Labour availability Information mobility
Political relations		Political cohesion Political stability	Political cohension Cooperation ability
Social relations	Social cohesion Interpersonal communication	Social cohesion Traditional cultural background	Entrepreneurship

12.5 TAXONOMY OF LOCAL SUCCESS FACTORS FOR METROPOLITAN AREAS

We conclude our analysis of Italian local success stories with the extremely interesting case of old industrial metropolitan areas like Milan and Turin. As in the case of the peripheral areas, the phenomenon of urban revitalization may be interpreted through the construction of a taxonomy of success factors. The literature in this case is scarce, as the phenomenon is more recent, although our previous model captured some signs of revitalization in these areas as early as in 1981 (section 12.3.1). We can define the success factors according to two dimensions: *the nature of the relations* involved (micro-, meso- and macro-economic relations; social and political relations)

Table 12.10 Taxonomy of local success factors in metropolitan areas

Spatial dimension	*Competence*	*Efficiency*	*Flexibility*	*Synergy*
Micro-economic relations	Concentration of R&S	Presence of private decision centres	Flexible automation technologies	New functional integration
	'Incubator' effect of advanced industries	Reduction in labour and floor-space intensity of production process	New managerial practices in industrial relations and labour organizations	
			Presence of small firms	
Meso-economic relations	Presence of qualified factors in industry and and in services	Inter-industry linkages	Production services as adjustment cost reducing factors	Interfirm integration (cooperation agreements, joint ventures)
		Sectoral diversification		
Macro-economic relations	Vocational training and managerial schools	Intermetropolitan communications networks	Downward flexibility of factor prices in a crisis condition	New industrial relations practices
Social relations	High education	Increasing awareness of being part of an urban group	Consumer services as personal cost reducing factors	New social solidarities/ class integration
Political relations	Ambitious cultural projects	Presence of public decision-making centres	Urban revitalization policies	Reduced political conflict/public and private partnership

and the specific *economic and spatial preconditions* for revitalization; these latter are classified, following some previous reflections (Camagni, 1986a), into competance, efficiency, flexibility and synergy (Table 12.10).

Regarding competence, the large town has always been recognized as a place with a high capacity for attracting and concentrating research units, advanced industrial and services production, as the right place for 'incubation' of small and medium-sized high-tech enterprises (Ciciotti, 1986) and for vocational training and managerial schools. A new awareness of the need for a broader knowledge of the role of the city has recently emerged among politicians and local governments, strengthening the local 'competence' through large research projects, such as the Progetto Milano, sponsored by the Lombardy regional government, and many local institutions, and the Progetto Torino, sponsored by the Turin city council.

The efficiency of urban structures may be interpreted as the outcome of the integration of different activities located in the area (inter-sectoral relations between industry and production services), of the widespread diffusion of communication networks, linking the individual city to an inter-continental urban structure, and of the presence of large public and private decision-making centres (Camagni and Predetti, 1988; Capello, 1988).

Flexibility is the most interesting and the most original of the success factors analysed so far. At the micro-economic level, this characteristic is linked to the efforts made by large firms to diminish the gap in flexibility with respect to small and medium-sized firms. These efforts may be summarized in the development of new managerial, organizational and technological models, such as flexible automation technologies, allowing greater product diversification and plant 'versatility', new organizational structures such as 'just-in-time' production, telecommunication systems, and remote-control systems allowing decentralization of routinized and standardized production phases (Camagni, 1987, 1988a). All these elements may be found in the revitalization of large industrial enterprises in the north-western Italian regions. Moreover, the presence of dynamic and flexible small and medium-sized units and the diffusion of differentiated production services makes it possible to reduce adjustment costs in large industrial units operating in the urban area.

In the context of macro and social elements, pronounced downward flexibility in relative factor prices in the north-western macro-region (both labour and floor-space prices) during the crisis period of the 1970s is an important element in the explanation of the revitalization process.

Another success factor in urban areas is the new tendency and capacity of the political authority to conceive and implement urban policy that supports revitalization. Salient examples are the new communication infrastructures (embodied in the Lombardia Cablata project for a modern fibre-optic network in the Lombardy region, or in the Passante Ferroviario, the underground link among the different railway stations in Milan), the rational use of derelict industrial lands and new attention to the development of science and technology centres (Gibelli, 1985; Camagni and Gibelli, 1986; Gibelli and Magnani, 1988). These tendencies are also present in other European towns with the development of international projects of urban revitalization and the creation of public agencies for the development of new strategies.

The synergy element is also part of the revitalization process for urban areas and, under different conditions, of the development of the north-eastern and central Italian regions.

At the micro level we find high-tech and innovative enterprises employing a new integration strategy: for example, the integration between production and strategic planning functions, between research and marketing, for the purpose of accelerating the innovation process (Camagni and Rabellotti, 1986; Camagni, 1988b). These strategies are seen in the 'barricentric' and therefore 'central' location of integrated 'mission units', committed to specific innovation projects.

At a macro-economic and social level, we find new industrial relations practices being adopted, together with new social solidarities and class integration. From the political point of view, we find reduced political conflict and the construction of a wide spectrum of public and private partnerships to run large projects ranging from advanced vocational training to infrastructure provision.

CONCLUSIONS

The main factors that allowed and caused the success of many local 'system' areas in the Italian experience, apart from the factor of labour costs lower than those in old industrial areas, may be summarized as the reduction of market transaction costs allowed by a spatially concentrated and specialized structure of small firms, and the existence of wide synergies between economic and social actors, arising from common cultural and political values.

The open problem now concerns the capability of these areas to secure continuing organizational and technological change under the new general conditions produced by past success and by the demographic generation transition: rising costs of land and labour, spatial congestion, new values among young people. Moreover, these same areas have to face the new competitive revitalization of many old industrial and metropolitan regions, which are adopting modern information technologies more quickly and using them more efficiently.

NOTES

[1] Though this paper is the result of a common research effort on the part of the two authors, R. Camagni has written sections 12.3 and 12.5 and R. Capello sections 12.2 and 12.4; the first section and the conclusions, as well as the regional indices of productivity and labour cost, have been jointly elaborated.

[2] Berardi, Fanti, Pacini, Romagnoli and Rondelli, 1986; Berardi and Romagnoli, 1984; Chiozzi and Gabbi, 1985; Cori and Cortesi, 1977; Fanti, 1984; Ganuggi and Romagnoli, 1982; Lorenzoni, 1979 and 1980; Lorenzoni and Sandri, 1982; Maccelli and Romagnoli, 1979.

[3] Anastasia and Rullani, 1981; Bagnasco and Trigilia, 1984; Brunetta and Segre, 1976; Zaninotto, 1978.

[4] Baldassarri, 1980; Balloni, 1979; Balloni and Vicarelli, 1979; Garofoli, 1981; Gaudeni and Vaciago, 1976; Jacobucci, 1985; Mazzoni, 1981; Niccoli, 1982 and 1984.

REFERENCES

Anastasia, B., and Rullani, E. (1981), *La nuova periferia industriale, saggio sul modello veneto*, Venice: Arsenale Cooperativa Editrice.

Aydalot, P. (ed.) (1986), *Milieux innovateurs en Europe*, Paris: Gremi.

Aydalot, P., and Keeble, D. (eds) (1988), *High Technology Industry and Innovative Environments: the European Experience*. London: Routledge.

Baccarani, C., and Panati, G. (1978), 'Prospettive di agevolazione finanziaria per le industrie del tessile-abbigliamento in una microarea di recente sviluppo', *Economia e politica industriale*, no. 19 (September), pp. 161–78.

Bagnasco, A. (1977), *Tre Italie*, Bologna: Il Mulino.

Bagnasco A. (1983), 'Il contesto sociale', in G. Fuà and C. Zacchia (eds), *Industrializzazione senza fratture*, Bologna: Il Mulino.

Bagnasco, A. (1988), *La costruzione sociale del mercato*, Bologna: Il Mulino.

Bagnasco, A., and Trigilia, C. (eds) (1984), *Società e politica nelle aree di piccola impresa: il caso di Bassano* Venice: Arsenale Editrice.

Baldassarri, M. (1980), 'Note sulla struttura industriale e sui rapporti tra banche e imprese nelle Marche', *Economia Marche*, no. 7, pp. 3–18.

Balloni, V. (1979), 'La direttrice adriatica allo sviluppo industriale del Mezzogiorno', *Economia Marche*, no. 6, pp. 7–53.

Balloni, V., and Vicarelli, R. (1979), 'Il sistema industriale marchigiano negli anni settanta', *Economia Marche*, no. 5, pp. 43–74.

Becattini, G. (ed) (1987), *Mercato e forze locali: il distretto industriale*, Bologna: Il Mulino.

Benedetti, E., and Camagni, R. (1983), 'Riflessioni sulla periferia', *Economia e politica industriale*, no. 39, pp. 19–37.

Berardi, D., Fanti, L., Pacini, S., Romagnoli, M., and Rondelli, O. (eds) (1986), *Relazione annuale sull'economia e l'occupazione nell'area tessile pratese*, Prato: Consorzio Centro Studi.

Berardi, D., and Romagnoli, M., (1984), *L'area pratese tra crisi e mutamento*, Prato: Consorzio Centro Studi.

Bianchi, G., and Magnani, I. (eds) (1986), *Sviluppo multiregionale, teorie, metodi e problemi*, Milan: Franco Angeli Editore.

Brunetta, R., and Segre G. (eds) (1976), *Struttura e crisi dell'economia veneta*, Venice: Marsilio Editori.

Cafiero, S. (1986), 'I termini attuali del problema dell'industrializzazione del Mezzogiorno', *Studi Svimez*, no. 2.

Camagni, R. (1984), 'Modèles de restructuration économique dans les régions européennes pendant les années '70', in P. Aydalot (ed.), *Crise et éspace*, Paris: Economica.

Camagni, R. (1986a), 'Robotique industrielle et revitalisation du Nord-ouest italien', in J. Federwisch and H. G. Zoller (eds), *Téchnologie nouvelle et ruptures régionales*, Paris: Economica.

Camagni, R. (1987), 'The spatial implications of technological diffusion and economic restructuring in Europe with special references of the Italian case', paper presented at the OECD meeting of experts on 'Technological Developments and Urban Change', Paris, 29–30 June.

Camagni, R. (1988a), 'Lo spazio nella planificazione', in Gibelli and Magnani, op. cit.

Camagni, R. (1988b), 'Functional integration and locational shifts in the new technology industry', in Aydalot and Keeble, op. cit.

Camagni, R., and Cappellin, R. (1985), *Sectoral Productivity and Regional Policy*, Document 92-825-5535-6, Brussels: Commission of the European Community.

Camagni, R., and Gibelli, M. C. (1986), 'Urban planning strategies in an era of deindustrialisation: the scheme for a technological pole in Milan', *Revue d'Economie régionale et urbaine*, no. 5, pp. 663–75.

Camagni, R., and Predetti, A. (eds) (1988), *La trasformazione economica della città*, Milan: Franco Angeli Editore.

Camagni, R., and Rabellotti, R. (1986) 'Innovation and territory, the Milan high-tech and innovation field', in Aydalot, op. cit.

Capello, R. (1988), 'La domanda di reti e servizi di comunicazione nell'area metropolitana milanese: vincoli e strategie', in Camagni and Predetti, op. cit.

Cappellin, R. (1983), 'Osservazioni sulla distribuzione inter e intra regionale delle attività produttive', in Fuà and Zacchia, op. cit.

Cappellin, R. (1983b), 'Productivity growth and technological change in a regional perspective', paper presented at the eighth Pacific Regional Science Conference, Tokyo, 17–19 August.

Chiozzi, P., and Gabbi, M. (1985), 'Indagine qualitativa sui residenti nel Comune di Montemurlo', in *Indagine socio-economica sul Comune di Montemurlo*, Prato: Consorzio Centro Studi.

Ciciotti, E. (1986), 'Aspetti spaziali nel processo di formazione di nuove imprese: il quadro di riferimento delle analisi e alcune verifiche empiriche', in R. Camagni and

L. Malfi (eds), *Innovazione e sviluppo nelle regioni mature*, Milan: Franco Angeli Editore.

Cori, B., and Cortesi, G. (1977), *Prato: frammentazione e integrazione di un bacino tessile*, Turin: Fondazione Giovanni Agnelli Editore.

Dei Ottati, G. (1987), 'Distretto industriale, problemi della transazione e mercato comunitario: prime considerazioni', *Economia e politica industriale*, no. 51 (September), pp. 93–122.

Del Monte, A. (1982), 'Dualismo e sviluppo economico in un'economia periferica: il caso italiano', in Goglio, op. cit.

Fanti, L. (1984), *Modello produttivo e figure professionali intermedie nel tessile pratese*, Prato: Consorzio Centro Studi.

Federwish, J., and Zoller, H. G. (eds) (1986), *Téchnologie nouvelle et ruptures régionales* Paris: Economica.

Fioravanti, P. (1982), 'Struttura familiare e organizzazione del lavoro', *L'immagine dell'uomo*, May–December, pp. 314–20.

Fuà, G., and Zacchia, C. (eds) (1983), *Industrializzazione senza fratture*, Bologna: Il Mulino.

Ganugi, P., and Romagnoli, M. (1982) *Aspetti della domanda e dell'offerta di lavoro nell'area pratese*, Prato: Consorzio Centro Studi.

Garofoli, G. (1981), 'Aree periferiche: analisi territoriale e mercato del lavoro', *Economia Marche*, December, pp. 3–13.

Garofoli, G. (1985), 'Sviluppo multiregionale e sviluppo industriale', in G. Bianchi and I. Magnani (eds), *Sviluppo multiregionale: teorie, metodi e problemi*, Milan: Franco Angeli Editore.

Gaudeni, G., and Vaciago, G. (1976), 'Studi sull'economia marchigiana: una sintesi di critica', *Economia Marche*, no. 1, pp. 13–38.

Gibelli, M. C. (ed.) (1985), *La rivitalizzazione delle aree metropolitane*, Milan: Clup Editore.

Gibelli, M. C., and Magnani, I. (eds) (1988), *La pianificazione urbanistica come strumento di politica economica*, Milan: Franco Angeli Editore.

Giovannelli, L. (1983), *Una politica innovativa nelle piccole e medie imprese*, Milan: Etas Libri.

Goglio, S. (ed.) (1982), *Italia: centri e periferie*, Bologna: Il Mulino.

GREMI (1987), *Innovation Policies at the Local Level*. Round Table, Paris, 14–15 December.

GREMI (1988), *Milieux locaux et dynamiques d'innovation*, atti del seminario, Ascona, 14–17 April.

Innocenti, R. (ed.) (1985), *Piccola città, piccola impresa*, Milan: Franco Angeli Editore.

Irpet (1976), *Lo sviluppo economico della Toscana: problemi e prospettive*, Florence: Irpet.

Jacobucci, D. (1986), 'Piccole e medie imprese marchigiane ed esportazione: aspetti finanziari, valutari ed assicurativi', *Economia Marche*, no. 2, pp. 177–92.

Lorenzoni, G. (1979), *Una politica innovativa nelle piccole e medie imprese*, Milan: Etas Libri.

Lorenzoni, G. (1980), 'Lo sviluppo industriale di Prato', in *Storia di Prato (secolo XVIII–XX)*, vol. III, Prato: Edizioni Cassa di Risparmio e Depositi.

Lorenzoni, G., and Sandri, S. (1982), *La strategia di sviluppo della banca locale*, Prato: Edizioni Il Palazzo.

Maccelli, A., and Romagnoli, M. (1979), 'Aspetti del decentramento produttivo nell'area tessile pratese', *Inchiesta*, January–February, pp. 83–9.

Mazzoni, R. (1981), 'Alcuni aspetti del recente sviluppo economico delle Marche', *Economia Marche*, December, pp. 15–47.

Niccoli, A. (1982), 'Economia marchigiana negli anni settanta', *Economia Marche*, no. 2, pp. 167–205.

Niccoli, A. (1984), 'Alle origini dello sviluppo economico marchigiano', *Economia Marche*, no. 1 (June), pp. 3–17.

Onida, F. (1980), 'Esportazioni e struttura industriale dell'Italia negli anni '70', *Economia Italiana*, no. 1 (February), pp. 97–139.

Perrin, J. C. (1988), 'Technologies nouvelles et synergies locales: elements de théorie et d'analyse', in Aydalot and Keeble, op. cit.

Pettenati, P. (1982), 'Lo sviluppo economico delle aree periferiche: considerazioni generali', *Economia Marche*, no. 2 (December), pp. 159–66.

Piore, M., and Sabel, C. (1984), *The Second Industrial Divide*, New York: Basic Books.

Saraceno, P. (1986), 'La questione meridionale a fine 1985', *Studi Svimez*, no. 2, pp. 117–30.

Stöhr, W. (1986), 'Territorial innovation complexes', in Aydalot, op. cit.

Stöhr,W., and Tödtling, F. (1977), 'Spatial equity – some anti-theses to current regional development doctrine', *Regional Science Association Papers*, vol. 38, pp. 33–53.

Trigilia, C. (1985), 'La regolazione localistica: economia e politica nelle aree di piccola impresa', *Stato e mercato*, no. 14 (August), pp. 181–228.

Zaninotto, E. (1978), 'Struttura tecnologica, professionalità, decentramento produttivo: ipotesi interpretative del caso Veneto', *Economia e politica industriale*, no. 18, pp. 147–65.

CHAPTER 13

Local development initiatives under incipient regional autonomy: the Spanish experience in the 1980s

Antonio Vazquez-Barquero

13.1 INTRODUCTION

The purpose of this chapter is to show how, during the 1980s, the concept of indigenous development became a development strategy of the non-metropolitan areas in Spain. Mobilization of the indigenous development potential has become a mechanism, used by local communities, for obtaining the restructuring of the local productive system and helping to adjust the local/regional economy to the economic and institutional changes of the past decade.

Until the late 1960s, local development in Spain took place thanks to the effect of national criteria operating on local economic and socio-cultural factors, and it was conditioned by the impact of protectionist economic policies. After the opening of the Spanish economy to the world market, and because of the economic crisis and the 1978 constitution (which provided for territorial autonomy), however, the conditions for development strategy in Spain changed and indigenous development becomes a viable strategy for the local/regional governments in an international context.

The combined effect of the industrial crisis and of the transfer of competences to the local/regional communities has led some regional/local governments to adopt a development-from-below strategy in order to solve the unemployment and structural adjustment problems of their local economies. Nevertheless, local development policies are still in an experimental phase and are being followed in only a limited number of regions and towns.

In order to exemplify the recent Spanish experience in the use of local development instruments, two different cases have been selected, an under-developed rural area and an 'old' industrial area. The first describes the integrated local development programme undertaken by the Lebrija Town Council, in Andalusia, in order to initiate a process of industrialization in an agricultural economy through the mobilization of local resources and initiatives. The second studies how the *industrialde* (industrial parks) have become useful instruments for industrial and spatial restructuring of certain areas of an old industrial region in decline, in this case the Basque country.

13.2 LOCAL DEVELOPMENT AND THE TERRITORY

For decades the functional approach to development has dominated scientific thought, and the urban-industrial concentration/diffusion model has been the basis for development policies and programmes. Nevertheless, in recent years the territorial approach to development has become more and more popular (Friedman and Weaver, 1979). A new paradigm has been outlined in which the territory is seen, not as the coincidental arena of economic and functional relations but rather as an agent of socio-economic change, and indigenous local development appears as a possible strategy for the restructuring of the productive system (Gore, 1984).

The economic space must be understood in relation with the spatial differences that each of the successive productive systems have introduced in a particular area, and thus with respect to the changes within its structure and in the social relations (Massey, 1984). Historically, each territorial community has been shaped by the interests and relations of its social groups, which form their own identity and culture so distinguishing their community from others. Thus the territory may be understood as the fabric of all kinds of interests of a territorial community, and this allows us to perceive it as an agent of local/regional development, as long as its integrity and territorial interests are maintained in the process of development and structural change.

Today, in spite of the strong international interdependency of most localities, this concept is significant not only in explaining the reality of some of the more or less marginal areas of the capitalistic system: Scott (1984) recognizes the importance of what is local even in advanced metropolitan areas. Its present reality, its very physical configuration, can only be explained by keeping in mind the role played by the local dimension of the productive system, of the labour market, of community life and its administrative and management system. Often, the local community's characteristics constitute the most relevant element of the identity of the social groups and, in any case, an active reference point in the political consciousness of the workers.

During the period of economic growth following the Second World War this local differentiation was used at first by national government and entrepreneurs and later by international actors (including multinational firms) to fit local economies into the 'new' (qualitative and structural) division of labour (Muegge and Stöhr, 1987). The spatial impact of this process has been more diffused in advanced economies than in recently industrialized countries, like Spain, where it only affected a few areas so far.

Of course, industrial diffusion had started in the 1960s and 1970s in Spain, but the process had not reached major momentum when the crisis of the late

1970s arose (Vazquez, 1986). The explaining factors are: that Spain is a latecomer to industrialization, that for decades it has been a relatively closed economy with strong national protectionism and that during the Franco period the labour market was politically constrained.

These conditions have given rise to a new conceptualization of territorial development in the last decade. Starting with work on indigenous potential, autonomous development and self-reliant development (Stöhr, 1981 and 1985; Bassand *et al.*, 1986) we can define territorial development as a process of internal growth and structural change that leads to an improved standard of living within the local/regional community. The economic and socio-cultural dimensions within the process can be identified. On the one hand, local entrepreneurs use local potentials productively, so that their products are competitive on the market, and on the other hand, the local values and institutions become the basis of the development processes. This is well exemplified by the Spanish experiences of industrialization in rural areas (Vazquez, 1983; Instituto del Territorio y Urbanismo, 1987a).

Local development processes in the past have taken place in a spontaneous manner, rather than as a result of direct intervention by local public managers. At present, however, it is possible to design and implement a policy for the promotion of local development. The measures that implement this policy should be geared towards creating a favourable economic environment for the development of these processes, towards utilizing the local potential productively, towards resolving the specific problems of each local economy, and, lastly, towards overcoming the obstacles so that the local communities can take advantage of the articulation of the local economy with the national/international economic system and state development policies.

Yet, how to specify a local development strategy? In recent years several proposals have been formulated. Gore (1984) mentions, among others, Johnson's (1970) strategy of market towns and rural growth centres, Sachs's (1980) eco-development strategy, Rondinelli and Ruddle's (1978) strategy of integrated regional development and Stöhr and Tödtling's (1979) strategy of selective spatial closure. All are most attractive, by being alternatives to the traditional policies of development from above. In general, however, there are important limitations hindering their implementation in countries like Spain. These strategies lack economic and institutional analysis of the territories for which they are proposed, they lack specificity in the intervention instruments they suggest, and in any case, they are excessively utopian.

In this author's opinion, a local development strategy only makes sense when defined in relation to the economic, socio-cultural and political characteristics of the territory in which it is to be adopted. Beyond this, in countries like Spain and Italy where industrial development still combines two models, namely the urban-industrial concentration/diffusion model and also the diffuse rural/industrial growth model (Fuà, 1988; Vazquez, 1988), there still exists a pattern that can serve as a guide for designing indigenous local development strategies. In any case, the most convenient way should be to propose measures that tend to restructure the local productive system little by little.

Each local/regional community has a productive system with its own peculiar characteristics that must be kept in mind when designing and implementing the development strategy. For the purpose of simplification we can define two extreme situations. There are areas in which the local industrialization process has not yet begun, but where an important development potential exists, the mobilization of which could transform its traditional productive system. On the other hand, there are other areas where development processes, based on indigenous and/or exogenous initiatives, have begun and which experience a period of crisis that impedes the continuity of the initiated development processes. The first applies to rural areas, the second to 'old' industrial areas.

In each case, the development strategy should specify different policy measures. But all strategies have some points in common. Thus, it is necessary to correct the dysfunctions created over a period of decades by the traditional development policies in the local productive systems, and which have impeded or conditioned the emergence and expansion of the indigenous development processes. In order to accomplish this, a necessary condition is that the local development strategy be assumed by all levels of public management as one of the development strategies, and any actions that block the local communities from taking advantage of their integral development potential must be impeded. Where indigenous development potential is found to exist, policy measures should be taken such as to allow its integrated mobilization. The public managers should pay particular attention to the preparation of the institutional and economic atmosphere that permits the indigenous development model to operate in the area, stimulating local entrepreneurship through social animation, information and financing (Coffey and Polese, 1984; Quevit, 1986).

Moreover, it is only possible to define and implement the indigenous development policy and strategy in a context where local/regional communities are capable of managing their territorial development policy. It would be convenient to give greater competences to the territorial communities and create a financial system sufficiently independent of the central administration that it will allow them to control the processes and functions (commercial, financial, technological, entrepreneurial) that take place within their territory. In other words, the devolution of responsibilities to the local/regional communities is a necessary condition for the success of the local development policy.

Lastly, the success of indigenous development policies is also conditioned by the policy formulation process. Given that the indigenous local development processes demand the combined action of the public and private agents and institutions, it is a necessary condition that in the design of the strategy, in the proposal of the measures and in each stage of the formulation of the local development policy all agents can participate in carrying out the projects (potential investor, research and training centres or financial institutions) and participate in the decision-making process. Were this not so, the actions foreseen in the development programme would run the risk of not being accomplished as desired. It will also be necessary to coordinate these actions through commitments (such as the 'contracts programme' in France or the stock companies in Spain) that give coherence and feasibility to the actions of the public and private agents.

13.3 INDIGENOUS INITIATIVES FOR LOCAL DEVELOPMENT

During the 1980s the local/regional governments began to assume an important role in the structural change of the Spanish economy, driven mainly by the need to solve the local communities' problems. Today there exists a growing recognition of the leading role reserved for the local public managers in economic development policy (Torres, 1987).

Confronted with rising unemployment, some local governments have adopted an active attitude towards the change of the local productive system and in the promotion of initiatives for the creation of jobs. Two types of case are the most common. On the one hand is that of production systems, mainly agricultural, which are unable to occupy the local labour force; in this type of case the enlargement of the productive structure through industrial initiatives is considered a solution to local problems. The other case is that of a production system, principally industrial, that has gone into crisis in the mid-1970s; in this case reconversion of the industrial fabric and reindustrialization have become one of the solutions to local problems.

In order to exemplify initiatives for local development in Spain, two cases were chosen: Lebrija (Andalusia), which is a traditionally agrarian area, and Oñate (Basque country), which is an area of industrial tradition within a region of advanced industrialization in decline. The selection has been made with the following criteria. First, smaller localities have been considered because they seem more relevant at the present time and more in need of such initiatives. Areas that are marked by special functional characteristics, as would be the case with mountain or border areas, have been excluded. Finally, not only were integrated local development actions considered, but also sectoral measures for the creation of local employment.

13.3.1 LEBRIJA, ANDALUSIA: FROM AGRICULTURAL CRISIS TO INDUSTRIAL DEVELOPMENT

Lebrija is a good example of the use of local initiatives in the transformation of the productive system. Through the promotion of industrial initiatives, the local government has made an attempt to initiate a process of local industrialization that will broaden job possibilities and improve the local community's income level (Champetier, 1986).

Rural economy to be industrialized

Lebrija is a small municipality of the province of Seville (Andalusia) located in the lower Guadalquivir river, 40 kilometres from Cadiz and 80 kilometres from Seville. Its surface area is 39,972 hectares. It has a resident population of 27,913 (in 1986) focused in the population nuclei of Lebrija and El Cuervo. The working-age population (between 15 and 65 years old) represents 61.7 per cent of the resident population. Its population is young, given that in 1981 37.2 per cent (national average 24.5 per cent) of the inhabitants were under 16 years of age.

Lebrija's productive system is centred, above all, in agriculture (69.9 per cent of total employment, compared to 15.9 per cent national average). Agricultural land occupies 37,086 ha, of which 15,000 ha has been irrigated after important desalination and drainage projects were undertaken. However, it should be mentioned that the land ownership structure is strongly concentrated (*latifundia*). This historical characteristic is common throughout Andalusia, especially in the province of Seville. The main agricultural products are cotton, wheat, wine, sunflowers and beets. The industrial sector is not yet of great importance and secondary activities occupy 19.3 per cent of total employment (national average, 31.8 per cent).

Unemployment is the crucial problem of this community. It is difficult to estimate its magnitude because hidden work situations as well as underemployment exist. Also, during certain times of the year, around 50 per cent of the *jornaleros* (day workers, constituting 44 per cent of agricultural employment) become unemployed. According to official figures, unemployment affects 20.9 per cent of the working-age resident population (in Andalusia the rate of unemployment was 31.3 per cent at the end of 1986 and in Spain as a whole it was 21.2 per cent). The issue is not so much its present size, but rather that if the productive system is not transformed, unemployment will continue to grow. The local labour supply will continue to increase, given the importance of the young population and the limited absorption capacity of large metropolitan areas in Spain or abroad for migrants during the next decade. On the other hand, labour demand will decrease because agriculture will continue to reduce occupation as more agricultural processes become mechanized and as a result of Spain's entry into the European Community.

Thus, the solution to Lebrija's present and future problems must lie in a transformation of its productive system and change of its economic base. The local community is aware that Lebrija must produce more industrial goods and services if it wants to create more jobs. An agreement exists between the local social groups and institutions on this issue, but some differences on local development strategy exist. In any case, mobilization of the local resources may well be one of the instruments for its achievement.

The local development programme

After the new constitution of 1978 and the first municipal elections of 1979, the local government confronted the unemployment issue. Thus, during the past years it has tried to launch an integrated local development strategy, capable of mobilizing the financial and human resources of the community and undertake solutions for specific problems. A local development programme has thus been worked out, geared to raising employment and to improving the standard of living, through industrial development and the improvement of socio-cultural activities.

The programme aims mainly to promote rural-based processing and industrial activities, as well as those linked to the Lebrija production system, mobilizing the community's technical and financial capacity, promoting all projects coming from the local community and giving support to local cooperatives, to local investors and to external entrepreneurs who wish to invest in useful local projects.

For this purpose, the local government created the Industrial Promotion Department (Departamento de Promoción Industrial) in 1984 and in February 1985 a study was finished on Lebrija's economic possibilities. The purpose was to identify projects that were technically and financially feasible, coordinate promotion activities, give technical assistance to those projects proposed by local agents, and serve as a centre for information and negotiation on loans, financial aid and subsidies.

Furthermore, with the purpose of facing some of the serious social problems (such as illiteracy) and making more efficient use of human resources, a community development initiative has been started. The aim is to stimulate the entire population's participation in all the projects through the creation of new associations, to push towards a global solution to the community's problems and to generate understanding and agreement between the local actors and develop new forms of solidarity.

Undoubtedly, both aspects of the local development strategy complement and strengthen each other. First of all, the increased solidarity favours the local community's financial participation in local projects, as occurs with the Translesa tomato-canning company in which 160 residents participate financially and which will be discussed below. On the other hand, the rather weak cooperative movement has gained more momentum, as is shown by the *jornaleros'* co-operatives, a mechanics' co-operative, another one of bricklayers, another of cotton producers (800 workers) and another of ceramics (fifteen young people). Lastly, the training of youths and adults through various programmes is improving the skilled labour force and makes it both productively and socially useful.

The programme contains a number of further projects that are slowly taking shape. Some have a social character, such as the construction of two industrial estates (one of them inaugurated in June, 1986), the creation of the Industrial Promotion Department and the creation of a Training and Cultural Centre (inaugurated in May 1987). At the same time, however, multiple entrepreneurial initiatives have appeared: manufacturing of kitchen cabinets, manufacturing of sanitary packaging for hospitals, manufacturing of goat cheese, construction of a slaughterhouse and the creation of a transportation co-operative.

The first project finished was the construction of a tomato-canning plant, Translesa, in the Marismas industrial estate. The enterprise was financed by the municipal government, regional government, private savings, and the cotton producers' co-operative. The ceramic co-operative is now functioning and its products are sold in the regional market. By the end of 1987 some fifty other projects were under way and another thirty under study.

Local products are mainly sold in the regional and national markets. New firm managers have come to Lebrija from other areas of Andalusia or Spain in general, frequently attracted by the mayor or the town council. In the case of external investments, the new local firms have selected the new managers.

The financing of the local development projects is one of the aspects that the local managers have been most careful about. Apart from the local government's budget funds, local managers have been able to obtain funds from the autonomous community of Andalusia and from the central government, and they have also tried to obtain funds from the European Social Fund (for training and education). Local private savings, however, play a strategic role in carrying out the key projects.

The local development actors

The design and implementation of the local development programme and the launching of the local projects have been made possible thanks to the co-operation of public and private agents and to the coordination of the policy measures of the various levels of the administration.

The heart of the programme is the local community and, more specifically, its mayor, who has managed to stimulate local activities, enthuse all types of interests with the programme and coordinate the policy measures. The town council services are also playing an important role in the creation of the two industrial estates, as well as in implementing the community development initiative.

Nevertheless, the Industrial Promotion Department is the most strategic unit in the whole process. Its major functions are to formulate (with the support of external consulting) and direct the local development programme, give the necessary technical assistance to the projects, stimulate the projects (Translesa and the ceramics co-operative), solve training problems and serve as social animator.

In order to achieve greater efficiency in the promotion of development and to increase flexibility in the management of the local projects, it was planned to transform the Industrial Promotion Department into a stock company. The town council launched the Lebrija Economic Development Society (SODELSA) in 1987, and owns 100 per cent of the shares.

Setting aside the local offices of the central government, the major actors are the network of co-operatives that have been developed in recent years and the residents, who have pushed through new initiatives. The labour unions, not very important in the area, joined the project through commissions, just as the social and cultural organizations do in the community development initiative. Finally, local entrepreneurs, initially sceptical of the town council initiative, have progressively been interested in the local projects.

Private regional entrepreneurs and institutions such as the chamber of commerce do not give specific support to the Lebrija development initiative. The same occurs with the university or the regional research centres, although some links are being established. Nevertheless, the regional administration gives great support to the programme, helping to carry out and finance specific projects.

The Agricultural Directorate of the autonomous region of Andalusia participates in rural development activities; the Regional Labour Directorate does so in the training of development agents and in the creation of units for job promotion; the Regional Economic Directorate finances project viability studies and the construction of some industrial buildings, such as that of the ceramics co-operative, and provides financial support for local development projects. Lastly, specialized agencies, such as IPIA (Industrial Promotion Institute of Andalusia) and SODIAN (Society for the Industrial Development of Andalusia) were ready to support the creation of firms and participate in their financing, whereas the Regional Institute of Agrarian Reform has already done so.

As far as state institutions are concerned, the support of the Ministry for Labour and Social Security towards the local development programme should be mentioned: financing the Lebrija economic study, subsidizing job creation and participating financially in the tomato cannery.

General comments

Although the Lebrija initiative for local development is still in the implementation phase and a final assessment cannot be made, some important results have been achieved. First of all, local, individual private financing could be mobilized and a number of local cooperatives formed. The Marismas Industrial Estate (8 ha.) was built and other industrial estates are under construction. Eleven firms, employing a total of seventy-five workers (plus thirty-five temporary workers) have been set up. Finally, fifty-one more firms are under way and 866 jobs should be created.

The Lebrija experience is of great interest for the rest of Andalusia, where analogous economic situations can be found. Its way of facing the problems of an underdeveloped rural area is a new approach that can be generalized towards other towns. Local government has managed to mobilize local human and financial resources, it has created a high local synergy, joining the efforts of local groups for solving community problems, and lastly it has shown a capacity for institutional coordination through specific projects.

13.3.2 OÑATE, BASQUE COUNTRY: FROM INDUSTRIAL CRISIS TO REINDUSTRIALIZATION

Oñate is a good example of how the utilization of an instrument, the *industrialde* (industrial park), primarily designed to solve a specific problem – the need for organized industrial land – may become a mechanism for promoting reindustrialization of an 'old' industrial economy, and therefore for increasing employment (Llorens and Larraya, 1987).

An industrial economy to be restructured

Oñate is a small town in the province of Guipuzcoa (Basque country), located within the Alto Deva district, well known for its local entrepreneurial capacity in promoting indigenous industrialization. Its surface area is 108.2 square kilometres and the population 10,460 (in 1986). During the period 1950–70 the population had cumulative annual growth of 2 per cent, but has stagnated since then, experiencing a slight fall in the last five years. Its labour-force participation rate (41 per cent of total population) is higher than in Guipuzcoa (39 per cent).

Its productive system centres mainly around industry, given that 61 per cent of total employment is in industrial activities. These activities are quite diverse: metal construction (28 per cent of industrial employment), foods, wood furniture and lumber, electrical material, and first transformation of metals and metal commodities. The co-operative movement is important (46 per cent of industrial employment) and is linked to the Mondragón group. Agricultural activity is not very relevant, given the fact that only 2 per cent of land surface is devoted to agriculture (forests are important) and only 1.5 per cent of local employment is in agriculture.

In the early 1980s, Oñate needed its productive system to be industrially and spatially restructured in order to solve the problems of unemployment and the lack of industrial land. Although the economic crisis has had less impact in Oñate than in the province of Guipuzcoa and in the rest of the Basque country, owing to the fact that the co-operatives have adapted better to the changes in production and in demand, in 1986 the unemployment rate was rather high (16.7 per cent of the working population), though less than in the Basque country as a whole (23.5 per cent in 1986). Nevertheless, unemployment among the young (81 per cent of the people between 16 and 20 years of age and 52 per cent of those between the ages of 21 and 25) and the stagnation of jobs (due to the rationalization of investments and to the introduction of capital-intensive equipment) present the local community with the need for creating jobs, and mainly for presently unskilled workers.

Furthermore, industrial land with adequate facilities is scarce. Certain old industrial plants located in the town centre need to move because of pollution problems, and in order to be able to expand under good conditions. On the other hand, if the flight of local firms to other municipalities, because of the lack of industrial land, was to be avoided, it was necessary to make available new land. Furthermore, it seemed convenient to promote and support new initiatives for job creation that could appeal among the young.

Thus, the solution to Oñate's problems seemed to revolve around the creation of new industrial land and so provide industrial facilities to present and potential entrepreneurs. The local community, and, above all, its mayor and town council, were aware of the strategic value of endowing the municipality with an industrial park for supporting the industrial and spatial restructuring process which has to take place.

The *industrialde* programme

The *industrialde* programme represented the interests of the Oñate Town Council and the Society for Industrial Promotion and Reconversion of the Basque government, as well as the need for new locations for the local firms. The *industrialde* programme seeks to create industrial parks equipped with common services (central offices, telex, telefax, library, meeting rooms, parking space etc.), when industrial land is lacking, or is very expensive. Its aim is to suit the needs of small firms and to serve as a stimulus for technological change and for job creation. New plant and firms have priority, particularly if they incorporate new technologies in their processes and/or products, but the relocation of plants is also acceptable, mainly when new investments and technological improvements are involved.

The Oñate *industrialde* was created in 1982, through a stock company, the Sociedad Oñatiko Industrialde SA. Its objectives are diverse, according to the specific interests of each shareholder, mentioned below. The major objective is to make urbanized industrial land available to those firms near the town centre that are planning to expand their business.

Secondly, the intention was to favour Oñate's industrial development through the fostering of technological change and new investments, to support the new entrepreneurs, to promote common services within the industrial park, to make management assistance available to small firms, and to activate the creation of service firms within the *industrialde*. Finally, the overriding objectives were to reduce unemployment and improve the population's standard of living.

The corporation has a capital of 215 million pesetas. The shareholders are the Sociedad de Promoción Industrial of the Basque regional government (51 per cent of the capital), the Guipuzcoa provincial government (24 per cent), Oñate Town Council (24 per cent) and the Mancomunidad (inter-municipal public service agency) for the Alto Deva municipalities (1 per cent). The *industrialde* is directed by a manager, who also acts as an adviser to small firms, and a board of administrators presided over by the Mayor of Oñate.

The firms located in the *industrialde* have signed a lease with a purchasing option after nine years. The price is established in relation to the number of square metres built up and it is paid through monthly quotas with 12 per cent annual interest (in 1986, the price per square metre was 360 pesetas, far below the free-market price). The common services and professional assistance are paid for separately and have low rates for the firms located in the *industrialde*.

The results have been, up to now, very stimulating. First of all, industrial land has been made available to the local firms. The initial plan of supplying 18,000 square metres, with a budget of 160 million pesetas, was surpassed by the demand. The project was enlarged, more land has been acquired, and two industrial estates, Benzano and Zubillaga, have been built.

Secondly, in 1986, twenty-four firms employing a total of 202 workers (of which thirteen were new firms generating sixty-four jobs) have been located in the two industrial parks. The type of activities reproduces the local industrial fabric, with firms devoted to the transformation of wood, metal products, mechanic shops and bakery. As far as the service sector is concerned, the most outstanding one is a firm devoted to the selling of computer services. With regard to further development, the endowment of common services in the *industrialde* is limited and must be strengthened in the future.

Local development actors

The *industrialde* programme formulation and its implementation has been possible thanks to the cooperation of the public agents and to the participation of local entrepreneurs. The basic idea of what would later be the *industrialde* programme surfaced during the mayor's and town council representatives' meetings of the Mancomunidad for the Alto Deva municipalities. Its implementation was possible thanks to the initiative of the Mayor of Oñate, who saw it as a way of solving the local problems: to supply industrial land, create jobs, promote new initiatives and prevent the local firms' flight towards other municipalities. The town council not only took the initiative of creating the society and sought other institutional support, but also participates with 24 per cent of the stock company's capital and made an office available for the manager while work on the industrial park was under way.

The participation of local entrepreneurs in the project was decisive. Prior to the project's launching, fifteen firms showed an interest in buying urbanized land for a plant location. Later, once the industrial estates had been built, the local entrepreneurs bought the land they needed. The Mondragón co-operative firms were supportive of the project, and are present in the industrial park (Fagor, a firm of the Mondragón co-operative group, occupies 70 per cent of the land built up at the Berezano industrial estate). The local community has also accepted and supported the *industrialde* project.

The provincial government (Diputación de Guipuzcoa) participates in the project with a 24 per cent share of the stock-company capital. At the time of the society's creation, the Diputación had not yet decided to participate, but it acquired half of the capital subscribed by the town council at the time of the first expansion. The Diputación's interests in the project are rather general in nature and boil down to stimulating initiatives for local job creation and provincial development.

The government of the autonomous Basque region is present in the project through the Society of Industrial Promotion, which has a 51 per cent share of the company owning the *industrialde*. From the beginning, the regional public managers were in favour of the project since it met some of their own objectives, such as the supply of industrial land, the support of industrial initiatives, the promotion of technological development, and the creation of firm services.

There is no central state participation in the *industrialde* programme.

General comments

Oñate's *industrialde* is undoubtedly a most interesting example for other industrial areas in the Basque country and in the rest of Spain. It shows how local development measures are useful for the industrial and spatial restructuring of an area and can solve problems that distorted the functioning of the local productive system. It also shows how an action designed to provide organized industrial land has a general positive effect in the local economy. Last of all, it has served as a model for regional policy given that what was at first an industrial promotion measure for Oñate became, in time, an industrial promotion programme of the Basque regional government, the provincial governments, and other town councils. At the end of 1986, five *industrialdes* were functioning in the province of Guipuzcoa and fourteen other *industrialde* projects were in different stages of realization in the provinces of Guipuzcoa (nine), Vizcaya (four) and Alava (one).

13.4 LOCAL DEVELOPMENT STRATEGY IN SPAIN

Having seen how local development policies play an active role in the restructuring of some local productive systems, a set of questions arise, which we will try to answer. Is locally initiated development a widely adopted strategy in Spain? Which agents have been the most dynamic in the formulation and implementation of the existing processes? Have important changes been produced in the local development strategy? If so, what are the new proposals?

To answer the first question, a number of cases of local indigenous development do exist in Spain, the majority of which are entrepreneurial experiences in industrial activities. They have developed by using local (mainly human and financial) resources, thanks to the initiative of private agents. They are located in rural and urban centres spread throughout the territory.

Recent studies (Instituto del Territorio y Urbanismo, 1987a) have evaluated the importance of indigenous development in Spain. By the early

1980s local indigenous industrialization had become a common phenomenon in Spain, affecting at least 10 per cent of the active industrial population. Local firms produce a wide range of mature industrial goods, of which between 10 and 20 per cent of total production is exported abroad, according to product and area.

Local industry has been one of the characteristics of the growth and structural change process of the Spanish economy. In fact, the industrial revolution has been completed and modern growth began thanks not only to the traditional urban/industrial polarized growth model, but also to the diffuse industrialization model in rural areas. In fact, although development from above has been the predominant strategy since the 1940s, development from below has always been a hidden but important characteristic of the growth and structural change process (Vazquez, 1988).

Before the late 1970s, the local indigenous development processes emerged, above all, as a result of the fact that local entrepreneurs assumed the role of using the local resources productively in order to adjust the productive system to the changes of national and international supply and demand. In other words, in each of the processes, the main mechanisms were a set of social, cultural and economic factors that combined in a spontaneous fashion. The role played by the local governments and by the state in general with regard to local initiatives was a rather passive one. This does not mean that the sectoral and regional policies did not influence (whether by favouring or by constraining) the indigenous processes, but rather that local indigenous development was not explicitly thought of as a growth and structural change strategy for the Spanish economy. This point has been explained in greater detail elsewhere (Vazquez, 1983).

Up to the 1970s the strategic agents in the local development processes were the private entrepreneurs and the public managers played, at best, a subordinate role. Nevertheless, a substantial change has taken place during the past decade with respect to the local development strategy, since the state, but more particularly certain local/regional governments, progressively assumed a greater role in the local development processes. In the case of the latter, this is due to the increasing financial and political autonomy granted under the new constitution.

As shown by the cases of Lebrija and Oñate it is not only that the local public managers assume a leading role in these processes and that public and private actions are coordinated for the solution of local problems. What counts above all is that the local development strategy begins for the first time to be defined and to be experimented with at the local level.

The locally initiated development policy is, therefore, not born as a result of a change in the state's development strategy but rather as a result of constitutional reform and the fact that some local/regional governments assume the task of defining and implementing it at a time when the institutional and economic conditions have greatly changed. Faced with the central government's lack of concern for territorial problems, the local governments are assuming this area of economic policy and begin to construct the local development strategy.

Yet, what are the facts that have favoured this important change in development strategy? Above all, the 1978 constitution has changed the institutional framework under which the economic system functions, and therefore has altered the development strategy's conditions. The transfer of competences to the regions abolished the most important restriction that prevented the local/regional governments from taking control of territorial development. In other words, the new constitution creates a favourable environment for local development policies to be worked out and implemented, at the local and regional level (Vazquez, 1987).

The crisis has also seriously affected the Spanish economy in the last ten years. Its impact has been felt by all of the territorial productive systems, though in different ways. State policies, however, have only been directed towards correcting the macro-economic imbalances (above all, the inflation and the foreign deficit), and towards making functional adjustments in the sectoral industrial system (industrial reconversion), but without paying too much attention to the territorial differences and differentiating between growth and structural change policies. Some regional governments (those of Catalonia, the Basque country, Andalusia or Valencia) and local governments (Lebrija or Oñate), however, have faced the unemployment problem by using instruments that allow for a better functioning of the local/regional productive system.

Last of all, some local governments have realized that indigenous development is a viable strategy, have adopted it and have attempted to employ actions aimed at stimulating local initiatives. Trying to implement the theories and technical expertise created during the last decade, they have begun to experiment with local development policies and to learn from their own experience how to design and carry them through.

This means a change in the development strategy, not only because local development policies are now possible, but also because increased complexity has been introduced by having a larger number of agents intervening in the processes. This is a new strategy slowly being assimilated by both public and private territorial actors. The more dynamic local entrepreneurs continue to be strategic agents in local development processes, since they are ready to take the risks that these actions entail, and they have shown flexibility and efficiency in carrying on the investment projects. But the entrepreneurial environment has changed profoundly in many localities thanks to the local public managers' new attitude.

It should also be recognized that co-operative firms are increasingly important in local development, owing to their capacity for gathering and using (human and financial) resources in these projects, and so reduce individual risks. The more successful, however, like Mondragón or Guisona, have adopted management forms, production methods and marketing techniques like those of private firms.

The other potential local actors at the moment are still playing a secondary role in local development policy. In general, the unions, without deep local roots, have not been particularly dynamic in the promotion of local initiatives, although they have usually finally accepted their implementation in view of their positive effects, such as maintaining local employment and improving the standard of living of the local community. Other local interest groups have been sceptical if not opposed to the changes in the productive system through local development initiatives, if they thought that their short-term interests were at stake. Among the external actors, as previously mentioned, central government has been rather passive towards the new development strategy. The 1985 Law of Regional Incentives thus recognizes as one of its objectives to 'reinforce the indigenous development potential of the regions', but among the promotion areas are not included those that have the greatest capacity for indigenous development. On the other hand, the adoption of the ILE-OCDE programme of local employment initiatives in Spain has not been very successful owing to insufficient information, to the scarce financial means applied to it, and to the lack of an efficient instrumentation by the Ministry of Labour and Social Security (Otero, 1984).

Despite the experimental character of the local development policies and their recent introduction among the state instruments for economic adjustment, certain characteristic features can already be observed. In the first place, integrated local development programmes have not yet been widely accepted among the local public managers. The main reason for this seems to be that they would imply a global transformation of the productive system in many cases and would require not only the availability of a considerable amount of human and financial resources, but, above all, a wide mobilization of the local community, which involves a major social and economic transformation over a short period of time. Only when local problems are severe and/or when a strong local leadership exists within the community have these profound changes been well received.

Second, the new regional governments have widely incorporated sectoral measures within their regional policy instruments. Concrete examples are the *industrialde* in the Basque country, or the technological institutes in the Valencia region (RICO, 1988). The Institute for the small and medium-sized industry of Valencia (IMPIVA) has begun a technological assistance programme for the local industrial firms. It financially supports research and development centres already under way (INESCOP in Elda), and gives technical and financial aid to those that function with difficulty (AICE in Castellon and AIDIMA in Alfalfar) or promotes their creation (AITEX in Alcoy and AIJU in Ibi). All these centres offer quality-control services, technological assistance, diffusion of information and training to local firms, as well as carrying out R&D projects. Given that the productive systems of all these local communities are usually specialized in one industrial activity, the research and development centres are also specialized by sectors (shoes, ceramics, furniture, textiles and toys).

The regional government of Valencia has recently broadened its technological policy and has promoted the creation of a technology park within the Valencia metropolitan area. As to the local development policy, this is a particularly interesting example, not only because the initiative surges from a regional government, but mainly because it is aimed at promoting technological development, and eventually, the creation of new products in those sectors in which the region's productive system is specialized. In other words, this is the last stage of a series of industrial policy instruments that begins with the creation of R&D centres by local firms (INESCOP in Elda, for example), that continues with the creation and promotion of various technological institutes and that culminates in the technology park of Valencia.

The Valencia technology park is seen as an initiative for concentrating public and private research and development centres, associated with entrepreneurial productive activities. Within it, technological institutes, technological firms and joint services will be located. The technological institutes are to be the key element of this park, and will be rooted within the Valencian industrial tradition. They will be specialized by product, from food production, machinery, metal products, ceramics to biomedicine and optics.

The technological firms are defined as firms with a high level of R&D in relation to the volume of sales. Research and development will be done by the firm alone, or in collaboration with the institutes, the university, or other specialized centres. The objective is to diversify the industrial fabric of the region and incorporate the more advanced technologies, both regarding processes and products. For this reason service firms geared towards the transfer of expertise, both organizational and entrepreneurial, or to improvement in the capacity of marketing techniques, design or quality control are also allowed to locate in the Valencia technology park.

Finally, what are the institutional forms used for the implementation of local development policies? On the basis of the information available through case studies, it seems that the local public managers have shown a certain scepticism towards using public institutions for the management of local development policies. The idea of choosing a stock company as the legal form of the agency for local development is widespread. This applies to the *industrialde* in the Basque country, the development agency of Lebrija, Ubrique's agency for the promotion and marketing of local products, or of Valencia's technology park.

It appears that public managers are aware of the convenience of building a flexible, creative and operative agency in order to manage local problems and that this is difficult through the present structure of the public administration. A stock company seems to be a better mechanism for the collaboration and coordination of the agents interested in the project, as well as for the efficient implementation of the measures. Beyond this, it allows for the financial participation of public and private agents and the control and supervision (even by institutions which do not finance the project) of the agency's and firm's actions.

13.5 CONCLUDING REMARKS

Having studied the most important features for the local development initiatives in Spain, an external observer could single out some points of interest within the new process.

First of all, the 1978 constitution, and the consequent devolution process, have created the conditions for local development to become a viable strategy of growth and structural change.

The 1960s and 1970s saw an important wave of industrialization, but the dominating efforts on the part of central government to design and implement sectoral and regional policies have had an uncalculated impact on local development. In fact, the measures have been mainly directed towards achieving a high aggregate industrial growth rate and not towards meeting the local demands or solving territorial problems. Local development has been possible thanks mainly to the spontaneous reaction of the local communities.

The 1978 constitution has opened the way towards a new sectoral and regional legislation, closer to the territorial reality of Spain. Local and regional governments increased their financial and institutional autonomy, and, for the first time, local development strategies can be worked out and implemented. As local/regional governments were democratically elected, the mayor (and the regional government) emerges as a 'new' local actor. At the same time, territorial (mainly local) governments have become an important trigger of local development initiatives, in so far as they create a better economic environment for local entrepreneurship.

Besides this, local initiatives are emerging, not only because the constitution has created a new legal and institutional environment, but also because the economic crisis confronted local communities with a very difficult situation of rising unemployment, and without migration as an escape valve. These facts have transformed the political and social behaviour of the local actors and brought about both the awareness that local community work is a necessary condition for solving local problems and the willingness of local actors to cooperate. In some cases, left-wing local groups have changed their attitude of strong criticism towards the functioning of the capitalist system and to the economic strategy of the central government. Where they have discovered that a local development strategy is viable under present legal and technical conditions, they try to press for the implementation of local development policies and they show themselves ready to cooperate and collaborate with other local groups. On the other hand, more conservative local groups have reacted in an analogous manner, when they realized that local entrepreneurship has been reinforced because of the change in the local, legal and social atmosphere.

Furthermore, the case studies indicate that different actors play different roles in the various stages of the formulation and implementation of local development policies. A recent report on the formulation process of local development policies in Spain (Instituto del Territorio y Urbanismo, 1987b) has identified three stages: initiative, formulation and implementation of the local development programme.

During the initiative stage of the local programme the local community and the local public managers obtain an awareness of the problems concerned and decide to overcome these through the initiatives for local development. The initiative usually begins with the local managers, but initiatives do not necessarily arise just anywhere because of local unemployment and/or the crisis of the productive system.

For the emergence of such initiatives, a set of factors must occur. Four major factors, among others, have been identified in the case studies: the existence of unfulfilled demands on the part of the local community in the social (employment) and economic (land, infrastructures etc.) areas; the willingness of mayors and town councillors to undertake economic development projects that were not traditionally their task; the availability of sufficient human resources (entrepreneurs and labour force) and financial (savings) resources in the local community; and a local institutional agreement through the tacit or express acceptance of the local development strategy by the local entrepreneurs and, if applicable, the social consensus between political groups and with the labour unions.

The formulation stage of the programme calls for agreement between the various agents, local leaders, regional government and financial institutions. Such questions as financing, viability of projects and competence for action require the support of the other actors. Consulting firms can play an important role by specifying the actions required to achieve the local development goals. But it is up to the local public managers to obtain the other agents' support.

The programme implementation stage requires the creation of an agency for the management of the programme, in which all the actors directly intervening in the implementation must be present.

Finally, as mentioned in the report of the Instituto del Territorio y Urbanismo (1987b), it must be pointed out that the local development policies have been worked out and implemented with an environment full of constraints. As previously mentioned, local development initiatives in Spain have arisen without the existence of a sufficient legal or institutional frame to ease their implementation. Some local leaders felt the need to find a solution to local problems and have tried to stimulate local initiatives and coordinate the actions of all the actors involved, but in the midst of a legislative vacuum on local development and on the fringe of the economic and sectoral policy of the central government.

In fact, we are witnessing in Spain for the first time the definition of local development strategies which condition the behaviour of the public and private, local and external actors. In any case the local policy formulation process is being carried out on a trial-and-error basis and all actors are learning how to take advantage of it and how to behave in each of the stages of the local policy formulation process.

ACKNOWLEDGEMENTS

I thank Antonio Torres, Mayor of Lebrija, and J. L. Llorens, Director of IKEI, for their help.

REFERENCES

Arrow, J. K. (ed.) (1988), *The Balance between Industry and Agriculture in Economic Development*, London: Macmillan.

Bassand, M., Brugger, E. A., Bryden, J. M., Friedmann, J., and Stuckey, H. (1986), *Self-Reliant Development in Europe. Theory, Problems, Actions*, Aldershot: Gower.

Champetier, Y. (1986), 'Lebrija. Creatividad y solidaridad', Mimeo, Brussels: Groupement Européen pour la Promotion des Initiatives Locales pour l'Emploi.

Coffey, J. W., and Polese, M. (1984), 'The concept of local development: a stages model of indigenous regional growth', *Papers of the Regional Science Association*, no. 54.

Folmer, H., and Oosterhoven, F. (eds) (1979), *Spatial Inequalities and Regional Development*, Leiden: Nijhoff.

Friedman, J., and Weaver, C. (1979), *Territory and Function*, London: Edward Arnold.

Fuà, G. (1988), 'Small-scale industry in rural areas: the Italian experience', in Arrow op. cit.

Gore, C. (1984), *Regions in Question*, London: Methuen.

Instituto del Territorio y Urbanismo (1987a), *Areas rurales con capacidad de desarrollo endógeno*, Madrid: MOPU (Ministerio de Obras Públicas y Urbanismo).

Instituto del Territorio y Urbanismo (1987b), 'The formulation process of local development policies. The Spanish experience', report submitted to the OECD meeting on 'The Rural Policy Formulation Process', 1–3 June, Paris.

Johnson, E. A. J. (1970), *The Organization of Space in Developing Countries*, Cambridge, Mass.: Harvard University Press.

Llorens, J. L., and Larraya, J. (1987), 'El proceso de formulación del programa de desarrollo local en Oñate: El Industrialde', mimeo, Madrid: Instituto del Territorio y Urbanismo, and San Sebastian: IKEI.

Massey, D. (1984), *Spatial Division of Labour*, London: MacMillan.

Muegge, H., and Stöhr, W. (1987), *International Economic Restructuring and the Regional Community*, Aldershot: Gower.

Otero Hidalgo, C. (1984), 'El programa ILE–OCDE en la estrategia de promoción de empleo', *Información Comercial Española*, no. 192.

Quevit, M. (1986), *Le Pari de l'industrialisation rurale*, Lausanne: Editions Regionales Européennes.

Rico Gil, A. (1988), 'La experiencia valenciana en la promoción de la innovación', *Papeles de Economía Española*, no. 35, pp. 142–52.

Rondinelli, D. A., and Ruddle, K. (1978), *Urbanization and Rural Development: A Spatial Policy for Equitable Growth*, New York: Praeger.

Sachs, I. (1980), *Stratégies de l'écodéveloppement*, Paris: Editions Ouvrières.

Scott, A. (1984), 'Production, work and territorial development: a theoretical synthesis', paper read at IGU Commission on Industrial Systems, 20–26 August, Nebian.

Stöhr, W. (1981), 'Development from below: the bottom-up and periphery-inward development paradigm', in Stöhr and Taylor, op cit.

Stöhr, W., and Taylor, D. R. F. (1981), *Development from Above or Below? The Dialectics of Regional Planning in Developing Countries*, Chichester: Wiley.

Stöhr, W., and Tödtling, F. (1979), 'Spatial equality – some anti-theses to current regional development doctrine', in Folmer and Oosterhoven, op. cit.

Torres Garcia, A. (1987), 'El papel de los ayuntamientos en la creación de empleo', *Ayuntamientos Democráticos*, no. 67, p. 22.

Vazquez-Barquero, A. (1983), 'Industrialisation in rural areas: the Spanish case', paper read at the OECD intergovernmental meeting, Senigalia, 7–10 June (CT/RUR/113/06, OECD).

Vazquez-Barquero, A. (1986), 'El cambio del modelo de desarrollo regional y los nuevos procesos de difusión en España', *Estudios Territoriales*, no. 20, pp. 87–110.

Vazquez-Barquero, A. (1987), 'Local development and regional state in Spain', *Papers of the Regional Science Association*, vol. 61, pp. 65–78.

Vazquez-Barquero, A. (1988), 'Small-scale industry in rural areas: the Spanish experience since the beginning of the century', in Arrow, op. cit.

PART V

INTERNATIONAL PROMOTION OF LOCAL DEVELOPMENT INITIATIVES

CHAPTER 14

Background and structure of a European programme for local development strategies

Wolfgang J. Steinle and Deborah J. Moya

14.1 INTRODUCTION

In one of his well-known 'green' sentences, Samuelson wrote that 'no-one should pay any appreciable insurance premium to be protected against the risk of a total breakdown in our banking system and of massive unemployment in which 25 per cent of the workers can find no jobs'. Whereas Samuelson may be right with regard to the banking system, the present economic situation raises some doubts with regard to the unemployment aspects of his argument. In a variety of the highly developed (and the developing) countries, registered unemployment is presently well in excess of 10 per cent, not to mention hidden unemployment, make-work or underemployment, which, in certain economies, may be more significant than registered unemployment *per se*.

Seen against the background of the present economic climate, the problems with which we are faced have dramatically changed within a few years time.

The difficult present economic situation - apart from global aspects of stagflation, technological change, international trade and division of labour - may be traced back to the factors inherent in the history of industrialization, public involvement in the private sector, the role and identity of public administration as well as the culture of the labour force.

14.2 BACKGROUND TO THE APPROACH

Without serious efforts to adapt our thinking to the new social and economic situation and the changing role and nature of the factors which bear impact upon economic and social developments, it seems unlikely that we shall be able to master the situation which has emerged in recent years.

14.2.1 THE HISTORY OF INDUSTRIALIZATION: TRADITIONS, ATTITUDES AND THE CULTURE OF DEPENDENCY

In the course of industrialization, traditions and attitudes developed which have considerable influence on the present economic situation. The history of advanced industrial countries at least up to the middle of the 1970s shows an almost continuous increase of large businesses in employment and output. With regard to the structure of the labour force, the implications of this tendency are large reductions in numbers of self-employed, and substantial increases of numbers of employees. In general terms, the history of industrialization is one of increasing dependency of labour.

What is of particular importance here is that this history implies a tradition of employment and economic culture in which education, training, work experience and policy instruments are geared mainly towards dependent employment. Educational and training systems do not generate managers or entrepreneurs but people with technical skills to fit into profiles of dependent employment. This 'culture of dependency' is an essential component of present economic and social structures, and plays a decisive part in the present unemployment situation.

14.2.2 INVOLVEMENT OF THE PUBLIC SECTOR IN THE PRIVATE ECONOMY

In the postwar period public expenditure has become an ever-increasing component of GDP. The public sector now occupies a dominant role in the economy and thus also determines to a large extent the performance of the private economy. Investment and innovations are no longer an exclusive function of private capital equipment and profit expectations, but also, and perhaps predominantly, of public programmes and incentives. Moreover, people's emotions in connection with public deficit spending should not be lightly disregarded, since they may also affect private investment.

Massive public programmes have an impact upon the situation and orientation of private-sector situations and strategies. They form an integral part of private-sector investment decisions. Before any decision in the private sector is made, the availability of public-sector incentives, potential tax reliefs, and other benefits are examined.

Thus, the orientation and structure of public policies have the power to make or break change and innovation and are decisive factors in the development of the entire economy. This facet of the economic environment forms an institutional and social framework in which both individual initiatives and expectations are tendentiously reduced to marginal activities and replaced by bureaucratic actions and reactions on the part of both the public and the private economic sectors. This, of course, serves to reinforce the pervasive culture of dependency.

14.2.3 THE JOB GENERATION GAP

In the course of industrialization a 'job generation gap' has arisen, in that the distribution of economic activities over the industrial life cycle has become increasingly biased towards mature enterprises or corporations, which do not create enough jobs to offset unemployment to a significant extent. In spite of these circumstances, alternative means are not being sufficiently utilized to create those jobs which are necessary to reduce unemployment.

The existing job generation gap is a result of the combined development of the culture of dependency and changing economic conditions. The present economic climate necessitates smaller flexible units instead of large organizations which by their very nature lack flexibility.

The trend towards smaller units of economic activity, however, involves managerial skills adapted to the specific situation in smaller companies and it implies additional demand for managers. However, managers, owing to the culture of dependency, are not generated in sufficient numbers. Large corporations regenerate their own supply but complementary managerial skills are not supplied.

14.2.4 LONG-TERM UNEMPLOYMENT AND ITS EFFECTS

At the beginning of the 1970s – before the recession set in in early 1974 – the interval between becoming redundant and finding new employment was a relatively short one. In the years following, however, this process slowed increasingly.

Those groups of the population which became redundant had dwindling chances of becoming re-employed. In this way, a group of 'hard-core' unemployed was formed which continues to grow.

Long-term unemployment is clearly on the rise and is now not only touching so-called problem groups (e.g. older persons, the handicapped, low-qualified persons etc.). It is increasingly becoming a very real threat to young people who have completed vocational training (including highly qualified university graduates) and are hunting for first employment.

Increasing length of unemployment leads not only to greater integration difficulties on the labour market, but also to a certain attitudinal, motivational and self-image change in those involved: during the first weeks of unemployment – if a new job is not already in the offing – unemployed persons are quite active in seeking new employment, making use of a number of different sources. Afterwards, job-hunting behaviour changes and the job market is watched only selectively. In the course of this process, a point of resignation and apathy may be reached which leads to a drop in motivation to continue to job-hunt actively. A person can come to experience feelings of low self-worth, powerlessness and hopelessness which can seriously affect his or her ability to self-assess strengths and weaknesses objectively and realistically appraise the opportunities available to him or her.

14.2.5 THE VALUE SHIFT AND LATENT POTENTIAL

While institutional structures, attitudes and expectations are still largely oriented to a culture of greater dependency, a shift in work values is apparent and gaining in momentum.

Recent years have seen the development of a changing understanding of the role of work and a fundamental re-examination of the basic assumptions behind the organization of work, the relation of work to leisure time and the relation of the individual to work. A growing desire is apparent to create work in a context which is personally meaningful and satisfying. This shifting of values, which had its origins among the younger generations over a decade ago, is spreading to a considerable cross-section of those of working age.

Indeed, under growing pressure and the unappealing prospects of long-term unemployment, the last few years have witnessed mounting numbers of start-ups of small businesses, often making use of newer forms of work organization. In addition to start-ups motivated by the shift of values and a search for work content and organization more in keeping with personal values, there have been start-ups by entrepreneurs wishing to start new but more traditional small enterprises. These numbers of business start-ups have occurred despite overall less-favourable circumstances – that is, despite the depressed economic climate and despite social and institutional undercurrents which are still predominantly oriented to a culture marked by dependency thus tending more to dampen and less to facilitate small business creation.

These developments would seem to indicate that a considerable latent potential for action and new economic activity exists in the population. It can be assumed that a greater proportion of the population would be more likely to take action if conditions for an initial start-up were more favourable and support services were available.

While numbers of start-ups are high, however, there are also high failure rates among young, small enterprises in their first years of business. The failures are not necessarily due to 'bad' business ideas but often to lack of experience and business expertise on the part of new entrepreneurs who are, at first, unable adequately to lay the ground for a business start-up on their own and are unprepared for the variety of tasks demanded in running a business.

This is not surprising in view of the social and institutional framework outlined above. The present educational system cannot provide the necessary basis; it is geared to teaching horizontal but not vertical skills (i.e. it trains welders, carpenters, lawyers or engineers but not managers, entrepreneurs or inventors). Existing institutions and services are not designed to provide financial and other support to potential entrepreneurs low in capital, securities and experience with often rather vague business plans.

Interest and potential exists in the population to take action and find alternatives – and personally satisfying alternatives – to under- or unemployment. If this latent potential is to be adequately tapped and self-employment is to offer to many a viable and longer-term way out of the present dismal economic situation, new services in a new context are needed which allow for specific support on various levels of potential entrepreneurs and personal initiative. It is important that new services are geared to the budding entrepreneur and to young businesses as well as 'before the fact' to stimulating the potential in those who have not yet taken action.

14.2.6 THE CHANGING ROLE OF POLICY MEASURES FOR BUSINESS CREATION AND EXPANSION

Traditionally policy measures to stimulate the creation or expansion of business worked on the basis of making available funds or tax relief. This type of intervention presupposes flexible potential entrepreneurs or existing companies; incentives work under the assumption of realistic business plans for growth and expansion existing in sufficient numbers together with the ability to implement them.

In looking at the process of business creation and stabilization which is displayed in Table 14.1, it becomes apparent that financial incentives only become relevant at the later stages of this process (not during the development process but only at the stage of operationalization or implementation). The existing shortage of entrepreneurial and of managerial skills indicates that – in order to stimulate the creation and expansion of businesses – it seems appropriate to apply measures which sustain the creation process at an earlier stage at which motivational components, information, advice etc. are crucial.

Table 14.1 'PICTIS' – formation phases and requirements

	Requirements				
	Motivation	*Information*	*Finance*	*Labour*	*Premises*
Phase 1: Idea generation	x	x			
Phase 2: Conceptualization	x	x	x		
Phase 3: Testing	x	x	x		
Phase 4: Implementation	x	x	x	x	x
Phase 5: Stabilization	x	x	x	x	x

Source: Adapated from *Empirica: Policy Instruments to Facilitate the Creation of Small and Medium Sized Companies*, Bonn, 1986.

14.3 THE LOCAL PROGRAMME

In the light of the factors outlined above, a programme aimed at the promotion of indigenous potential and oriented to intervening in the present economic and social situation must develop and work within a new conceptual framework involving extremely flexible training structures.

The training offered must work with and complement traditional services introducing new elements and bridging existing gaps. To do so it must cover a broad scope, at the same time remaining responsive to individual needs. Training should range from an initial phase of stimulating the awareness of local residents and local and regional authorities to assistance by providing support in the development of business ideas matching product ideas with potential markets, to long-term on-the-job training, and assistance for groups of people establishing or stabilizing co-operatives or small businesses.

The main emphasis must be on actually training people and not on incentives of whatever nature. Incentives are only useful to the extent that they facilitate the implementation of development projects; they cannot, however, generate such projects. It is imperative that such training differ substantially from traditional concepts. The function of training must be to stimulate and motivate local residents to take the initiative if the existing problems of development are to be overcome. In this framework, training is no longer to be used as an instrument to match people with professions and the globally perceived demand for certain skills. The function of training should rather be first to identify concrete local development potentials and then match them with the available human resources. This should include the considerable economic potential of the increasingly qualified unemployed population as well as the growing numbers of females entering the labour market and the new job profiles demanded by this segment of population.

Using training as a vehicle for local social and economic development, the projects initiated under the auspices of the European Social Fund (ESF) operated along the following guiding principles of the programme:

(a) to harness local economic and social energies through the stimulation and creation of additional economic activities. This entails:

- activating and motivating people for local development;
- bringing about positive effects on local development by training people to help themselves;
- using training in a proactive instead of a reactive mode;

(b) to collaborate with and upgrade local institutions to make them self-sustaining and enable them to continue providing the training services needed for endogenous local development, once the programme is terminated. This involves:

- filling the existing gaps in provisions for endogenous local job creation;
- seeking and encouraging the collaboration of local authorities in this initiative.

14.3.1 STRUCTURE OF THE PROGRAMME

The set of eleven local pilot projects which make up the experimental programme each operated over a period of three years. The projects were selected by ESF staff in collaboration with national, regional and local authorities. The first seven projects began in the first half of 1982; four new projects were added to the programme in 1983. All projects except one were granted a one-year extension of their running time so that the original seven projects terminated their ESF project period at the end of 1985 while the four later projects terminated in December 1986.

In each project, previously existing or newly established private non-profit organizations were responsible for formulating local-development plans and carrying out the administration and organization of programme activities in their own local area.

Due to their varying contexts, the various local areas are marked by economic and social problems which are quite different in nature. They range from areas which are highly dependent on agriculture with excessive hidden unemployment, to highly congested urban areas with problems of industrial reconversion in which underemployment and high registered unemployment is typical.

Table 14.2 displays some of the main characteristics of the local programme. As can be seen, the pilot actions that form part of the programme are located in declining industrial areas, in semi-industrialized zones as well as in rural and underdeveloped areas of the Community. Each pilot action makes use of a selected array of goals, measures and criteria, according to the local socio-economic environment.

Table 14.2 General characteristics of the pilot actions forming part of the programme

Project organizers	*Pilot region and structure*	*Beginning of project*
Enterprise Carrickfergus Ltd Carrickfergus Northern Ireland United Kingdom	Carrickfergus: small town, traditionally dependent upon large declining companies	1962
Creuse-Expansion-Tourisme Gueret France	Creuse: rural, partly tourism	1983
Centre d'Etudes et de Formations Rurales Appliquées Bron France	Haute-Loire; rural, high underemployment	1962
Association de Développement des Pyrénées par la Formation Toulouse France	Pyrenees: rural, mountainous, some tourism	1982

Table 14.2 *Cont'd*

Project organizers	*Pilot region and structure*	*Beginning of project*
Istituto per la promozione dello sviluppo economico e sociale Rome Italy	Cilento: typical under-developed region, lack of indigenous development potential	1982
Tyne & Wear Enterprise Trust Ltd Newcastle upon Tyne United Kingdom	Newcastle upon Tyne: urban, high unemployment, characterized by declining industries	1982
Itinéraire pour le Développement et l'Emploi dans les Voges Remiremont France	Vosges: partly rural, partly manufacturing oriented (textiles etc.)	1982
Wigan New Enterprise Ltd Wigan United Kingdom	Wigan: urban, high unemployment, within community area of Manchester and Liverpool	1982
Association pour la Promotion et la Protection des Iles du Ponant Auray France	Iles du Ponant: small islands, fishery, some tourism	1983
L'Association pour le Développement par la Formation des Pays Aveyronnais et Tarnais Albi France	Tarn and Aveyron: rural, some tourism, area near Toulouse	1983
Dundee Training for Employment of Enterprise Project Dundee Scotland United Kingdom	Dundee: urban, peripheral declining industrial area	1983

Source: *Empirica: Assessment of Locally Initiated Training and Employment Pilot Actions. Final Report*, Pt I, Bonn, May 1987, p. 9.

It is clear that the differing socio-economic situations in the pilot areas profoundly influence the general direction of the programme. In rural areas the emphasis is on improvements to the local social fabric; the activities focus on revitalization, the maintenance of present population or existing economic activities. The projects in rural zones include beside economic goals demographic, social and psychological goals, such as breaking down the social and psychological isolation of residents in rural areas. In the industrialized areas, the creation of additional jobs, upgrading entrepreneurial skills and maintaining or establishing new small businesses is stressed. The programme aims at the stimulation of local entities as diverse as entrepreneurs, small and medium-sized businesses, agricultural or manufacturing co-operatives, family businesses, part-time jobs, or complementary and voluntary activities from which economic benefits may accrue.

While the various project areas of necessity must pursue differing goals and strategies depending on the given local circumstances, they all follow the same general goals which involve the corresponding activities shown in Figure 14.1.

The local programme works to promote a new role of training in economic development. The function of training is not just to provide people with professional skills as they are globally demanded but rather to use training in a larger sense to overcome specific local development problems. This orientation profoundly influences the nature and contents of training. In the pilot programme, people are not trained for professions independently from where and in which industry the skills acquired may serve them. Rather the training offered is goal-oriented and specifically tailored to concrete projects serving perceived local development needs or potentials.

14.3.2 THE COMPLEMENTARITY OF THE LOCAL TRAINING PROGRAMME TO EXISTING STRUCTURES

The function of the local training programme is not to replace traditional institutions but rather to operate and provide training measures in the gaps left by existing structures.

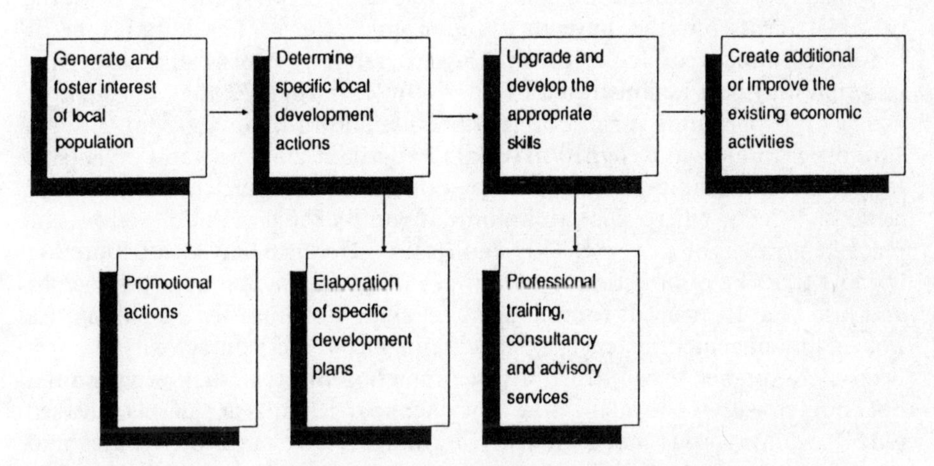

Figure 14.1 General goals and activities of the pilot actions. Source: *Empirica: Assessment of Locally Initiated Training and Employment Pilot Actions. Final Report*, Pt I, Bonn, May 1987, p. 10.

During the period of its functioning, the local training programme has proved to be largely complementary to existing services and has been able to introduce a number of elements which had not been available before. Some of these elements include the use of active recruitment, collaboration with various other organizations, the information channels used, the use of training to match concrete local needs for certain skills with local supply, the way in which services are offered and the use of integrated measures. The use of integrated measures, the information channels used as well as the way in which services are offered have particular relevance to the creation of jobs and new economic activities and accordingly will be discussed in this chapter.

One of the most remarkable characteristics of the local programme is the conceptual framework on which it is based. While traditional measures are largely segmented, the programme follows a holistic 'integrated approach'. This means that, instead of planning and implementing independent measures for certain groups, such as young people, unemployed executives, women, long-term unemployed etc., local labour markets are seen and approached as a whole. This approach makes it possible to take given constellations specific to the local labour market into consideration.

Such a holistic approach is only possible if there is a central coordinating point which is well informed of the activities, needs, problems and potential of a given local area. All of the programme projects, in fact, function to a greater or lesser degree on the basis of local information networks. In this way, local accountants, politicians, agencies etc. are aware of programme activities and refer potential entrepreneurs to the pilot projects. The projects, in turn, are aware of where funding is available for what, where business premises are vacant, which enterprises are looking for people with which skills, and so on.

A local flow of information among all relevant parties is not only desirable in social terms but can have direct economic effects. The importance of the establishment of local information-exchange networks and how they function may best be illustrated by an example. In the UK, the management extension programme is run both nationwide and through the pilot actions. This programme was designed to retrain redundant managers and executives for smaller companies. The overall placement rate of participants is 30 per cent. In Wigan, where this programme is run by the local pilot action, the placement rate is 80 per cent. This significant difference may be attributed to the fact that the pilot action is well aware of where smaller enterprises exist and just what their needs for management skills are. It is then a small matter to bring management extension participants into direct contact with the respective companies. In contrast, after completion of other such programmes not coupled with the local training approach, participants are simply released onto the anonymous labour market and have many more difficulties and much less of a chance of finding employment.

It is true that, with few exceptions, all individual components of training and assistance (e.g. vocational training, financial sourcing, market analysis etc.) made available through the local programme to potential entrepreneurs are generally offered by various existing institutions or private consultants.

While there is some overlap with traditional structures as to the kinds of services offered, there are few or no duplications with regard to preconditions for the use of traditional institutions and the local programme or the way in which services are offered. The use of traditional services puts the potential entrepreneur in the position of having to 'sell' his ideas to various agencies, banks etc. or to be willing - and financially able - to employ a private consultant to do this work for him.

The necessity of having to 'sell one's idea' and 'to go it alone' can have undesirable consequences. Potential entrepreneurs often do not have the skills themselves and do not know where to get the assistance needed to cope with the many kinds of problem emerging on the way from an idea to the actual setting up of a business. This lack of problem-solving strategies is often compounded by the fact that, instead of being dealt with as they crop up, difficulties are sometimes covered up in order that an idea might appear more worthwhile and gain in 'marketability'.

In addition, most traditional services go on the premise that those seeking assistance already have a solid idea of what they are going to do and how they are going to do it and that they need mainly marginal help in locating sources of finance. This means that, at the time he begins to approach traditional services, the potential entrepreneur must already have a concrete and well-formulated business idea as well as the strong personal motivation necessary to see the various procedural steps through more or less alone. No provisions are made for those with just budding ideas or for those who want to 'do something' but do not yet know exactly what.

By contrast, aside from 'having heard of the programme' and 'interest in taking a look', there are no rigid preconditions for the use of the local training programme services. The local programme is unique in the way in which it offers its services. All assistance and support services necessary to facilitate the creation of businesses are available from one source and do not have to be painstakingly ferreted out in a multitude of institutions.

Another unique feature of the programme is the basic procedure it has adopted (see Figure 14.2 for an overview of the programme's general task structure). Services offered go far beyond those generally offered elsewhere and individuals or groups are given support from the moment interest is expressed in finding a workable business idea to the actual birth of a business and beyond.

In fact, the task of the programme often begins even before active interest is expressed in 'doing something'. At this stage, motivation seminars are held to reorient participants from seeing only the inevitability of their situation and passively waiting for someone to offer them jobs to accepting the idea that they can take an active role in their future.

At the next stage, group and individual training is available to assist partici-

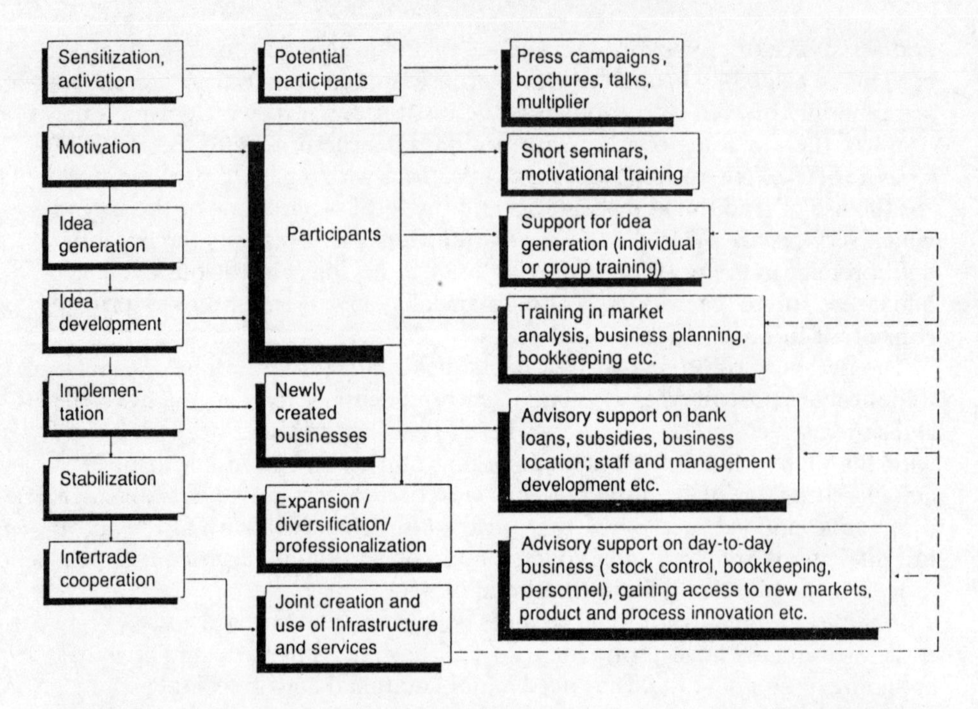

Figure 14.2 General task structure of training for development. Source: *Empirica; Assessment of Locally Initiated Training and Employment Pilot Actions*, Final Report, Pt I, Bonn, May 1987, p. 13.

pants to develop and formulate business ideas on the basis of given skills and interests. Potential entrepreneurs are then given support in setting the stage for their future business, i.e. they are assisted in sourcing grants and aids, carrying out a market analysis, developing marketing strategies, estimating necessary skills, capital, equipment, forward planning etc. When the point of implementation – the actual business set-up – has been reached, new entrepreneurs are given on-the-job training as to how to run a business and gain the necessary technical and managerial/administrative skills as they go.

With this procedure, many potential difficulties can be recognized at an early stage and nipped in the bud. This procedure also enables emerging entrepreneurs to gain the experience and learn the skills and strategies necessary to deal successfully with difficulties which may arise in the future. With the help of this conception-to-birth-to-weaning support offered by the local programme several enterprises have been created in the pilot areas which would certainly not exist had such assistance not been available.

This multifaceted training programme offers a number of advantages and is certainly integral to the programme concept. Its very extensiveness, however, and the fact that it offers a great deal more support than existing services, involves a latent danger to which the local training programme should be constantly alert: namely, the danger of overshooting its goal and coddling participants by providing well-meaning though too much 'assistance'.

Not only the amount of assistance but also the structures within which assistance is provided will greatly influence the actual effects achieved by training. Traditionally, training involves one or more persons who 'know' (the teacher(s)) and one or more persons who 'don't know' (the student(s)). This constellation is based on a particular form of relationship which can easily fall into and serve to reinforce old dependency patterns.

One of the programme's focal points of orientation is its 'help people to help themselves philosophy'. This and the procedures it adopts in pursuing this objective further distinguish the training programme from most existing educational provisions and consultancy services.

The purpose of local programme training and support is to equip people with the appropriate base of knowledge and the skills necessary to enable them to master their situation actively and independently. The training and assistance made available to the local population must focus on the intrinsic problems of all participating parties and support personal involvement and initiative. It should *not* provide substitutes for the latter.

It is essential that training serves to activate individual abilities and skills to the point that participants in training become able to analyse their own situations, assess where strengths and weaknesses lie, collect the necessary information and make their own decisions. It is equally essential that this often lengthy and protracted process begins with the first contact a participant has with a project; it cannot be postponed until after a business start-up.

In the long run, neither the programme, the local area nor the trainee is served if an individual is put on his feet by project staff before he has learned the elementary skills of crawling and walking on his own; the first time the going gets difficult, he may fall and be in exactly the same position he was in before.

The 'coddling approach' presents a very real danger since it is so near to structures both instructors and participants have all grown up with and have come to accept – and perhaps expect – as 'normal' in many areas of education, training and consultancy (for example, in traditional business consultancy, a contractor pays for the consultant to do something for him and not for the consultant to assist him in doing that something himself). It is important that this danger remains in the awareness of all internal and external project staff in order that this self-defeating approach may be avoided. To illustrate the 'coddling approach', preparing business plans for participants, submitting them to banks or development agencies, raising funds for participants etc. fall in this category. The category also includes telling participants what to do and exactly what step to take next. Instead of promoting personal involvement and initiative they rather serve to reinforce those passive expectations the programme aims to overcome. An approach more likely to prepare trainees for independent coping and personal competence would be to make participants aware of different sources and channels of information, leaving it to each trainee to experiment with the results each source produces; assisting participants to set up a preliminary business plan, helping and encouraging them to find and close the gaps themselves and strengthen weak points; making participants aware of the various factors to be considered in raising funds, allowing trainees themselves to decide on the procedures they wish to use, etc. It is true that this second approach is more difficult and time-consuming in a short run. It is, however, the approach which will provide individuals with the expertise and skills to be able to deal adequately and flexibly with varying business situations in the future.

14.3.3 OBJECTIVES AND CONTENTS OF TRAINING ACTIONS

Table 14.3 displays the objectives of training courses by project for the entire period of project operation. As can be seen from this table, most of the French and Italian seminars and courses have focused mainly on vocational training and to a lesser extent on the management and creation of businesses, whereas the main emphasis in Creuse and Tarn and Aveyron has been on the acquisition or improvement of skills. With regard to the United Kingdom projects, the main emphasis in Wigan has been on the management and creation of businesses followed in close second and third position by the acquisition or improvement of specific skills and the development of business ideas; Tyne and Wear has clearly focused on the management and creation of enterprises; in Carrickfergus, the main concern has been the acquisition or the improvement of specific skills; whereas in Dundee, the main focus has been on self-improvement and motivation.

A separate consideration of courses organized in 1982/3, 1984 and 1985/6 indicates that some interesting developments have taken place with regard to course objectives. Courses in Cilento and the Pyrenees continue to deal primarily with vocational training but, while Cilento initially paid little attention to the creation of businesses, a larger proportion of its courses now cover this area. The opposite seems to be true for the Pyrenees.

With regard to the United Kingdom projects, while Tyne and Wear courses remained true to their original central focus, namely the creation of businesses, both Carrickfergus and Wigan courses have changed their main points of emphasis. In both projects, the initial emphasis was on vocational training. In Carrickfergus, the focus then changed to the acquisition and improvement of new skills and the development of business ideas and in the last years to training for project collaborators. In contrast, the main concern in Wigan progressed from vocational training to the development of business ideas to the creation and management of businesses.

As Table 14.4 shows, overall, the majority of participants begin training with the intention of creating their own business or participating in the creation of cooperatives. The greatest number of participants attended

Table 14.3 Objectives of group training (number of activities) (1982-6) (%)

General objective of the action	*Cilento*	*Haute-Loire*	*Pyrenees*	*Vosges*	*Creuse*	*Iles du Ponant*	*Tarn and Aveyron*	*Carrick-fergus*	*Tyne and Wear*	*Wigan*	*Dundee*
Discussion of local development problems and potentials	15.1	7.3	5.4	10.4	14.9	0.0	14.3	5.3	4.8	2.6	4.1
Vocational training course	39.5	14.6	25.0	9.0	6.4	77.8	14.3	15.8	7.3	4.3	4.1
Technical, legal, fiscal and statutory information	7.0	5.2	7.6	4.5	21.3	11.1	0.0	5.3	0.8	0.4	0.0
Management and creation of enterprises, cooperatives and associations	14.0	1.0	15.8	13.4	4.8	0.0	4.8	5.3	28.2	25.0	2.0
Search for new activities and new revenues	3.5	0.0	8.7	16.4	9.6	0.0	14.3	0.0	2.4	2.2	2.0
Acquisition or improvement of specific skills	3.5	3.1	4.3	4.5	35.6	0.0	33.3	26.3	4.8	22.8	16.3
Development of business ideas	2.3	2.1	5.4	1.5	4.3	0.0	0.0	21.1	5.6	19.8	22.4
Training for project collaborators	1.2	0.0	0.0	3.0	0.5	0.0	0.0	10.5	7.3	0.9	0.0
Self-improvement, positive reinforcement, motivation	7.0	7.3	8.7	14.9	1.6	0.0	14.3	5.3	10.5	3.0	36.7
Other	3.5	25.0	6.0	16.4	1.0	0.0	4.8	5.3	26.6	18.5	12.2
NA	3.5	34.4	13.0	6.0	0.0	11.1	0.0	0.0	1.6	0.4	0.0
Number of responses	86	96	184	67	188	9	21	19	124	232	49

Source: *Empirica: Assessment of Locally Initiated Training and Employment Pilot Actions, Final Report*, Pt V, Bonn, May 1987, p. 4.

courses for this initial reason in Tyne and Wear followed by Wigan, whereas participants in Dundee were most interested in acquiring self-confidence followed by acquiring skills to find a job. The greatest number of participants in Haute-Loire came to training courses wishing to expand their present activity. In the Iles du Ponant, Creuse, Vosges, Aveyron and the Pyrenees, most participants entered training to become more efficient in their present activity while in Carrickfergus, the largest proportion of participants attended to acquire skills to find a job. In Cilento, an equal proportion of participants entered training to start their own business as to acquire skills to find a job.

Table 14.5 gives an overview of the main subjects dealt with in group training over the course of the programme. On the whole, relatively high emphasis was placed on such topics as financing and marketing across projects. While on-the-job training and the evaluation of problems or needs have received considerable attention in the Italian and the French projects, the former seems to have been of only minimal importance in the United Kingdom projects with the exception of Dundee. In this regard, however, it must be noted that much of the on-the-job training carried out by the United Kingdom projects is done on a one-to-one basis and is therefore not included in Table 14.5. The United Kingdom projects have tended to place higher emphasis on more traditional subjects throughout and in the later project years on the development of business ideas.

14.3.4 STRUCTURE OF PARTICIPANTS

Table 14.6 gives an overview of the socio-demographic characteristics of participants in the individual projects. On the whole, participation rates reflect the objectives of the projects as well as the structure of the regions' labour markets.

Over all projects, there were fairly balanced proportions of employed and unemployed participants (46.9 per cent and 53.1 per cent respectively) with a slight predominance of the unemployed.

In general, almost twice as many men as women took part in the local programme. Participants tended to be over 25 years of age (over 70 per cent) and to have over five years of professional experience (61.5 per cent).

A closer look, however, reveals systematic differences between the French and Italian projects on the one hand and the United Kingdom projects on the other. The majority of the French and Italian projects evidence higher participation rates for women and young people; the results with regard to women are particularly striking as these regions have significantly lower activity rates for women. Given the high participation of young people, women and the relatively low female activity rates in these regions, it is not surprising that large proportions of course participants in these projects have no learned vocation or under five years of working experience.

With regard to work status, the French projects exhibit higher participation rates of employed persons while Cilento and the United Kingdom

Table 14.4 Objectives of participants in training courses (1982-86) (%)

	Start own business/ co-operative	*Acquire skills to find job*	*Be more efficient in present activity*	*Expansion of present activity*	*To acquire self-confidence*[1]	*Communication*[2]	*Number of responses*
Cilento	34.1	34.1	32.3	16.8	—	30.7	375
Haute-Loire	15.7	16.6	29.8	31.9	—	31.0	477
Pyrenees	33.3	21.5	49.3	25.7	—	24.9	840
Vosges	41.2	16.4	48.5	25.2	—	17.9	330
Creuse	36.2	18.1	55.4	22.6	—	24.9	177
Iles du Ponant	18.2	10.0	69.1	10.9	—	14.5	110
Tarn and Aveyron	27.2	8.6	75.7	32.1	—	19.8	81
Carrickfergus	25.7	43.2	37.8	13.5	32.4	—	74
Tyne and Wear	69.7	14.8	17.4	14.4	35.6	—	264
Wigan	46.9	32.9	28.9	23.2	35.8	—	827
Dundee	40.2	48.5	11.3	19.1	63.4	—	194

[1] Only for English sites.
[2] Only for French and Italian sites.

Source: *Empirica: Assessment of Locally Initiated Training and Employment Pilot Actions*, Final Report, Pt V., Bonn, May 1987, p. 5.

Table 14.5 Contents of seminars and training courses (1982-6) (and percentage of trainers who named topics, multiple response)

Cilento	*Haute-Loire*	*Pyrenees*	*Vosges*	*Creuse*	*Iles du Ponant*	*Tarn and Aveyron*	*Carrick-fergus*	*Tyne and Wear*	*Wigan*	*Dundee*
General adminis-tration (38.4%)	Financing (15.6%)	On-the-job training (35.3%)	Evaluation of problems and needs (43.3%)	Evaluation of problems and needs (34.0%)	Financing (77.8%)	On-the-job training (42.9%)	Financing (26.3%)	Marketing (51.6%)	Marketing (47.0%)	Evaluation of problems and needs (34.7%)
On-the-job training (38.4%)	Prototype develop-ment (13.5%)	Financing (27.7%)	Assistance to indivi-dual proj. (43.3%)	Financing (32.4%)	Prototype develop-ment (33.3%)	Production techniques (38.1%)	Marketing (26.3%)	Market research (44.4%)	Financing (42.7%)	Verbal, written and presenta-tion skills (28.6%)
Marketing (31.4%)	Market research (8.3%)	Technical training (theory) (25.5%)	On-the-job training (35.8%)	On-the-job training (19.7%)	Market research (22.2%)	Marketing (19.0%)	Accounting (15.8%)	Financing (43.5%)	Market research (42.7%)	Financing (28.6%)
Technical training (theory) (31.4%)	Accounting (7.3%)	Production techniques (23.9%)	Financing (32.8%)	Prototype techniques (19.7%)	Production techniques (22.2%)	Technical training (theory) (19.0%)	Grants and aids (15.8%)	Preparation of a busi-ness plan (41.9%)	Development of business ideas (35.8%)	Market research (24.5%)
N=86	N=96	N=184	N=67	N=188	N=9	N=21	N=19	N=124	N=232	N=49

Source: *Empirica: Assessment of Locally Initiated Training and Employment Pilot Actions*, Final Report, Pt V, Bonn, May 1987, p. 6.

Table 14.6 Socio-demographic characteristics of participants by project site

	Cilento	*Haute-Loire*	*Pyrenees*	*Vosges*	*Creuse*	*Iles du Ponant*	*Tarn and Aveyron*	*Carrick-fergus*	*Tyne and Wear*	*Wigan*	*Dundee*	*Total*
Sex												
male	51.3	47.4	57.4	62.9	74.1	64.1	82.5	83.3	76.5	57.9	68.8	60.1
female	48.7	52.6	42.6	37.1	25.9	35.9	17.5	16.7	23.5	42.1	31.2	39.9
Age												
under 25	59.7	29.9	37.7	16.9	15.0	50.0	13.8	31.4	15.1	28.3	28.5	28.8
25-64	40.3	70.1	62.3	83.1	85.0	50.0	76.2	67.2	83.7	71.6	71.5	70.8
65 and over	—	—	—	—	—	—	10.0	1.4	1.2	0.1	—	0.4
Education												
16 and younger	53.0	24.0	26.0	48.0	22.9	30.0	35.5	33.9	66.5	78.3	76.9	47.0
after 16	47.0	76.0	74.0	52.0	77.1	70.0	64.5	66.1	33.5	21.7	23.1	53.0
Professional experience												
no	52.4	9.3	12.2	3.5	6.6	15.8	4.8	7.7	1.2	4.9	6.8	8.9
5 years or less	40.8	36.6	37.7	33.7	26.4	34.2	28.6	30.8	16.1	16.9	20.5	29.6
more than 5 years	6.8	54.1	50.1	62.8	67.0	50.0	66.6	61.5	82.7	78.2	72.7	61.5
Marital status												
single	73.5	39.0	56.9	26.3	32.4	49.5	37.2	28.2	27.6	30.4	37.9	42.5
married	23.0	55.2	37.7	67.2	64.7	45.7	61.5	67.6	65.4	61.4	51.6	51.6
widowed/divorced	0.5	5.8	5.4	6.5	2.9	4.8	1.3	4.2	7.0	8.2	10.5	5.9
Income												
low	67.1	36.7	53.9	28.5	31.9	40.4	26.7	38.6	52.6	62.4	77.9	51.4
average	29.6	59.8	45.3	58.1	58.4	57.7	56.6	54.4	42.2	31.3	21.4	43.5
high	3.3	3.5	0.8	13.4	9.7	1.9	16.7	7.0	5.2	6.3	0.7	5.1
employed	32.8	67.6*	58.4*	73.3	54.2	83.6	70.4	58.1	37.1	29.1	12.4	46.9
unemployed	67.2	32.4	41.6	26.7	45.8	16.4	29.6	41.9	62.9	70.9	87.6	53.1
No. of responses	375	477	840	330	177	110	81	74	264	827	194	3749

* This question did not exist in the first questionnaire.

Source: *Empirica: Assessment of Locally Initiated Training and Employment Pilot Actions*, Final Report, Pt V, Bonn, May 1987, p. 7.

projects drew larger proportions of the unemployed. It must be noted, however, that the category 'employed' also includes underemployed persons.

Tables 14.7–14.9 show the professional background and distribution of the major groups of participants for each project. Seen as a whole, with few exceptions, the main groups represented reflect the basic structure of the pilot areas and have remained relatively stable throughout the life of the projects.

A comparison of 1982/3, 1984 and 1985/6, however, reveals the developments which have occurred in this area during the course of the projects. With regard to the United Kingdom projects, Tyne and Wear has continued to reach more qualified groups throughout the project. However, while Entrust attracted almost exclusively more qualified participants in the earlier project years, it has gone on to reach larger proportions of housewives and unskilled manual workers in the last years. Wigan, too, began with a more highly qualified group but progressed to include a broader base of participants, while Carrickfergus originally attracted particularly large proportions of business owners and skilled workers, but as this project neared its close semi-skilled workers and trainees made up the largest proportion of participants. Dundee has been able to reach larger proportions of lower-qualified groups. The French and Italian projects seem to have reached a relatively broad base throughout the four years although there have been more shifts in the specific groups represented.

14.4 EXAMPLES OF NEW BUSINESS CREATION

The main and ultimate objective of the training actions is to create viable new jobs or economic activities and to stabilize others.

As these goals are similarly pursued by many existing services and traditional institutions, it is of particular interest to consider the procedures followed by local programme projects and the ways in which the services they offer differ from and complement already existing services. In this context, a number of questions arise. Just what is the contribution being made by the local programme with respect to the creation of jobs or new economic activities? In what ways do programme services represent something new? Are they able to reach those for whom traditional measures are somehow inadequate or inappropriate?

This section examines those training actions explicitly geared to the creation of new economic activities, their procedures and results. Section 14.4.1 gives an overview of relevant training actions, and section 14.4.2 presents several examples focusing on the particular contribution made in this area by each of the pilot projects, the jobs and economic activities created as a result of training actions and the problems encountered in the implementation process.

Table 14.7 Professional background of participants in training courses (1982/3): main groups represented

Cilento	*Haute-Loire*	*Pyrenees*	*Vosges*	*Creuse*	*Iles du Ponant*	*Tarn and Aveyron*	*Carrick-fergus*	*Tyne and Wear*	*Wigan*	*Dundee*
Students	Housewives	Farmers	Farmers				Owners of businesses	Skilled manual workers	Managers and executives	
(33.3%)	(22.4%)	(16.7%)	(27.9%)	—	—	—	(30.2%)	(18.7%)	(22.0%)	—
Skilled manual workers	Clerical workers	Self-empl. profess.	Semi-skilled manual workers				Skilled manual workers	Clerical workers	Clerical workers	
(8.7%)	(12.1%)	(9.0%)	(11.7%)	—	—	—	(18.6%)	(13.5%)	(19.8%)	—
Civil servants	Civil servants	Housewives	Managers and executives				Semi-skd. manual workers	Managers and executives	Skilled manual workers	
(8.7%)	(9.4%)	(7.8%)	(10.4%)	—	—	—	(14.0%)	(11.6%)	(15.4%)	—
Self-empl. profess.	Family help	Clerical workers	Clerical workers				Unskilled manual workers	Self-empl. profess.	Owners of businesses	
(7.2%)	(4.6%)	(7.5%)	(8.4%)	—	—	—	(7.0%)	(10.3%)	(11.0%)	—
N=69	N=415	N=335	N=154	—	—	—	N=43	N=155	N=91	—

Source: *Empirica: Assessment of Locally Initiated Training and Employment Pilot Actions*, Final Report, Pt V, Bonn, May 1987, p. 8.

Table 14.8 Professional background of participants in training courses (1984): main groups represented

Cilento	*Haute-Loire*	*Pyrenees*	*Vosges*	*Creuse*	*Iles du Ponant*	*Tarn and Aveyron*	*Carrick-fergus*	*Tyne and Wear*	*Wigan*	*Dundee*
Housewives	Civil servants	Farmers	Skilled manual workers	Owners of businesses	Farmers	Farmers	Owners of busi-nesses	Skilled manual workers	Managers and executives	
(17.3%)	(29.0%)	(14.7%)	(14.8%)	(38.3%)	(50.0%)	(37.3%)	(50.0%)	(19.6%)	(13.3%)	—
Agricul-tural workers	Clerical workers	Clerical workers	Managers and executives	Farmers	Agricult. workers	Owners of businesses	Managers and executives	Managers and executives	Skilled manual workers	
(11.7%)	(22.6%)	(13.5%)	(13.9%)	(25.5%)	(10.5%)	(11.9%)	(21.4%)	(15.2%)	(12.8%)	—
Students	Spouses of business owners/ farmers	Skilled manual workers	Clerical workers	Self-empl. profess.	Family help	Civil servants	Skilled manual workers	Clerical workers	Owners of businesses	
(11.2%)	(22.6%)	(11.6%)	(13.0%)	(12.8%)	(10.5%)	(10.4%)	(14.3%)	(10.9%)	(9.6%)	—
Trainees or appren-tices	Farmers	Housewives	Farmers	Family help	Skilled manual workers	Clerical workers	Housewives	Owners of businesses	Clerical workers	
(9.8%)	(8.1%)	(9.7%)	(10.4%)	(4.3%)	(5.3%)	(9.0%)	(7.1%)	(6.5%)	(8.0%)	—
N=214	N=62	N=259	N=115	N=47	N=38	N=67	N=14	N=92	N=188	—

Source: *Empirica: Assessment of Locally Initiated Training and Employment Pilot Actions*, Final Report, Pt V, Bonn, May 1987, p. 9.

Table 14.9 Professional background of participants in training courses (1985/6): main groups represented

Cilento	*Pyrenees*	*Vosges*	*Creuse*	*Iles du Ponant*	*Tarn and Aveyron*	*Carrick-fergus*	*Tyne and Wear*	*Wigan*	*Dundee*
Housewives (20.7%)	Farmers (18.7%)	Farmers (32.8%)	Skilled manual workers (25.4%)	Farmers (25.0%)	Farmers (35.7%)	Semi-skd. manual workers (47.1%)	Managers and executives (23.5%)	Clerical workers (14.2%)	Skilled manual workers (15.5%)
Students (19.6%)	Clerical workers (16.3%)	Spouses of business owners/ farmers (13.1%)	Farmers (23.8%)	Spouses of business owners/ farmers (15.3%)	Agricult. workers (28.6%)	Trainees or appren-tices (11.8%)	Self-employed professionals (11.8%)	Managers and executives (12.2%)	Semi-skd. manual workers (14.4%)
Skilled manual workers (8.7%)	Housewives (9.8%)	Clerical workers (9.8%)	Clerical workers (10.8%)	Family help (11.1%)	Family help (7.1%)	Managers and executives (5.9%)	Housewives (11.8%)	Housewives (12.2%)	Unskilled manual workers (11.9%)
Clerical workers (4.3%)	Skilled manual workers (9.3%)	Skilled manual workers (6.6%)	Housewives (8.5%)	Skilled manual workers (6.9%)	Housewives (7.1%)	Students (5.9%)	Unskilled manual workers (11.8%)	Skilled manual workers (11.1%)	Housewives (8.8%)
N=92	N=246	N=61	N=130	N=72	N=14	N=17	N=17	N=548	N=194

Source: *Empirica: Assessment of Locally Initiated Training and Employment Pilot Actions*, Final Report, Pt V, Bonn, May 1987, p. 9.

14.4.1 OVERVIEW OF TRAINING ACTIONS GEARED TOWARDS THE CREATION OF NEW ECONOMIC ACTIVITIES

To lay the ground for subsequent, more specific analyses of the creation of new economic activities, this section gives an overview of those training activities (group and individual) which were planned and executed by the pilot actions with the explicit purpose of creating new economic activities.

Table 14.10 shows the proportion of training courses dedicated to the creation of new economic activities in each pilot action. As can be seen from this table, about 30–40 per cent of participants in training courses indicate planning to set up a business or co-operative. In Wigan and Newcastle, this proportion is significantly above average, whereas it is very small in Haute-Loire.

The professional characteristics of those participants wanting to start their own business or planning to participate in the creation of a cooperative as well as those participants who have other objectives are displayed in Table 14.11. As can be seen, participants planning to create businesses or co-operatives differ from those who want to achieve other goals in several aspects. In most of the projects (with the exceptions of Haute-Loire, the Pyrenees and Tarn and Aveyron), the group of potential 'creators' consists mainly of workers; in Tarn and Aveyron, 'creators' are predominantly farmers. The Pyrenees and Haute-Loire are in an exceptional situation, with family aids, spouses of business owners and housewives heading the list of potential creators of businesses. In many cases participants with other objectives have another professional background.

14.4.2 SELECTED EXAMPLES OF TRAINING ACTIVITIES

In the previous section, we discussed the general significance and some of the main characteristics of individual and group training devoted to the creation of new economic activities. In order to analyse the specific contribution made by the pilot actions in the job-creation process as well as the nature of the local development problems faced by the pilot actions, this section presents a set of actual cases and the measures used to deal with them by the projects.

The examples are presented in Figures 14.3–14.9 according to the following standard format:

Phase 1: initiation of contacts between participant(s) and pilot actions, preparatory work to formulate project ideas.

Phase 2: training and support phase to lay grounds for establishment of new economic activities.

Phase 3: start-up phase and services made available during implementation.

Comments: The examples are commented to illustrate specific problems which arose as well as to highlight areas in which project activities were complementary to programmes of existing structures.

Table 14.10 Participants in group and individual training planning to create new economic activities

	Cilento	*Haute-Loire*	*Pyrenees*	*Vosges*	*Creuse*	*Iles du Ponant*	*Tarn and Aveyron*	*Carrick-fergus*	*Tyne and Wear*	*Wigan*	*Dundee*
At the beginning of the course											
Plan to create a business	19.2%	14.3%	23.7%	35.8%	33.3%	15.5%	16.0%	23.0%	63.3%	45.7%	37.1%
Plan to create a cooperative	18.9%	1.5%	11.7%	6.7%	2.8%	3.6%	11.1%	2.7%	8.0%	1.6%	6.2%

Source: *Empirica: Assessment of Locally Initiated Training and Employment Pilot Actions. Final Report*, Pt VII, Bonn, May 1987, p. 2.

Table 14.11 Professional background of participants in group training planning to start a business or co-operative

	Cilento	*Haute-Loire*	*Pyrenees*	*Vosges*	*Creuse*	*Iles du Ponant*	*Tarn and Aveyron*	*Carrick-fergus*	*Tyne and Wear*	*Wigan*	*Dundee*
Professional background of persons who plan to start their own business or want to participate in the creation of a co-operative											
Workers	25.0	8.0	12.5	25.7	26.6	40.0	27.3	47.4	29.3	30.2	42.3
Executive manager, owner of business, self-employed	6.3	0.0	7.1	19.9	20.3	5.0	9.1	21.1	19.6	20.9	16.7
Clerical worker	5.5	16.0	12.1	11.8	12.5	5.0	9.1	5.3	12.0	8.5	6.4
Farmer	7.0	6.7	18.9	9.6	4.7	0.0	36.4	0.0	0.0	0.0	0.0
Housewife, family help, spouse of a business owner	14.8	40.0	20.4	11.0	17.2	15.0	9.1	5.3	4.9	9.3	12.8
Student, trainee or apprentice	24.2	5.3	5.7	2.2	1.6	0.0	4.5	0.0	7.1	8.2	5.1
Civil servant	1.6	2.7	2.5	3.7	3.1	0.0	0.0	0.0	3.3	4.4	2.6

Professional background of persons who do not want to start their own business or want to participate in creation of a co-operative											
Workers	21.9	4.2	12.3	13.9	17.7	6.7	6.8	34.5	12.5	11.2	41.4
Executive manager, owner of business, self-employed	9.3	5.2	11.6	8.8	20.4	5.6	23.7	43.6	41.3	32.6	6.9
Clerical worker	6.1	12.9	11.8	9.3	6.2	3.3	6.8	1.8	11.3	17.8	3.4
Farmer	5.3	4.7	15.5	32.0	35.4	41.1	37.3	0.0	0.0	0.0	0.0
Housewife, family help, spouse of a business owner	16.6	35.3	13.6	12.4	5.3	12.2	16.9	3.6	11.3	15.3	10.3
Student, trainee or apprentice	23.5	3.5	7.3	1.0	0.9	3.3	0.0	7.3	1.3	8.4	16.4
Civil servant	2.4	13.7	1.8	6.2	0.9	0.0	11.9	3.6	2.5	0.5	2.6

Source: *Empirica: Assessment of Locally Initiated Training and Employment Pilot Actions. Final Report*, Pt VII, Bonn, May 1987, p. 2.

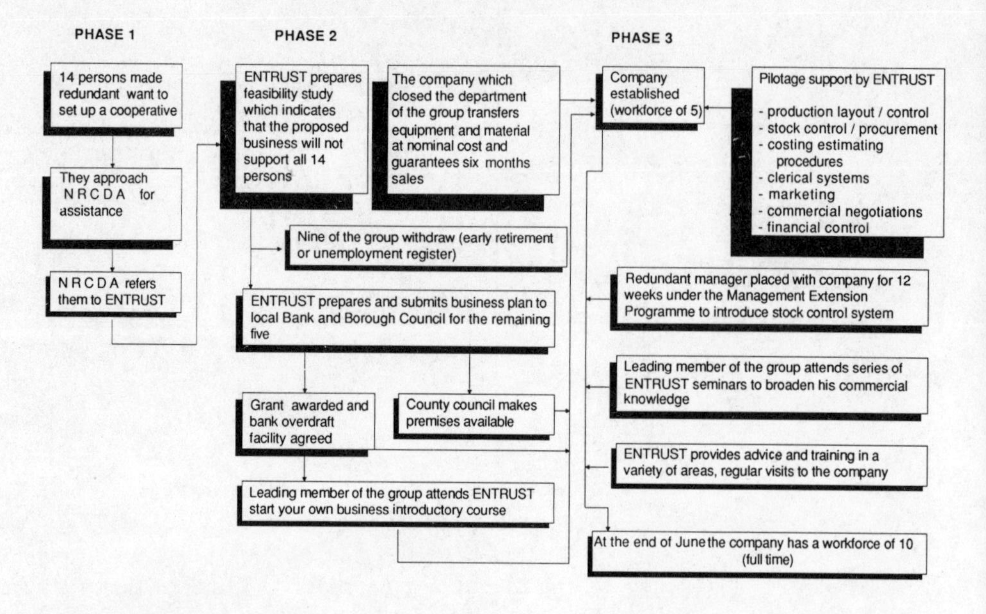

Figure 14.3 (Example 1) Creation of a company for the manufacture of insulation joinery (ENTRUST, Newcastle upon Tyne). NRCDA is the national organization for the development of co-operatives. Source: *Empirica: Assessment of Locally Initiated Training and Employment Pilot Actions, Final Report*, Pt VII, Bonn, May 1987, p. 4.

Comments: The workforce had no managerial knowledge. Moreover, the leading member of the group had been the union representative at the company which closed down.

The NRCDA referred the group to ENTRUST because it could not give the in-depth assistance necessary. At ENTRUST, the current manager of the new company took advantage of the whole range of services for new starters. He continues to make use of services for existing businesses to gain the skills and competence necessary to be able to continue without assistance.

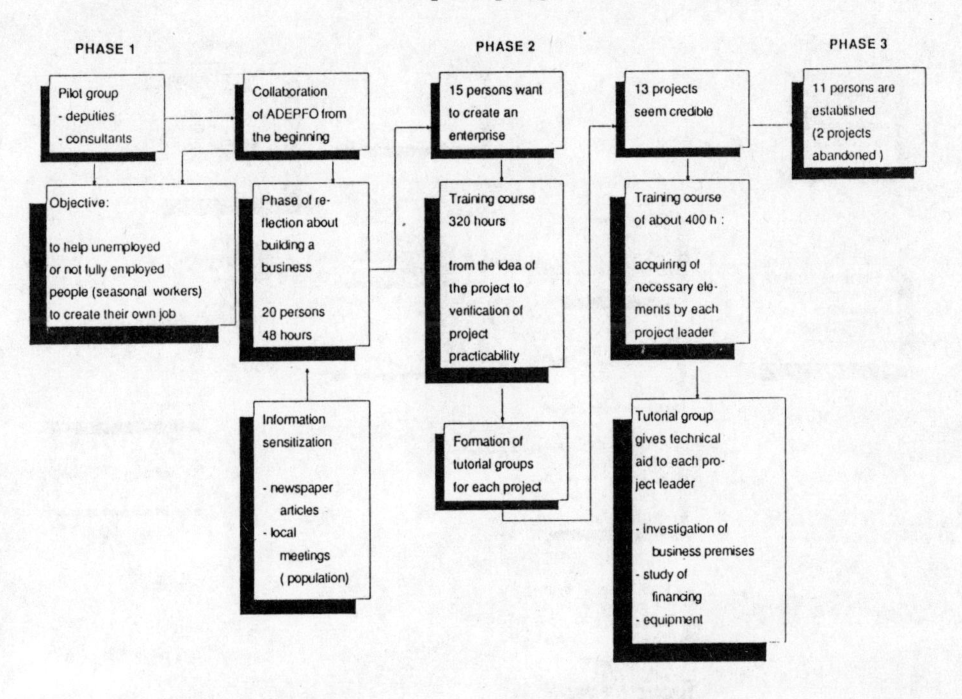

Figure 14.4 (Example 2) Creation of different enterprises and independent jobs (ADEPFO, Pyrenees). Source: *Empirica: Assessment of Locally Initiated Training and Employment Pilot Actions, Final Report*, Pt VII, Bonn, May 1987, p. 7.

Comments: The involvement of all concerned partners makes possible the integration in a previously determined field. Procedure: information/determination/orientation; the content of the training course is tailored to individual needs; technically based training and implementation; each individual project falls under the responsibility of a special tutorial group.

It must be noted that traditional services do not allow for the financing of this type of procedure.

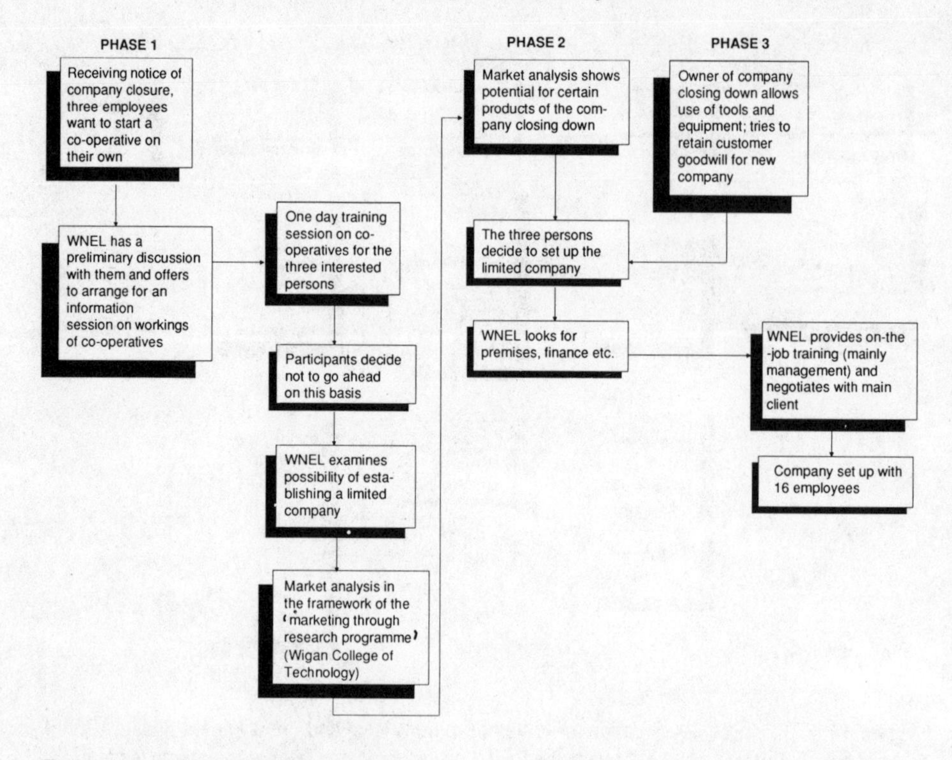

Figure 14.5 (above) (Example 3) Creation of a company for the manufacture of layette boxes etc. (Wigan New Enterprise). The three participants invested £9,000. WNEL arranged a bank loan of £10,000. Source: *Empirica: Assessment of Locally Initiated Training and Employment Pilot Actions, Final Report*, Pt VII, Bonn, May 1987, p. 8.

Comments: MBL Products Limited concerned the impending closure of a local manufacturing company. WNEL advised the workers on the viability of retaining the Co-operative Development Association regarding the possibility of trading as a co-operative. The workers decided against this, however, and WNEL then put together a business plan prior to seeking overdraft facilities from the bank.

The local authority assisted in the search for premises, and WNEL liaised with central government departments and training agencies regarding financial support. Management training was given to the new owners of the business and WNEL seconded one of the trainees from the Marketing Through Research programme to help with the marketing aspect.

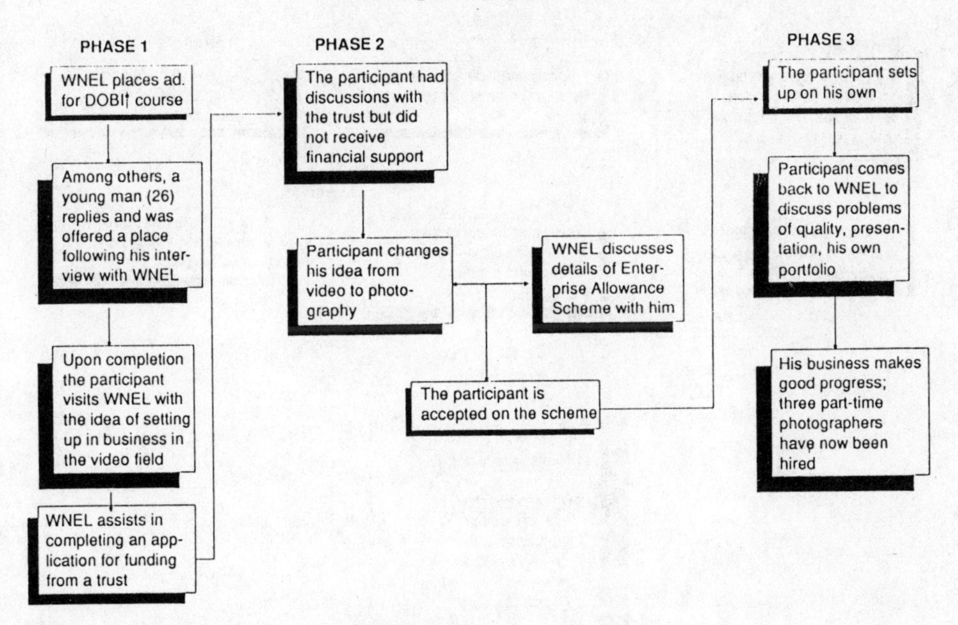

Figure 14.6 (Example 4) Creation of a photographer's business (Wigan New Enterprise). DOBI stands for 'Development of a Business Idea', and the course is 150 hours of group training. Source: *Empirica: Assessment of Locally Initiated Training and Employment Pilot Actions, Final Report*, Pt VII, Bonn, May 1987, p. 10.

Comments: The participant was referred to WNEL by a national programme for youth support – The Prince's Trust. The Trust asked WNEL to examine the business idea and give some training and support. The participant attended one of the WNEL courses run at Wigan College for Start and Manage Your Own Business, for which MSC (Manpower Services Commission) training allowances were paid. Following completion of the course, individual training was given and WNEL was able to introduce some new work to the participant. The process of developing and implementing the business idea took about two years during which WNEL provided intensive training and related support.

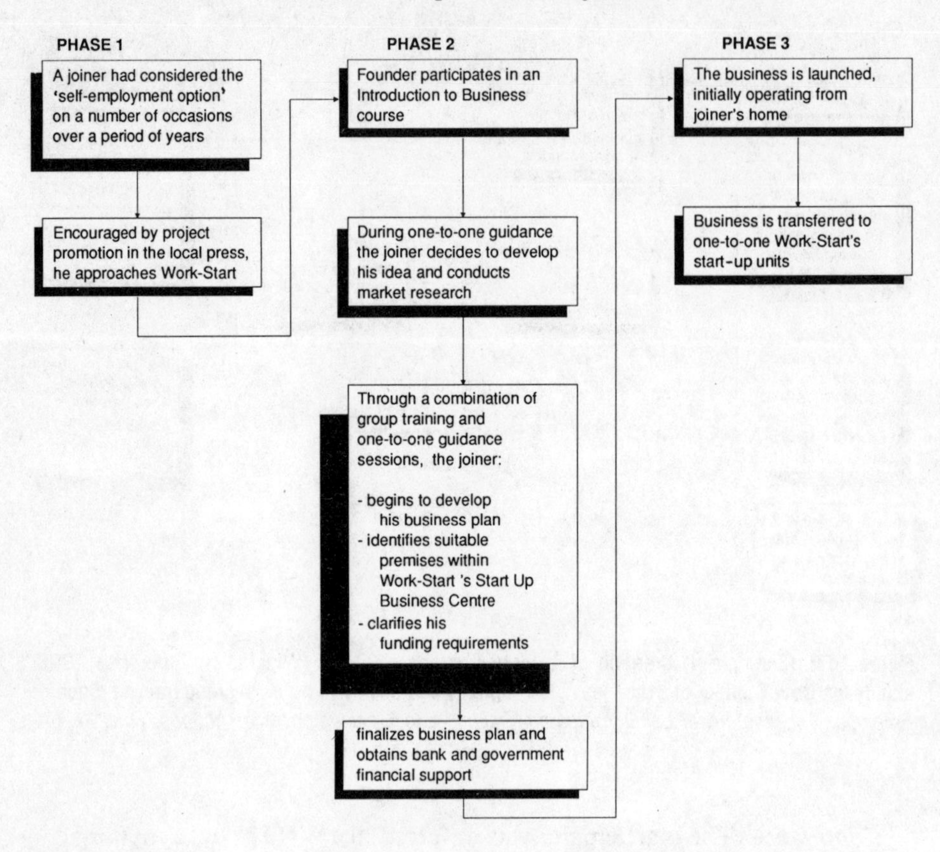

Figure 14.7 (Example 5) Creation of a joiner's business (Work-Start, Dundee). Source: *Empirica: Assessment of Locally Initiated Training and Employment Pilot Actions, Final Report*, Pt VII, Bonn, May 1987, p. 11.

Comments: It is too early to know how secure this business will be but to date the owner has been able to maintain a healthy work load. He continues to receive a certain amount of advice and guidance from Work-Start. As time progresses and his finances improve he will be able to obtain consultancy services from some external business consultant.

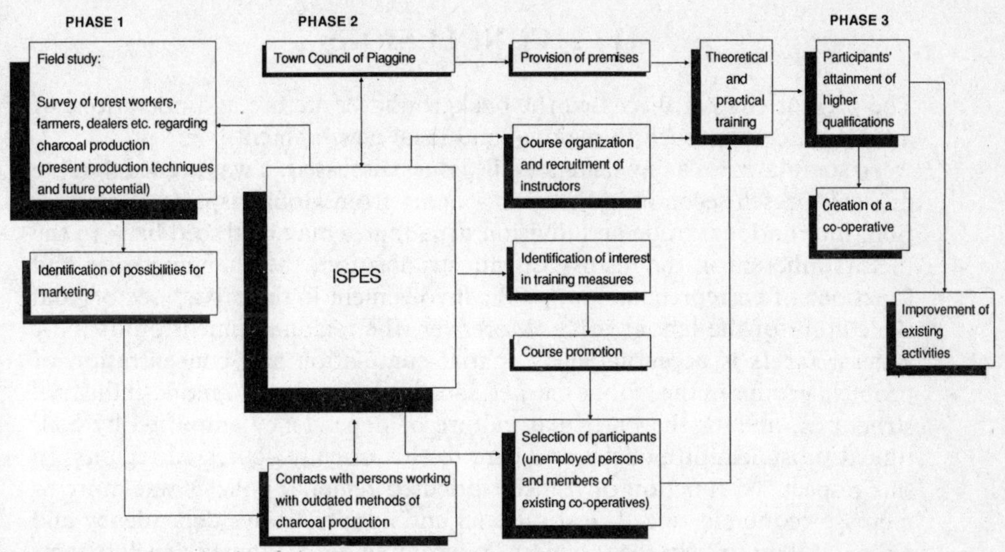

Figure 14.8 (Example 6) Creation of a co-operative in charcoal production (ISPES, Cilento). Source: *Empirica: Assessment of Locally Initiated Training and Employment Pilot Actions, Final Report*, Pt VII, Bonn, May 1987, p. 12.

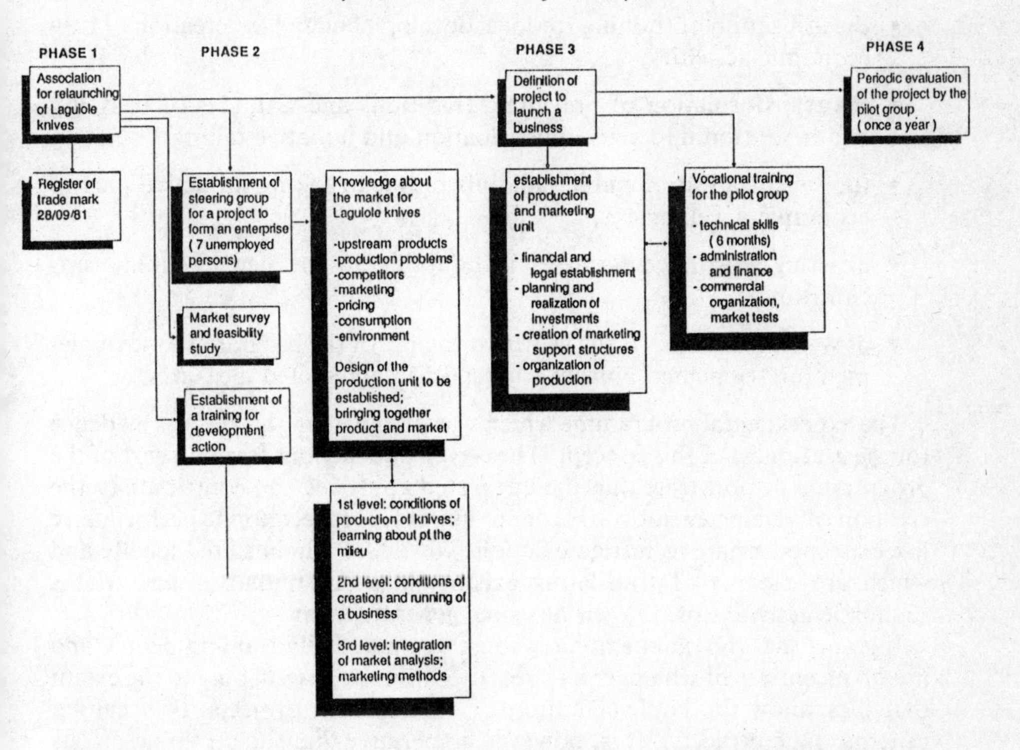

Figure 14.9 (Example 7) Unit for the production and marketing of Laguiole knives (ADEFPAT, Tarn and Aveyron).

14.5 CONCLUSIONS

The present chapter discussed the background, structure and operations of an integrated approach to regional and local development.

To summarize in a few sentences the issues discussed, it was argued that the difficult present economic situation - apart from global aspects of stagflation, international trade and division of labour - may be traced back to the factors inherent in the history of industrialization, the changing roles and functions of entrepreneurship, public involvement in the private sector, and the culture of the labour force. Moreover, the regional dimension of these general facets is accentuated by spatial cumulation and concentration of problem groups in the labour market, localized traditions of mono-industrial structures, and/or the particular culture of dependency amplified by continued subsidization policies and care for historically depressed regions. In this respect the function of transfer-oriented regional policies was more to increase economic, social, institutional and administrative dependency and passivity than to help these regions in attaining self-sustained development.

To overcome this situation requires new approaches. The integrated approach presented in this chapter is based upon the following aspects:

- the utilization of training for local development and the creation of new economic activities;
- the transformation of prevailing traditions and attitudes of passivity and hesitation into active participation and initiative-taking;
- the sensitization of public administration to taking an active part in economic development;
- assisting existing enterprises in adapting to the new economic circumstances;
- as well as - and this is the central principle of the programme - complementing segmented policies with specific, integrated measures.

The experimental programme which was designed on this basis is evidence for the usefulness of this concept. The eleven pilot actions forming part of the programme demonstrate that the integrated approach can contribute to the creation of viable new jobs. In order to upgrade local economic performance it seems appropriate to instigate actions which are administered locally and which are capable of translating existing ideas or initiatives into viable economic activities or, in their absence, generate them.

It seems that the main emphasis must be on actually training people and not on incentives of whatever nature. (Incentives are useful only to the extent that they allow the implementation of development projects; they cannot generate such projects). It is, however, imperative that such training differs substantially from traditional concepts. The function of training should be to stimulate and motivate local residents to take the initiative in an effort to overcome the existing problems of development. In this framework training is no longer to be used as an instrument to match people with professions and globally perceived demand for certain skills. Rather, the function of training should be to match concrete local-development potentials with the available human resources.

The training thus offered covers a broad scope. It ranges from an initial phase of stimulating the awareness of local residents and local and regional authorities to assistance to individuals willing to establish an economic activity by matching product ideas with potential markets, to long-term on-the-job training, and assistance for groups of people establishing co-operatives or small businesses.

Logically, the local training programme cannot change a deeply rooted situation overnight. Whereas this paper indicates positive medium-term effects, it remains to be seen what results will be achieved in the longer run.

CHAPTER 15

Learning experiences from OECD and EC reviews of local employment initiatives

Andreas Novy

15.1 INTRODUCTION

Today unemployment is considered a national and Europe-wide problem; but the place where it is daily experienced is at the local level. 'That is where the stress and strain - and the practical solidarity - actually occur'. For this reason it is not surprising that there is 'increasing pressure from local communities and interest groups on local elected representatives and other individuals and groups' (OECD, 1986d, p. 10).

Unemployment is now not perceived as cyclical and temporary, as it was in the 1960s, but as *structural* and *persisting*. As a result of the dwindling faith in centrally controlled demand management, more flexible, often supply-side, instruments are being applied. In this situation local employment initiatives (ILEs: Initiative Locale de Création d'Emploi; Initiative for Local Employment Creation) are being promoted by the OECD and EC and presented as one remedy to reduce unemployment starting from the grassroots (see also Chapter 2).

> In July 1982, the OECD Council authorized a new Programme of Co-operation and Action concentrating on Local Employment Initiatives. Its main objectives are threefold. First, to promote the exchange of experience and information on the development of local employment and enterprises. Secondly, to design the methods for and to undertake the social and economic evaluation of new employment and enterprise initiatives. Thirdly, to provide technical support and assistance for participants in the design, implementation and assessment of national programmes. (OECD, 1987)

The EC programme was also started in 1982 following a series of 'Consultations' with existing local initiatives in the framework of its 'Programme of Actions on the Development of the Labour Market (European Commission, 1985b). More recently an information network on local initiatives, ELISE (the European Information Exchange and Resource Unit based in Brussels), was established in 1985 with the joint support of the EC and OECD (McEldowney and Aitken, 1987).

Probably thousands of ILEs exist all over Europe, several hundred of them described in the OECD and EC reviews quoted. They vary widely in their objectives, organization and activities ranging from the St Helen's Trust, which was held up as an example to British firms by the Prime Minister, Mrs Thatcher (OECD, 1986d, p. 44), to non-conventional ILEs that grew out of the alternative movement and seek to overcome capitalism (European Commission, 1986, p. 5). Nevertheless, there is common ground: the objective of the local community to become the master of its destiny instead of staying dependent on central and foreign interests.

This chapter aims to give an overview of ILEs described in OECD and EC reviews. To achieve this purpose a new typology of ILEs is introduced and local initiatives are analysed in a wide framework. The respective national framework of ILEs is given special attention as well as structural challenges at the global level. We shall see that these environment aspects are of crucial importance to ILEs and influence the genesis of ILEs as well as their respective success or failure. In the final section learning experiences are elaborated and ILEs as an indigenous policy instrument are analysed within a framework that takes global economic restructuring into account.

The reviews discussed in this chapter are - as will be seen - mainly concerned with initiatives for local employment and firm creation and only partly with broader local employment initiatives as defined by Stöhr in Chapter 1. As they summarize the work of the OECD and EC they only include case studies of West European market economies.

15.2 LOCAL EMPLOYMENT INITIATIVES

ILEs are defined by the OECD as 'local in origin.... [They] respond to local needs ... and are created and controlled by individuals and groups in the community'. Furthermore, they have as a central objective the creation of viable and worthwhile employment and only survive on a basis of long-term economic viability (OECD, 1985g, p. 15). The EC characterizes ILEs as 'enterprises' that support viable and sustainable employment, as of local origin and solving local problems (European Commission, 1985). This description is similar to but somewhat more narrow than the definition of local development initiatives (LDIs) given by Stöhr in Chapter 1: 'we need to consider four aspects of local development: the origin of the initiative, of the resource inputs and of the control mechanisms, and the destination of the benefits'. The EC description focuses primarily on enterprise and job creation.

A decisive distinction, however, exists between ILEs as community businesses and ILEs as traditional entrepreneurial activity. The EC programme deals to a great extent with alternative and communally oriented ILEs which try to achieve social and economic objectives at the same time. The OECD programme, on the other hand, concentrates on economic initiatives and does not clearly distinguish ILEs from small and medium-sized enterprises (SMEs). But as ILEs encompass a wide variety of initiatives there are also activities included that do not satisfy the characteristics of LDIs, as resources or initiatives do not come out of the local community, are not controlled locally or do not benefit the local community.

The descriptions in the OECD reviews are more detailed whereas the EC consultations, which usually last for one to four days, only give a short overview of local initiatives. (This overview does, nevertheless, demonstrate the impressive variety of existing initiatives.)

First, a short description of the types of area where ILEs are mainly located is given. Then the decisive distinction between ILEs as 'community businesses' and ILEs as 'businesses in the community' will be introduced.

15.2.1 AREA TYPES OF LOCAL EMPLOYMENT INITIATIVES

Usually, ILEs dealt with in the OECD and EC reviews develop in communities which are hard-hit by global changes manifesting themselves in high unemployment rates. But sometimes these initiatives also spring out of an enterprise culture like that of north-central Italy (OECD, 1986b, p. 13). Creative communities can profit from the new international division of labour and thereby enforce spatial polarization.

ILEs can be found in all types of area and in all countries of the OECD, in neglected inner cities as well as in peripheral rural areas. All European nations of the OECD and all member states of the EC have contributed more or less extensive case studies. Irrespective of the diversity of locations and the rapid shift of problem areas there are two main types of area where ILEs are important; old industrial and rural peripheral areas.

Old industrial areas

ILEs 'have a special relevance in areas which have previously housed one or a few industries ... which have been hard hit by the changing trade conditions and where, therefore, there have been *large-scale redundancies*' (European Commission, 1986, p. 45; my italics). These old industrial areas are frequently dominated by single industries at the end of their product life-cycle (Tichy, 1987) and are increasingly in fierce competition with newly industrializing countries (NICs). But 'capital has ... shown a marked proclivity eventually to abandon that which it has engendered in the first instance, namely dense communities of workers' (Scott and Storper, 1986, p. 309). A characteristic example is Bradford, until the 1960s 'the world leader of the textile industry' (Martinos, 1986, p. 7). One may call these former flourishing places the losers in the new international division of labour because until recently prosperity prevented them from diversifying their local production structure. In these cases ILEs emerged as a reaction to actual or imminent mass unemployment.

These industries are often controlled by the state and more or less nationalized. The political manoeuvring space in defending these industries, however, is shrinking and firms - both nationalized and private - face two defensive alternatives: they can reduce and *rationalize* production or they can *relocate* production to regions with lower labour costs and no workers' resistance (Fröbel *et al.*, 1986).

In old industrial areas local community animation and mobilization by voluntary or public initiatives are of prime importance. The main handicap of ILEs in these regions represents the lack of entrepreneurship, resulting from decades of 'normative acceptance of people working for a wage' (Hudson and Sadler, 1986, p. 185). Workers, who have got used to selling their labour power, can no longer exercise this routine activity and, therefore, severe problems arise. Their potential advantage is the external shock that might promote local solidarity across class boundaries and strengthen regional alliances.

Rural peripheral areas

The second type of regions where ILEs frequently emerged are peripheral rural areas. The slowdown of worldwide growth rates led to rising unemployment in former labour-absorbing industrial centres. The dwindling chances of finding a job in urban industry increased the willingness of potential young outmigrants to create ILEs in their home communities.

In Italy, for example, 'contracting labour markets of the north have ceased to offer outlets for the needs of southern immigrants' (European Commission, 1986, p. 23). The reaction of the potential outmigrants frequently was the setting up of agricultural co-operatives, training courses in traditional handicraft or community mobilization.

15.2.2 FUNCTION, ORGANIZATION AND OBJECTIVES OF LOCAL EMPLOYMENT INITIATIVES

In this section a typology is introduced that reflects the observable differences between ILEs. First, there are '*community businesses*' as base initiatives, and might be compared to what Johannisson in Chapter 3 calls social entrepreneurs. Second, there are traditional entrepreneurs that will be summarized as '*businesses in the community*', following a British organization covering 220 enterprise agencies (OECD, 1986c, p. 13). They resemble what Johannisson calls autonomous entrepreneurs.

This classification resembles the British distinction between the Conservatives' strategy of 'restructuring for capital' and the response of local Labour-dominated authorities of 'restructuring for labour' (Boddy and Fudge, 1984).

Local employment initiatives as 'community businesses'
In this type of ILEs are included what the OECD called 'community companies and co-operatives', 'local community development initiatives', 'ILEs for the social integration of different groups into the workforce' and the 'alternative entrepreneur' (OECD, 1985g). They can all be considered as LDIs and as part of the *voluntary sector*.

Community businesses often *emerge out of the alternative movement*, try to mobilize the grassroots and to create a common awareness within the community. They consider the present economic crisis as a symptom of the decline of industrial capitalism, and, in concrete terms, perceive the consequences of industrial restructuring as individual unemployment, marginalization and alienated work for the sake of efficiency. Usually they experiment with new forms of enterprise as regards their ownership, management and types of organization (Martinos, 1986, p. 57) or with co-operatives.

Community businesses are characterized by their *mixed objectives* as they aim at social and economic goals. Economically they aim at viability, and 'profit is not necessarily the prime objective' (European Commission, 1986, p. 20, my translation). Often they have particular social or environmental objectives concerning the utilization of local natural resources or productive skills. These ILEs satisfy the needs of a new middle class emerging out of the social movements which the dominant economic system cannot or does not want to satisfy. These needs include primarily biologically grown agricultural products, handicraft and ecologically sane production in general. Community businesses cover all types of activities with a *bias towards the service sector*.

> [In Germany] only 40 per cent of these projects are able to finance themselves ... only 20 per cent have the means to pay all their workers ... Many of these projects are subsidized by public authorities or by a network of sympathizers ... and it is very difficult to distinguish whether this situation is a matter of principle or due to the commercial non-viability of the project. (European Commission, 1986, pp. 130, 227, my translation)

As they have even more difficulties in obtaining financial support than traditional start-ups, their greatest problem is undercapitalization. Another problem represents their neglect of business and market orientation. But 'if one wants to stand one's ground on the market one has to play the game of the market. Growth and profit or loss and bankruptcy. That's a fact, not a critique' (European Commission, 1986, p. 131, own translation). This double dependency on the market and public goodwill indicates that there does not exist an 'alternative economy' *per se*, independent of the outside world.

Community businesses are often oriented towards marginalized social groups, trying especially to promote the underprivileged. Besides often highly qualified individuals and professional or voluntary social workers these ILEs are frequently founded by 'problem groups' of the labour market, such as women, people from ethnic minority groups, handicapped people, young people or long-term unemployed. Sometimes workers start a co-operative to take over their former employer's enterprise and, mainly in Italy and France, traditional co-operatives are founded (European Commission, 1986).

Unfortunately, high unemployment often makes job creation *the* urgent need. Therefore, one can observe that more and more ILEs lose other broader objectives and are reduced to being mere job creators. Therefore there emerges a 'predominance of direct help [to create jobs] over the richness that is represented in these experiences in social, economic and cultural respect' (European Commission, 1986, p. 226, own translation). This, however, reduces the broad variety of ILEs to small-scale businesses uniformly oriented towards profit maximization.

Local employment initiatives as 'businesses in the community'

More and more communities realize that they need a broad variety of SMEs so as to diversify the local economy. In this way it is hoped that the crash of a large firm will have less serious consequences for the local economy than otherwise. Local economic policy is no longer oriented to short-term welfare maximization but to long-term sustainability which is more easily achieved by means of viable SMEs. Therefore, local decision-takers increasingly promote traditional local businesses and start-ups, whose objective is the ordinary business goal of individual profit maximization.

'Businesses in the community' are not necessarily LDIs as one or more of their four elements of local initiatives and inputs, local control and benefits are often not fulfilled. They have no social or community-oriented objectives. But through their activity they frequently also help to soften the social problems of unemployment and economic decline. In some countries such as Italy they are reported to occur spontaneously and are suspicious of central government. In other cases they are induced by public authorities, as in Scotland (OECD, 1986b).

In Italy 'the number of industrial enterprises rose from 650,000 to 958,000 between 1971 and 1981 and their average size fell from 9.3 to 7.4 [employees]' (OECD, 1986b, p. 13). The increasing number of SMEs face strong world-wide competition and are, therefore, forced to innovate. Owing to favourable legislation and weaker union organization than in large firms, SMEs are able to increase the flexibility of production and enable unbureaucratic technology transfer from the universities. In recent years central government has realized its failure in regional and labour-market policy, which mainly promoted 'cathedrals in the desert' (OECD, 1985f, 1986b). Between 1985 and 1988 the government, therefore, wanted to spend $1.5 billion to promote youth entrepreneurship in southern Italy.

Science parks are nowadays seen as instruments enabling local sustainability and fostering local technological competitiveness (Nijkamp and

Table 15.1 Typology of local employment initiatives (based on OECD and EC reviews)

	Business in the community	*Community businesses*
Main actors	autonomous entrepreneur	social entrepreneur
Objectives	entrepreneurial economic profit maximization	mixed social and economic objectives
Positive externalities	increasing local economic sustainability	integration of problem groups, social innovation; increasing local economic sustainability
Organization	individual enterprises	co-operatives, self-managed firms, non-profit organization
Activities	all types of economic activities	bias towards the service sector

Stöhr, 1988). It is hoped that once a critical mass of scientists and innovators is reached positive cumulative effects will spread all over the region. A successful local technology policy has to take into account local skills and other traditional activities, improve and adapt their special skills within the university and other research establishments and inject these into local firms (OECD, 1986e, p. 11). This will enable science parks to become 'catalysts for regional development through the creation of a scientific and technical culture' (OECD, 1986e, p. 13).

15.3 THE RESPECTIVE NATIONAL FRAMEWORKS

Time and again the reviews refer to national events, activities, rules and organizations, i.e. the national framework in general, as being crucial for all actions at the local level and especially for ILEs. These respective national conditions influence the actors at the local level in diverse aspects as they limit and enable activities.

15.3.1 POLTICAL, SOCIO-ECONOMIC AND LEGAL SYSTEMS

'The consensus on the welfare state ... is breaking up without any clear indication of what it is going to be replaced with' (Hettne, 1987, pp. 11f.). For the nation state, to solve its legitimation crisis regarding local development would require it to pursue an equitable and just regional development policy, whereas the rationality crisis was believed to make it necessary to foster unbalanced growth for the sake of efficiency and international competitiveness. Regional and local problem areas would in the latter case have to be neglected for the benefit of national wellbeing (Johnston, 1986,

p. 277). The legitimation crisis of today is mainly the result of the success of yesterday, when state bureaucracies were expanded to grant justice to all. Today, however, it is becoming more and more obvious that bureaucracies exercise unaccountable power and foster a mentality of dependency and inflexibility.

The resistance to the modern welfare state has two different causes. First, it tends to restrict individual liberties. The restrictions range from the permissions needed for very simple economic and even non-economic activities in social security legislation in many countries, which undermine creative responses to unemployment (OECD, 1985b). The so-called value shift towards postmaterialism, which is partly politically instrumentalized, reinforces this resistance. Second, many citizens are sceptical about state power and intervention, 'partly because of the mixed record of nationalized industries, partly because of resistance to high levels of taxation' (OECD, 1986b, p. 76).

The fiscal consolidation policy practised in all Western countries has led to a cut in resources for social security and labour market policy which strikes the declining regions and cities hardest and leads to increased social and spatial polarization. It has to be stressed that in 'countries where the welfare state is being more or less dismantled a wave of criminality, political violence, racism and neofascism has become the worry of increasingly feeble national politics' (OECD, 1987, p. 11).

The question 'federal versus unitary state' is an important one, as ILEs are more easily encouraged in federal states. In the reviews, however, one cannot find clear evidence that unitary states are an obstacle to the creation of ILEs. In Sweden, for example, the fact of being a unitary state has not prevented the Swedish Parliament from enlarging local power. Municipalities in Sweden are often considered a model of local economic policy-making in Europe (OECD, 1985c). In the United Kingdom, on the other hand, central government has actually cut local political power. In this case local authorities strongly experience their dependence on the political goodwill of the national government. Because they are only statutory creations it is easier for central government to erode the financial base of ideologically disliked local authorities (Waters, 1987, pp. 14f.).

Independent of the federal v. unitary state dichotomy, *long-term political decentralization* and *the disposition of sufficient financial resources* are prerequisities of a forward-looking involvement of local authorities in economic development. The new local economic policies are different from the outward-looking job and investment hunting of previous years. 'The emphasis ... is now firmly on indigenous development – creating or preserving jobs on the foundation of an area's own resources – and this has inevitably led to a more decentralised area based approach to economic development' (Waters, 1985, p. 47). Nowadays city representatives self-consciously and clearly express their claims: 'If central government can't solve our problems then it should give us the freedom and means to do so ourselves' (OECD, 1986d, p. 42).

15.3.2 FINANCIAL SYSTEMS

Community businesses and support agencies are often not commercially viable, either because they are not familiar with business life, do not have good connections to the business world, or are engaged in activities that cannot survive in a profit-maximizing economy. It remains a political question whether the public should subsidize these activities. Obviously they do not only create employment but also take over public social, environmental and communal functions. The subsidies can come from the local or national government or from private supporters of these initiatives. For self-managed enterprises there are additional difficulties, as 'the variable capital and the principle "one man, one vote" certainly do not contribute to gaining bankers' confidence' (European Commission, 1986, p. 18, own translation).

Traditional entrepreneurs, who want to start a business and have a marketable product, can usually obtain money from the banking system or are supported by newly created public or private funds. Sometimes, however, financial circles are conservative in lending money to risky new initiatives, a fact which obviously discriminates against projects requiring venture capital. This difficulty can be overcome more easily at the local level by local public guarantees. Enterprise boards in Britain have utilized pension funds for promoting ILEs (McKean and Coulson, 1987).

15.3.3 SOCIAL AND CULTURAL SYSTEMS

It is increasingly realized that the socio-cultural system has a decisive impact on ILEs. In this respect two kinds of barriers in specific types of areas render the creation of ILEs more difficult.

In cities dominated by a mono-industrial structure a *strong trade-union movement* has usually been established. The individual worker has got used to delegate solutions of economic problems to trade-union leaders and management. In these regions it is very difficult in case of mass redundancies to overcome lobbying for outside help and – if it is not forthcoming – despair. In peripheral rural areas, on the other hand, there usually exists a deeply rooted *resistance to change*, as a result of unhappy earlier experiences with modernization projects and an inherent traditionalism (Waters, 1985, p. 40), both powerful barriers against ILEs.

In both cases, however, grassroots initiatives are of utmost importance for the viability and sustainability of local communities. They help to 'mobilise the community and to create a common awareness of local problems and opportunities' (Martinos, 1986, p. 54). Furthermore, employment and mobilization policy have to be combined to create a climate favourable for private initiative and self-help movements. Successful communities have shown that an open-minded culture is favourable for ILEs. In economic respects, a longstanding commercial and craft tradition promotes initiatives as in the case of north-east and central Italy (OECD, 1986d, p. 41). In political respects, the Swedish democratic and participatory culture encouraged ILEs (OECD, 1985c).

15.3.4 EDUCATIONAL AND VOCATIONAL SYSTEMS

Although state-school systems and employment services try to adapt people to the needs of the business world they often do not prepare them for taking initiatives. Künstler emphasizes that 'the training ... was not well adapted to the needs of those who would have to improvise and innovate in co-operative and other settings which make a considerable demand on management and human relationship as well as organisational skills' (Künstler, 1985, p. 14). In addition these programmes are often very inflexible and not well adapted to the local labour market. Furthermore, 'it was suggested that occupational training should *prepare workers not only for dependent employment but also for self-employment*' (Leve, 1985, p. 84). The economic involvement of local government makes it necessary to train and retrain government officials to give them basic economic knowledge and to achieve a 'more sympathetic attitude' towards local private business and 'greater flexibility' (Waters, 1985, p. 42).

Alternative enterprises which normally start with much dedication and social and cultural commitment realize after a short time that 'most of them have no competence at all for management and are strongly embarrassed facing problems of bookkeeping, stocks, the choice of business strategies and financing plans' (European Commission, 1986, p. 19, my translation). Specialized support agencies have to face this problem and help to make social and economic objectives compatible.

15.4 POTENTIAL AND ACTUAL AGENTS AT THE LOCAL LEVEL

The following main agents for local development are dealt with in the OECD and EC reviews of ILEs.

15.4.1 MEMBERS OF THE LOCAL COMMUNITY

It is the main advantage of democratic countries that in principle everyone who wishes has the right to participate in economic and political life irrespective of his or her race, sex or wealth. Innovative individuals are increasingly recognized as the wealth of a nation and 'identified as the key factor and at the same time, the limiting factor in the world market economy' (Hesp and Stuckey, 1987, p. 32). The growth of ILEs has seen the emergence of individual, public and community entrepreneurs whose action is aimed at job creation and other social goals (OECD, 1986d, p. 40).

Two types of creative citizen can be identified: the one, acting for his private profit, can be called the '*bourgeois*' or pure entrepreneur; the other type, looking for the common good, can be called '*citoyen*' (Kirsch, 1987). She or he is 'entrepreneur from necessity' (OECD, 1986a, p. 41) and is part of what IFDA calls the *Third System*. This is a 'system of power represented by people acting individually or collectively through *voluntary organisations and associations*. It is the main bearer of new values and visions' (Hettne, 1985, p. 51). Both types of innovators often start an employment initiative without even being recognized as such by public authorities.

Within a favourable framework that encourages private and communal activities *female agents* play an increasingly important role in these initiatives. In Sweden and the United Kingdom, for example, the percentage of self-employed women is rising much more quickly than that of salaried women (OECD, 1986a, p. 8).

15.4.2 LOCAL GOVERNMENT

Three categories of local economic policy activities are distinguished: '*Preaction* stands out for all forms of preparatory activities or, in other words, the building of capacity.... *Reaction* covers activities that are primarily undertaken as a result of more or less direct demands from business.... Finally, *proaction* ... refers to those activities where the municipality is the taker of the initiative' (OECD, 1986c, pp. 49f.). Until recently the economic role of local authorities has usually been reduced to preactive support of an adequate infrastructure (OECD, 1985c, p. 50), the hunting for large inward investment, the lobbying for jobs and money (OECD, 1986d, p. 8) and the control and/or implementation of centrally funded programmes (OECD, 1986d, p. 12).

The worsening of the international economic situation has put into question the traditional regional development policy of 'trickle down' because large firms are now very conservative in their investment decisions and public funds are reduced because of budgetary constraints (Stöhr, 1985). 'Such trends increasingly diminish the credibility of external solutions and lead to greater emphasis being placed on locally-generated development and local entrepreneurship, that is, job and business creation at the local level matching local needs and resources' (OECD, 1986d, p. 13). In Norway the municipality's involvement in economic affairs rose considerably (OECD, 1985a, p. 2). Often, mayors 'must now become entrepreneurs' (OECD, 1986d, p. 40).

This shift to indigenous development strategies was enforced by the political trend towards decentralization (OECD, 1986d, pp. 14f.). Decentralization *per se*, however, does not automatically increase the responsibilities of the municipalities as the Netherlands' reform shows (Waters, 1985, p. 5).

'In this new context, local government has found itself in a position of being expected to respond to serious and pressing social and economic problems without being adequately equipped in terms of specific legal power, financial resources and expertise' (Martinos, 1986, p. 65). Within a very short time the communal staff has to acquire general knowledge concerning business life and more specific knowledge relating to the status of local companies (OECD, 1985c, p. 29). Unluckily, the shortcomings of public statistics which are not adapted to the needs of local economic policy-makers render economic planning even more difficult.

Nevertheless, 'local government has acted on the whole in a positive and imaginative way' (Martinos, 1986, p. 63). Their activities shifted from preaction to *proaction*. Instead of capacity-building and *ad hoc* crisis management local authorities tried to increase their information about the local production structure in order to help 'to reduce the strong dependence of business life upon forces that lie outside the geographical borders of the municipality' (OECD, 1985c, p. 53). The real success stories are those ILEs which proactively promote change in time. They do not wait until they are adversely affected by the new division of labour but try actively to anticipate necessary changes by innovation and diversification. Unfortunately, this type of ILE is rarer than the reactive one. The city of Castanheira de Pera in Portugal represents an example of the proactive type (Martinos, 1986, pp. 19ff.).

15.4.3 CENTRAL GOVERNMENT

Traditionally central government interferes heavily from the outside in the local labour market using several instruments and organizations. Usually it is responsible for putting into operation the constitutional framework that distributes financial resources and allocates responsibilities. Secondly, it gives the local and regional authorities money to implement its policy at the local level. Thirdly, it has often created specialized regional development organizations like the Scottish Development Agency (which has only economic objectives) and the Highland and Islands Development Board (which aims at both economic and social development) (OECD, 1986b). Fourthly, the centrally financed public employment services are a main agent at the local labour market level. Examples cited in the reports are the British Manpower Service Commission (MSC), which was formerly responsible for manpower planning and training (Waters, 1985, p. 13) and the German Arbeitsbeschaffungsmassnahmen (ABM) (Stingl, 1985, p. 5), which assists projects that are in the public interest. In Italy, on the other hand, labour market policy was until recently very centralized (OECD, 1985f).

Furthermore, central government departments play an important role in local development (Waters, 1985, pp. 12f.; OECD, 1985a, p. 4) and, finally, the national location and spending policy has an impact on the local community (OECD, 1985c, p. 11). The 'central government monopoly in spending public money at the local level' (OECD, 1985c, p. 7) is, however, subject to heavy criticism. 'First, it is most unlikely that central agencies can innovate to a greater degree and can respond to problems quicker than local agencies.... Second, there is practically nothing that central government can show in connection with the mobilisation of resources outside the public sector.... Third, there is little coming from the national level in terms of coordination' and, finally, it is very doubtful whether subsidized large firms will stay longer than the duration of the subsidy (OECD, 1985e, p. 7).

15.4.4 SOCIAL PARTNERS

Entrepreneurs

Traditional local employers and large, usually multinational, firms are becoming increasingly involved in ILEs and participating more and more in local small-scale enterprise and job creation. Concerning their motives, however, the reviews show a certain ambivalence. On the one hand, one has to be aware of the fact that 'the business of business is business, and levels of employment are not in themselves a particular consideration' (OECD, 1986c, p. 2). 'Thus, especially for those bodies which have their origins and main interests in traditional SMEs, the "E" in ILE is more likely to stand for Enterprise than for Employment'. Therefore, it is 'necessary to point out that SMEs and ILEs are not quite the same thing' (European Commission, 1986, p. 52). On the other hand, a recent OECD review states that frequently the motives for large firms' involvement are social responsibility and enlightened self-interest (OECD, 1986c, p. 3).

This enlightened self-interest can be described as follows. Unlike financial capital, productive capital is in the short run immobile and a company which has made a considerable investment in one environment cannot afford to abandon it within a short time and would much prefer to develop the region in a positive way. Therefore, it subsidizes initiatives which promote local development. Sometimes 'positive efforts at job creation can help to ease the process of cutting back their workforce' (OECD, 1986c, p. 3). The support of non-profit bodies created by large firms to promote ILEs consists in financing, secondment of middle and senior managers, management and marketing advice. Spare premises and land are sometimes handed over to new entrepreneurs. However, for these organizations 'it is virtually impossible to reach specific target populations among the unemployed' (OECD, 1986c, p. 14).

Trade unions

In all industrialized countries trade unions are facing tremendous problems resulting from the industrial restructuring process and subsequent mass redundancies. For some time ILEs were attacked by unions as a form of self-exploitation that 'undermines the hard-struggled-for and successfully defended achievements of trade unions' (European Commission, 1986, p. 82, own translation). Following this line of thought, 'ILEs, by confusing policies that really create employment with a theory that advocates keeping unemployed people busy, are said to aggravate the polarisation and fragmentation of the labour market' (*Feedback ILE*, no. 6, 1987, p. 6). Trade unions tend to stress that structural problems such as mass unemployment have to be tackled at the national level. But the above-mentioned value shift and economic necessities to increase efficiency reinforce tendencies towards flexible and decentralized production, which undermines collective bargaining. This, however, represents a dangerous challenge to traditional centralized trade unions.

As they increasingly realize that class consciousness can only develop at the communal level (Walker, 1985), many trade-union activists are becoming involved in the creation of ILEs. 'Facing the threat of losing contact with a lot of workers, [unions are now trying] to do something directly at the local level, instead of demonstrating against unemployment, organizing marches, and proposing alternative economic and employment policies to the government' (European Commission, 1986, p. 83, my translation). In many countries ILEs are now welcomed even by the national trade unions, but the reservations against 'spreading the new entrepreneurial spirit or fostering self-employment' are still prevalent (*Feedback ILE*, no. 6, 1987, p. 6). Their activities centre around creating support agencies, alone or together with other organizations, promoting self-management and workers' co-operatives and publishing information material for their members.

15.4.5 UNIVERSITY AND RESEARCH INSTITUTES

'The speed of social and economic mutation has led universities to build new and better bridges with local and regional interests' (*Feedback ILE*, no. 4, 1986, p. 1). The new Finnish universities 'were founded specifically to adapt traditional skills to regional needs, and to encourage industry to modernise and take on board the latest methods of producing and working' (*Feedback ILE*, no. 4, 1986, p. 3). The possibilities of a university–industry interface are considered one of the main preconditions for founding technology parks. The Jamtland Project between Sweden and Norway is an interesting example of cooperation between two countries to promote remote rural areas (OECD, 1986b).

15.4.6 THE EUROPEAN COMMUNITY AND OECD

It needs to be stressed that EC bodies increasingly cooperate with regional and local authorities instead of national agencies. While the European Social Fund sponsored ILEs successfully (cf. Chapter 14), the European Investment Bank and the Regional Development Fund have been said to tend to discourage 'more innovative use for small scale projects' (Waters, 1985, pp. 13f.).

The OECD, as has been mentioned, collects information on ILEs and promotes the implementation of new initiatives. The ELISE project concentrates on 'the development of a documentation centre and a database of initiatives and support structures' (McEldowney and Aitken, 1987, p. 43).

15.4.7 INTERACTION BETWEEN LOCAL AGENTS AND SUPPORT SYSTEMS

It might appear as though there are a number of local agents acting in a vacuum without contact or interaction with one another. In reality, however, interrelations between individuals, groups and institutions are of crucial importance. To appreciate the complexities and development potential of local communities one has to examine these interactions carefully.

Science parks represent a good possibility for cooperation between industry, university and local authorities (OECD, 1986e). A balance of power or a partnership approach between local and national government facilitates the possibilities of broad action on behalf of the local community (OECD, 1985c, p. 6).

At the local level it is easier to reach cross-party agreements and to concentrate on the consensual objective of reducing unemployment, which has become a major threat to the community as a whole. 'One of the characteristics of recent local economic initiatives ... is the growing incidence of *joint ventures*, involving public, private and voluntary sectors' (Waters, 1985, p. 6). Social innovations at the local level, especially new forms of cooperation, try to 'alleviate the worst effects of the crisis' (Martinos, 1986, p. 63).

Everybody concerned with ILEs stresses that (informal) grassroots cooperation represents a necessary and important contribution to local development. But to the same degree (formal) institutional cooperation is needed to coordinate activities. For this reason specialized agencies 'with the aim of giving support to people wanting to create their own jobs' are set up in many countries (European Commission, 1985c, p. 320). Usually comprehensive support structures are preferred. A '*one stop support centre* available to all ... can offer advantages in terms of availability of expertise, combined with local knowledge of resources, markets and contacts' (European Commission, 1985a, p. 67). In countries where the 'alternative movement' is strong, however, as in the Netherlands and Germany, there exists a clear dichotomy between the two types of ILE that is also reflected in separate support structures (European Commission, 1986, p. 46; cf. also Chapter 2). MEMO and Netzwerk Selbsthilfe are examples of support organizations of this kind. These organizations are opposed to giving up their social and political objectives and become mere instruments of local job creation if they amalgamate with mainstream business-support structures.

A different type of development is observable within the French Boutiques de Gestion, which have now chosen a non-ideological approach because of the urgent problem of unemployment and a certain disillusion concerning self-managed firms (European Commission, 1985c, pp. 328f.).

Other types of support agency are financed solely by public or public/private ventures. Large firms sometimes themselves establish support agencies. Thus the British Steel Corporation in the UK, Elf-Aquitaine in France and Montedison in Italy have been successful in creating jobs in this way. In contrast to them the support agencies of the co-operative movements in France and Italy do not try to maximize employment creation but 'see their job much more as looking after their members' (European Commission, 1985c, p. 330).

The services these agencies offer to ILEs are varied and they 'distinguish themselves from traditional advice centres aiding small firms [which] ... are only equipped to give quick on-the-spot advice on particular technological problems.... The support agencies offer instead a *more global view of a business*' (European Commission, 1985c, p. 324). This hand-holding role ranges from promotional and wider development work and training to finance and monitoring. Large firms sometimes help through the secondment of staff or by permitting ILEs to use their private expertise, premises or equipment. Strong emphasis is placed on follow-up work because ILEs need more support in the everyday life of a business than established enterprises (European Commission, 1985a, c).

Independent private or public/private organizations are usually quicker and more flexible than public authorities. Within non-official support agencies which are financed by grants, subsidies and fees, the quantitative short-term success of industry-sponsored initiatives is greatest; but they are usually least preoccupied with long-term encouragement and continuing practical assistance. Community mobilization and the integration of problem groups in the labour market are only peripheral aims for them.

More holistic agencies with a bias towards community objectives usually remain dependent on public funds. Their disadvantage is the 'pressure to create the maximum number of jobs but with the risk that a political change will lead to a smaller budget or even closure' (European Commission, 1985c, p. 341). This leads to a more pragmatic approach towards ILEs. 'As one can observe, for example in the CDA (UK) or the Boutiques de Gestion (France), these services increasingly lose their ideological content and are more and more utilized as instruments stabilizing the capitalist system' (European Commission, 1986, pp. 225f., own translation).

15.5 RESULTS AND EVALUATION

The OECD's and EC's concern about ILEs centres around employment creation. In this respect successes are poor compared with macro-unemployment rates. But in some local communities results are obvious. Quantitative evaluations for certain cities have shown that ILEs 'have achieved the direct replacement of about one third of jobs lost' (Martinos, 1986, p. 62). 'Business in the Community' created 50,000 jobs a year and saved another 25,000. Elf-Aquitaine in France created 15,000 jobs and saved 6,500 others (OECD, 1986c, p. 13). Compared with these figures the number of jobs created in alternative enterprises seems minimal. It is estimated that there exist some 80,000 alternative jobs in Germany (European Commission, 1986, p. 125). But job creation and preservation are not the only quantitative criteria. Others are 'creation of business, wealth, added value and income, welfare for disadvantaged groups, economic regeneration of an area, community revival' (OECD, 1986d, p. 49).

Besides quantitative criteria, qualitative ones seem to be of utmost importance for all these ILEs which pursue social *and* economic goals. Social mobilization, the creation of an entrepreneurial climate, social peace and a proper environment are achievements of ILEs that are only politically appraisable. The social innovations taking place in many community businesses might indicate the work organization of the future, which will no longer be based on hierarchic and authoritarian but on symmetric relations, based on dialogue between managers and managed.

In more and more countries, therefore, 'local authorities are exposing the need for, and in some cases attempting, a true *social audit* which takes into account all the costs and benefits to public and private sectors alike', especially the 'unproductive cost of unemployment benefits and social security payments' (Waters, 1985, pp. 46, 51). By taking external costs into account political decisions are no longer solely based on narrow individual economic rationality.

15.6 LEARNING EXPERIENCES AND CRITICISM

15.6.1 GLOBAL CHALLENGES AND INDIGENOUS RESPONSES: STRUCTURE VERSUS AGENCY

As structural adjustment issues do not come into the explicit mandate of the OECD ILE programme, OECD reviews show a predominant concern with the micro-level of individual and communal activities. This reflects the division of labour in the mainstream social sciences betweeen structuralists and voluntarists based on an either/or assumption: *either* one analyses structural conditions, global developments and the international restructuring process *or* one concentrates on the individual and local actor who makes her or his own history.

A more adequate framework for analysing ILEs which overcomes this either/or assumption is given by structuration theory (Giddens, 1985).

> For without falling into the despair of thinking that there is a 'logic of history' which is always working behind our backs with inexorable necessity of falling into a simplistic voluntaristic illusion that we can be complete masters of our fate, Giddens enables us to understand both the limits and opportunities of shaping our destinies. (Bernstein, 1986, p. 242)

Local actors have to be seen as knowledgeable subjects who shape their own destiny but, nevertheless, have to take into account their structural constraints and possibilities. 'It has been said that structure is enabling as well as limiting to human action; it might just as well be said that agency is enabling as well as limiting to structure' (Walker, 1985, p. 187). *Structure is not something independent of agency*. This means, according to a famous remark of Karl Marx, that men make their own history but not under circumstances of their own choosing.

> This sophisticated theoretical approach is required in order to begin to comprehend more satisfactorily the links between people's knowledge of and feelings about space, their patterns of behavior and social practices, and the spatially uneven development of capitalist societies, and so reveal rather more about the processes of uneven development themselves. (Hudson and Sadler, 1986, p. 173)

An analysis of ILEs as local actors has to take into account structural conditions even if it is not interested in global changes at all. Local events can only be understood by taking the global restructuring process into account.

New international division of labour

Unemployment, although most strongly felt at the local level, often has its causes in changes that are external to the local community. The new international division of labour (Fröbel *et al.*, 1986) and its negative effects for specific local communities seems to be the crucial challenge that leads to ILEs. Some developing countries no longer only specialize in selling raw materials but, because of the possibility of multinational firms of worldwide scanning, they have redeployed manufacturing plants to NICs, which offer advantages in labour costs. This means that the production of more and more manufactured goods of increasingly sophisticated technology is being transferred to non-European countries. For the first time in the postwar period Europe is negatively affected by global economic restructuring, because of this competition from NICs. Obviously local communities are also affected in a negative way as they lose their bargaining power *vis-à-vis* multinational firms. 'With its wide array of locational choices, but with often very little dependence on any particular locality, the multi-establishment firm is able to make widely separated hinterlands compete with each other for development; in recent years an interregional and international bidding war for plant locations has been effectively ignited' (Scott and Storper, 1986, p. 306).

There is a fierce struggle between concrete place and abstract space, which is of increased importance. *Concrete place* is where members of ILEs live, have their friends and all other social relations that are vital for every woman or man. This place as a web of social relations inherited from history constitutes the life space where people become socialized, develop their personality and find their role in society. Unlike some capital (finance), which is perfectly mobile, workers, especially unskilled ones, are much less mobile for reasons other than economic ones. The often unsuccessful anti-closure campaigns are intended by workers as defending 'the right to "live, learn and work" in particular places' (Hudson and Sadler, 1986, p. 173).

Global capitalism, on the other hand, emphasizes abstract space, which is concrete place solely seen under the necessities of the needs for capital accumulation. It is a contradiction of capitalism that, owing to its need for realization of abstract value, it has to concretize itself in space because concrete places within abstract global space have to be chosen by multi-regional firms. Facing global capitalism every decision-taker, individual or collective, has to 'think globally and act locally' (OECD, 1986e, p. 2). If multilocational firms do this it is, however, usually to the detriment of a concrete locality as it is forced into a war between regions.

To clarify this, two examples will be given in which the neglect of global structural conditions might lead to incorrect conclusions. It is argued that one has to analyse science parks and SMEs more carefully than the OECD reviews do.

If science parks are spatially very concentrated they may increase rather than reduce spatial disparities. Furthermore, the fact that a science park is localized in a problem area does not *per se* represent local development. If it is financed by the central state (OECD, 1986e) it often closely resembles a traditional growth centre with a high-tech label. As the inputs usually come from the outside, the chances for local development consist in retaining as large as possible a part of the benefits in the respective locality. Of special importance in this respect is the upgrading of local labour power. Furthermore, local development policies have to aim at increasing local control and involvement in these initiatives (Stöhr, 1988). As this is a very difficult task science parks represent more often than not primarily a central-government response to world-market and multi-locational enterprises' needs rather than to autonomous local policy decisions of 'history-makers'.

Second, the fact that local SMEs flourish does not *per se* mean increased local self-reliance. The tremendous rise in self-employment in Italy, for instance, has to be analysed much more cautiously. One has to differentiate between those SMEs which work independently from large firms, those subcontractors that perform specialized high-tech or highly skilled work for them, and those that are strongly dependent on a large parent firm for routine functions. The rise in subcontracting and self-employment, therefore, might not contribute to increased freedom, independence and decentralization as the reviews implicitly suggest but, on the contrary, might lead to a deterioration of working conditions and increase the dependence of SMEs on the parent firm and serve mainly as an employment buffer (Holmes, 1986). This may be to the detriment of the region as a whole, rather than to its long-term advantage.

Local employment initiatives as an indigenous local policy instrument

The situation described above is the framework within which ILEs have come into being. More and more communities realize that they cannot compete in this 'war for enterprise location' and rediscover the indigenous development potential of their local economy. They foster their local community as a viable economic unit. This indigenous local policy requires the predominance of local actors, local decision-making, local benefits and local control of resources and innovation (cf. Chapter 1).

ILEs are *one* creative and innovative policy response to the above-mentioned global challenges that have been developed within local civil society and local public authorities. Most ILEs are explicitly or implicitly created as a response to external shocks or out of the willingness to avoid such shocks. In all cases creative individuals and groups have begun to look for solutions to these new and challenging changes. Through their initiatives they have often forced local officials to make the fight against local unemployment the main community concern.

Irrespective of the broad variety of ILEs all try to mobilize the indigenous human and natural potential of a region. In this they differ from traditional regional-policy approaches, which concentrate on the hoped-for 'trickle-down' effects from the centres, but do not take the concrete socio-cultural, political and ecological environment as their basis.

ILEs, usually deeply rooted in their local territory, obscure the neoclassical assumption of a plain, ahistoric space. Normally they will not take local support but afterwards following the neoclassical calculus, change their location with a view to profit maximization. In this way they help to sustain the local community. In this respect, ILEs are part of local capital, which generally is more dependent on a concrete locality, thereby differing markedly from multi-regional capital. Innovations in indigenous local businesses in general will be to the advantage of the local community. That is why the OECD and EC stress local innovation and entrepreneurship as crucial for sustaining a local community and implementing an indigenous development strategy, which is elaborated in detail in the following section.

15.6.2 INNOVATION AND ENTREPRENEURSHIP

Innovation and sustainability

'Unfortunately more jobs, income, GNP growth and regional exports are not enough to sustain a local economy in the long run' (OECD, 1986b, p. 8). Therefore, low vulnerability and high sustainability are what every local community needs to achieve. Strategies that promote territorial instead of mainly functional integration (Friedmann and Weaver, 1979) are better able to contribute to these goals. 'Current interest [in innovation] basically arises from the recognition that to survive in a competitive world there is no choice other than to set the regions on a technology course' (OECD, 1986e, p. 1). Science parks and other research and development efforts are, therefore, strategies forced upon local communities by international challenges. 'The region came to be seen either as a successful or unsuccessful adapter ... to the demands of the world market economy.... [However] in most countries political power is in the hands of those to whom global networks ... and/or their own power matter more than local communities' (Hesp and Stuckey, 1987, pp. 37ff.). Within the conflict between the 'integrated, multifaceted lifespace' which is 'home' to people and the 'monoculture' of the trans-nationalized economy ILEs try to strengthen local life space. The local community has to develop forces from within itself, including indigenous technological innovations to subsist in the ever-accelerating changes of the world system.

Entrepreneurship, diversity and creativity

> Diversity offers a local economy some degree of protection to unforeseen events – market changes, exchange rate changes, new competition, technological advances. With diversity there is always part of the economy relatively unaffected by changes in a single industry. (OECD, 1986b, p. 8)

The flourishing of ILEs does not only depend on 'usable' knowledge in business activities and technology. 'Creativity is something different from directly applicable (profitable) innovation. It heavily depends on "open ended childlike curiosity" ... therefore, innovative thinking cannot be guaranteed or directly achieved – it can be encouraged' (Hesp and Stuckey, 1987, p. 33). A culture of entrepreneurship will be additionally promoted by revising the educational system to bridge the 'gap between public domain and social development on the one hand and business world on the other' (European Commission, 1985, p. 73, my translation). 'Education will have to be a continuous process, *regular re-training* will have to become a rule, while people will have to be prepared to change their work during their lifetime' (OECD, 1985b, p. 64).

A healthy local development has to appreciate 'businesses in the community' as well as 'community businesses'. Acting in an entrepreneurial way represents a process, a mode of operation and an attitude and is a key feature of both 'risk-for-profit' actions by individuals and otherwise inspired community action (Martinos, 1986, p. 58).

> *Community entrepreneurship* is essential and means both a large number of individuals acting in an entrepreneurial manner, not only in the private but also in the voluntary and public sector, as well as an entrepreneurial opportunity seeking attitude in community processes, local organisation and within the broader local community. Community entrepreneurship is the motor that activates and utilises the indigenous potential of a locality and can bring about locally generated development. (OECD, 1986d, p. 4)

For alternative enterprises and trade unions a more positive attitude towards business knowledge and traditional entrepreneurial virtues would be helpful. Employers, on the other hand, should take enlightened self-interest as their watchword and try to maximize their long-term benefit within a local community.

Social and organizational innovation

New social and organizational innovations emerging within and around ILEs represent a very interesting learning experience. The European Commission stresses these meta-economic goals by arguing that

> LDIs are important not merely for the jobs they create but also as a way of preparing the ground for the development of areas suffering from high unemployment - promoting self-help, cooperation, local and regional regeneration - and integrating women, young people and disadvantaged groups into the labour market. They rebuild confidence in the area, maintain or develop skill usage and restore the capacity for enterprise. (European Commission, 1985c, p. 344)

This statement, however, is mainly valid for ILEs as community businesses.

The promoters of traditional enterprise frequently have also created - alone or together with other actors at the local level - new organizations, institutions and forms of cooperation as support structures. A variety of often antagonistic interests is represented in these support agencies. But, being rooted at the local level, they aim at activities that have *plus sum character*, meaning that everybody benefits. As mentioned above, it is likely that job and enterprise creation is in the interest of all if unemployment remains high. The Norwegian experience shows that if we are to achieve a positive process in a local community, it seems fundamental that a platform is established whereby the various local interest groups come into contact with each other, i.e. that the local community becomes a platform for solving local problems' (OECD, 1985a, p. 20).

With regard to alternative enterprises, public authorities have to pursue a careful strategy which is acceptable to both partners. Public authorities are primarily interested in utilizing the job potential of ILEs and their function of calming radical protest. Therefore, public authorities have to reward the meta-economic utility of integrating the marginalized population groups which alternative enterprises offer. The latter, on the other hand, have to accept that the state as donor wants to participate in deciding how its money is spent.

There are actually signs of convergence of different types of ILE (Martinos, 1986, p. 56). But whether there should be one or a few support agencies has to be answered by investigating in the respective local and national framework. It seems that a network of support agencies cooperating formally but staying independent may be the best construct. This pluralistic approach takes into consideration the variety of grassroots initiatives and at the same time maximizes general access to support agencies.

The 'public versus private' antagonism has to be de-ideologized and replaced by a pragmatic partnership model that is most easily realizable at the local level. 'What comes through loud and clear is that Government agencies now understand their limitations while the private sector is more ready to recognise the help it needs from the public sector' (OECD, 1986b, p. 73). The success of the state capitalist South-east Asian NICs such as Taiwan and South Korea shows the possibilities of creative interaction between public and private decision-takers (Moore, 1987).

Finally, it has to be stressed again that ILEs need to be promoted in time to diversify local economies that are heavily dependent on the world market. *Ad hoc* reaction can only ease the worst effects, whereas proactive planning can mitigate negative repercussions on a much broader basis.

15.6.3 NEW DIVISION OF ROLES BETWEEN LOCAL AND CENTRAL GOVERNMENT

We are living in a time of important institutional changes and it becomes more and more evident that local authorities have to play a decisive role. A consensus and partnership approach between local and national government has to be constitutionally fixed to avoid 'serious and broad based disagreement about local autonomy and discretion which has become a major issue in the UK since 1979' (Waters, 1985, p. 14). On the following pages this new relationship will be analysed in more detail.

Towards a flexible local government

ILEs flourish in communities that pursue an indigenous development strategy.

> The usual approach of assessing the needs of the periphery from the centre (top-down approach) should be reversed (bottom-up approach) without damaging central power but giving the periphery a freer hand to be more efficient in operating a policy that has been previously agreed at various levels of government. The decentralisation trend is now too strong to be opposed by central government. (OECD, 1985b, p. 4)

There are certain preconditions for municipal self-government illustrated by the Swedish example (OECD, 1985c). Of utmost importance is the political precondition of municipal democracy. Secondly, municipalities need to exceed a minimum size in order to execute their new activities. Furthermore, the right to levy an adequate amount of taxes, to have the planning monopoly or at least a right of veto, and a certain legislative power are prerequisites for self-government.

'*Local problems demand local solutions*' (OECD, 1985c, p. 19) cannot mean complete independence from central government. But local government should be allowed to act more flexibly and discreetly. Decisions should be

> taken at the national or provincial levels, but the direction of the path has to be indicated by the local communities who have to face daily the problems that arise.... What is not yet probably fully understood is that it is the role of the local community to implement – using a considerable amount of flexibility and resilience, together with a great freedom to make decisions – all the policies to combat unemployment. (OECD, 1985b, p. 64)

The above-mentioned restructuring processes are only possible because of technological progress. At the local level it should be up to the local community to direct technological development as far as possible in a favourable way. The local policy response must not concentrate on technological innovation but should search for new ways of cooperation and promotion of a culture of creativity, initiative and entrepreneurship. However, local agents are limited in their activities, as R&D is highly concentrated in a few regions (Malecki, 1986).

The national employment services in some countries are overprotective of some unemployed (e.g. youth in Denmark) while neglecting others (elderly people, immigrants). At the local level effective needs can be checked more easily and abuses avoided. Employment and training schemes executed by the local authority are likely to be more needs-oriented in at least two respects. First, the needy can be recognized more easily and the training courses can be better adapted to local and personal needs. Second, 'at the level of each local community it may be feasible to offer jobs paid below trade union rates, made available only to young and elderly people, who had suffered from prolonged unemployment' (OECD, 1985b, p. 57). This approach, however, undermines the collective bargaining power of trade unions.

Local government should try to cooperate with central authorities. 'But it is significant that when a national/local consensus is missing, a *local consensus* is enough to let the authority carry through initiatives ... experience shows that is is easier to reach a consensus at the very local level and build it up from there, mainly due to the visible common interest among local actors, especially on immediate and concrete issues' (OECD, 1986d, pp. 46.4). Similarly, mobilization is also easier at the local level with which local people can identify.

A 'like-minded' and active central government

A widely encountered opinion in the reviews is that 'even among those most in favour of an active employment development role for the local authorities, many consider that local authorities cannot solve structural problems, that "*one cannot shift unemployment from the national to the local level*" and that without national programmes and funds local authority action would be ineffective' (OECD, 1986d, p. 3). A positive environment for ILEs does not presuppose a powerless central government although certain activities have to be delegated to the local or private sector.

As there is no alternative to the welfare state, the existing system of social security has to be reformed instead of replacing it with a nineteenth-century watchdog state. This reform should promote private and communal initiative, flexibility and innovation to be able to compete internationally. On the other hand, it should enable everybody to live with dignity.

Indigenous development presupposes that the state takes over new responsibilities, which are described in detail in Chapter 2. 'Increased power at the district and regional level requires a growth of power at the centre' (Friedman and Weaver, 1979, p. 203). Examples of these new responsibilities are a better coordination of different national and national/local activities, improving international competitiveness facing international restructuring, and global long-term economic planning. This is of special importance because a democratically legitimized central government is the only body which derives its power from the obligation to take into account the wishes of all citizens. Traditional public responsibilities, on the other hand, like research and development, and effective regional and personal redistribution, also demand a strong and like-minded state for promoting local development.

Central government, furthermore, has to change its internal organization so as to become more flexible. Today, frequently, 'stability prevails over ability' (OECD, 1985b, p. 2). Debureaucratization will help to respond quicker to private-sector and local needs. But one has to bear in mind that real debureaucratization has to be accompanied by decentralization.

15.6.4 ON COOPERATION AND CONFLICT

The OECD and EC promote the broad variety of activities above described by stressing the mutual advantages of ILEs. This approach concentrates on partnership and cooperation and shows interesting perspectives for local development. Indeed; plus-sum games are to be promoted whenever possible! However, this approach contains a crucial weakness because it does not at all deal with power relations and structural constraints. Free agreements between individuals and groups always face the danger of domination by commanding unequal power. A cat and a mouse do have mutual interests: they both profit from unpolluted air and clear water and both appreciate it if there is no other cat. But obviously there are conflicts too.... A multinational corporation and a medium-sized local community are usually unequal partners. For this reason democratic bodies like the central state or even supranational organizations like the European Community have to become countervailing powers to develop a legal framework which strengthens as much as possible local communities and their respective citizens facing multinational firms.

Another problem not explicitly dealt with in the OECD and EC reviews is that of conflicts within the community. Do ILEs create 'marginalized jobs for marginalized people' while, on the other hand, the official economy flourishes? The decline of a region does not affect all members equally. Some might even benefit! One cannot neglect these problems, especially if one is interested in the wellbeing of the marginalized and disadvantaged, as the OECD and EC stress (European Communities, 1985; OECD, 1986a).

One of the politically most interesting points concerning ILEs is that criticism and support equally come from the Left and the Right. Neo-liberals do not want local authorities to 'interfere with the market mechanism' and, therefore, 'advocate deregulation at local level and a reduction in tax pressure'. Radicals, on the other hand, 'consider that it is pointless to seek local level solutions to structural unemployment' (OECD, 1986d, p. 17). As has been shown, ILEs are in obvious contradiction to the ideologies of the free market and of unaccountable central planning. The problems of the future demand new concepts (see also Chapter 2). In corresponding unanimity,

> support for small businesses and community based employment initiatives in particular has been welcomed and promoted by authorities of all political persuasions, albeit with different objectives in mind. To some, such activities are encouraging self-help and enterprises and reducing dependence on the state; while to others, community and co-operative enterprise are a desirable alternative to traditional employment in large private companies. (Waters, 1985, p. 48)

Within this group of promoters of ILEs one has to be cautious as parts of the Right try to instrumentalize ILEs for achieving old objectives. They often only take new words (self-help, self-reliance, flexibility) for old remedies (unrestricted market economy and watchdog state). Rights without means, flexibility without security is not freedom! On the other hand, all those elements of ILEs that encourage the full realization of all members of a local community should be encouraged. By this is meant the importance given to concrete space, the communal fight against unemployment, the delegation of competences to the community or individuals, the preferential interest for the underprivileged, and so forth.

This demonstrates that ILEs offer an interesting though ambivalent regional policy instrument. The concrete outcome of the respective initiatives depends, as has been said, on the structural conditions that enable and limit their actions as well as on the imaginative action of individuals and groups at the local level. Those ILEs that are aware of their structural limitations and possibilities are, therefore, in a privileged position: they are able to take them into account in their activities, which obviously increases their chances of being successful. Knowledge is power!

ACKNOWLEDGEMENTS

The author would like to acknowledge the helpful advice given on previous drafts of this paper by Herwig Palme, Franz Tödtling and especially to the project leader Walter B. Stöhr from the Interdisciplinary Institute for Urban and Regional Studies at the University of Economics in Vienna, J. B. Pellegrin from the ILE Programme of the OECD in Paris, and Peter Roberts from Leeds.

REFERENCES

Bernstein, Richard J. (1986), 'Structuration theory as critical theory', *Praxis International*, vol. 6, no. 2 (July), pp. 235–49.

Boddy, Martin, and Fudge, Colin (1984), *Local Socialism?*, London: Macmillan.

Commission of the European Communities (1985a), *Programme of Research and Action on the Development of the Labour Market – Local Employment Initiatives, A Manual on Intermediary and Support Organisations, Main Report by The Centre for Employment Initiatives, London*, Luxembourg: Office for Official Publications of the European Communities.

Commission of the European Communities (1985b), *Forschungs- und Aktionsprogramm zur Entwicklung des Arbeitsmarkts. Örtliche Beschäftigungsinitiativen.* Luxembourg: Office for Official Publications of the European Communities.

Commission of the European Communities (1985c), *Programme of Research and Actions on the Development of the Labour Market – Local Employment Initiatives. An Evaluation of Support Agencies by Centre for Research on European Women – CREW, Brussels*, Luxembourg: Office for Official Publications of the European Communities.

Commission of the European Communities (1986a), *Cooperation in the Field of Employment – Local Employment Initiatives. Report on a Second Series of Local Consultations held in European Countries 1984–85. Final Report by the Centre for Employment Initiatives, London*, Luxembourg: Office for Official Publications of the European Communities.

Commission of the European Communities (1986b), *Initiative locales pour l'emploi. Relève des expériences de création d'emploi non conventionelles. Programme de recherche et actions sur l'évolution du marché du travail*, Luxembourg: Office for Official Publications of the European Communities.

Feedback ILE, no. 4 (1986) and no. 6 (1987), *Newsletter of the Co-operative Action Programme on Local Initiatives for Employment Creation (ILE)*, Paris: OECD.

Friedmann, John, and Weaver, Clyde (1979), *Territory and Function*, London: Edward Arnold.

Fröbel, F., Heinrichs, J., and Kreye, O. (1986), *Umbruch in der Weltwirtschaft*, Hamburg: Rowohlt.

Giddens, Anthony (1985), 'Time, space and regionalisation', in Gregory and Urry, op. cit.

Gregory, Derek, and Urry, John (1985), *Social Relations and Spatial Structures*, London: Macmillan.

Hesp, P., Stöhr, W., Stuckey, B., and UNIDO secretariat (1987), 'Introduction', in Muegge and Stöhr, op. cit.

Hesp, Paul, and Stuckey, Barbara (1987), 'Local development in the global network – the role of individual creativity and social entrepreneurship', *Journal für Entwicklungspolitik*, no. 1/87, Mattersburger Kreis für Entwicklungspolitik an den österreichischen Hochschulen, Vienna, pp. 26–42.

Hettne, Björn (1985), 'Development theory and the European crisis', in Stefan Musto (ed.), *Endogenous Development: a Myth or a Path? Problems of Economic Self-Reliance in the European Periphery*, Tilburg: EADI.

Hettne, Björn (1987), 'The crisis in development theory and in the world', *Journal für Entwicklungspolitik*, no. 1/87, Mattersburger Kreis für Entwicklungspolitik an den österreichischen Hochschulen, Vienna, pp. 5–25.

Holmes, John (1986), 'The organisation and locational structure of production subcontracting', in Scott and Storper, op. cit.

Hudson, Ray, and Sadler, David (1986), 'Contesting works closures in Western Europe's old industrial regions: defending place or betraying class?', in Scott and Storper, op. cit.

Johnston, R. J. (1986), 'The state, the region and the division of labor', in Scott and Storper, op. cit.

Kirsch, Guy (1987), 'Bourgeois oder Citoyen? Die politische Zukunft des Wirtschaftsliberalismus', *Die Zeit*, no. 17/87, pp. 41f.

Künstler, Peter (1985), 'Conclusions from local consultations held by the Commission of the European Communities on ILE', in OECD, 1985d.

Leve, Manfred (1985), 'Final remarks', in OECD, 1985d.

McEldowney, James, and Aitken, Peter (1987), 'Small firm and local employment initiative policies in the European Community: evolution and contribution', in Sutton (ed.) (1987), *Local Initiatives: Alternative Path for Development*, Maastricht: Presses Interuniversitaires Européennes.

McKean, B., and Coulson, A. (1987), 'Enterprise boards and some issues raised by taking equity and loan stock in major companies. Policy Review Section', *Regional Studies* vol. 21, no. 4, pp. 373–84.

Malecki, Edward J. (1986), 'Technological imperatives and moderate corporate strategy', in Scott and Storper, op. cit.

Martinos, Haris (1986), *Employment Creation in Local Labour Markets. Four Case Studies*, ILE Notebooks, Paris: OECD.

Moore, Mick (1987), 'Miracles and mysteries in the economic take-off of Taiwan and South Korea', *Journal für Entwicklungspolitik*', no. 2/87, Mattersburger Kreis für Entwicklungspolitik an den österreichischen Hochschulen, Vienna, pp. 2–12.

Muegge, Herman, and Stöhr, Walter (eds) (1987), *International Economic Restructuring and the Regional Community*, Aldershot: Gower.

Nijkamp, Peter, and Stöhr, Walter (1988), 'Technology policy at the crossroads of economic policy and physical planning', in Peter Nijkamp and Walter Stöhr (eds), *Special Issue on Technology Policy of Environment and Planning, Government and Policy, C* (forthcoming).

OECD (1985a), *Local Employment Initiatives in Norway – Experiences and Organisation of the Work*, Paris: OECD.

OECD (1985b), *Employment Creation in Local Labour Markets – A Comparative Review of Four Cities*, Paris: OECD.

OECD (1985c), *Municipal Economic Policy in Sweden*, Paris: OECD.

OECD (1985d), *ILEs: A Challenge to Public Employment Services*, ILE Notebooks No. 4, Paris: OECD.

OECD (1985e), *Possibilities for Action by Cities and Towns in Promoting Local Employment Initiatives and Economic Development*, Paris: OECD.

OECD (1985f), *Rapport sur les initiatives locales de création d'emplois en Italie*, Paris: OECD.

OECD (1985g), *Creating Jobs at the Local Level: Local Initiatives for Employment Creation*, Paris: OECD.

OECD (1986a), *Women – Local Initiatives – Job Creation*, ILE Notebooks no. 6, Paris: OECD.

OECD (1986b), *The Impact of Innovation and Technology Diffusion on Job Creation at Local Level. Report of a Review Team of Selected European Case Studies under the Chairmanship of the Rt. Hon. Shirley Williams*, Paris: OECD.

OECD (1986c), *Large Firms and Job Creation*, Paris: OECD.

OECD (1986d), *Employment Creation Policies – New Roles for Cities and Towns*, Paris: OECD.

OECD (1986e), *Experience in OECD Countries with Regard to Science Parks as a Means to Regional and Local Economic Development*, Paris: OECD.

OECD (1987), *ILE Programme*, Paris: OECD.

Scott, Allen J., and Storper, Michael (eds) (1986a), *Production, Work, Territory. The Geographical Anatomy of Industrial Capitalism*, Boston: Allen & Unwin.

Scott, Allen J., and Storper, Michael (1986b), 'Industrial change and territorial organisation: a summing up', In Scott and Storper, op. cit.

Stöhr, Walter B. (1985) 'Changing external conditions and a paradigm shift in regional development strategies?' in Stefan A. Musto and Carl F. Pinkele (eds), *Europe at the Crossroads. Agendas of the Crisis*, New York: Praeger.

Stöhr, Walter B. (1988), 'Regional policy, technology complexes and research/science parks', in Maria Giaoutzi and Peter Nijkamp (eds), *Informatics and Regional Development*, Aldershot: Gower.

Storper, Michael, and Scott, Allen J. (1986), 'Production, work, territory: contemporary realities and theoretical tasks', in Scott and Storper, op. cit.

Stingl, Josef (1985), 'Employment policy and local initiatives in Germany', in OECD, 1985d.

Tichy, Günther (1987), 'A sketch of a probabilistic modification of the product cycle – hypotheses to explain the problems of old industrial areas', in Muegge and Stöhr, op. cit.

Walker, Richard A. (1985), 'Class, division of labour and employment in space', in Gregory and Urry, op. cit.

Waters, Nigel (1985), *The Role of Local Government Authorities in Economic and Employment Development*, ILE Notebooks, Paris: OECD.

Select annotated bibliography on local development initiatives and related issues

Andreas Novy

This bibliography aims at giving all those interested in local development initiatives (LDIs) an overview of publications covering this topic. Since this topic has no strict border lines separating one field of investigation from another, the bibliography covers a relatively broad range of subject matter more or less closely related to LDIs.

Section 1 deals with European and section 2 with non-European case studies of LDIs. These sections contain examples of LDIs comparable to those described in this volume. Section 3 contains other material on areas closely related to LDIs, namely local development, self-reliance, socio-economic alternatives and innovation and technology. Section 4 introduces journals that regularly contain articles on related subjects.

Multiple mention of some books in different sections is intentional and is aimed at increasing the ease of handling. In sections 1, 2 and 3 there are cross-references for works that cover subject matter dealt with in other sections, according to keywords explained below. More specifically, the following cross-references are distinguished:

LOCAL DEVELOPMENT: policy and theory, public and private activities, indigenous development at the local level.

SOCIO-ECONOMIC ALTERNATIVES: third system approach, third sector (cooperatives), self-management, alternative life-style and new social movements.

SELF-RELIANCE: especially at regional or national levels.

INNOVATION AND TECHNOLOGY: social, organizational and technological innovation and entrepreneurship; economic aspects of technological progress, especially its local and regional implications.

REGIONAL SCIENCE: those aspects concerning local development initiatives not covered by other subject categories, e.g. functional and spatial division of labour, international restructuring process, uneven development.

EUROPEAN CASE-STUDIES: case studies of local development initiatives in Europe (section 1).

NON-EUROPEAN CASE-STUDIES: case studies of local development initiatives in non-European countries (section 2).

The author is grateful to Professor Gruchman, who supplied information concerning Eastern Europe.

1 CASE STUDIES OF LOCAL DEVELOPMENT INITIATIVES IN EUROPE

Albrechts, Louis (ed.) (1987), *Bedrijvencentra. Exponenten van lokale tewerkstellingsinitiatieven. Lokale Initiatieven 3*, Brussels: Koning Boudewijnstichting.

This mainly Dutch volume analyses local employment initiatives generally and with an empirical emphasis on the Flemish part of Belgium. The English contributions concentrate on business centres and innovation networks in Europe.

Aydalot, Philippe (ed.) (1986), *Milieux innovateurs en Europe. Innovative Environments in Europe*, Paris: GREMI (Groupe de Recherche Européen sur les Milieux Innovateurs).

• INNOVATION AND TECHNOLOGY

This volume focuses on the environment, which is crucial for the innovation process. Individual firms have to be seen as part of an environment that is favourable or hostile to innovation. It examines this general framework in case studies about metropolitan and old industrial regions.

Bullman, Udo, Cooley, Mike, and Einemann, Edgar (eds) (1986), *Lokale Beschäftigungsinitiativen. Konzepte – Praxis – Probleme*, Marburg: Norbert Schüren.

• LOCAL DEVELOPMENT

This book aims at deepening theoretical work on alternatives in local employment policy. A special section deals with local employment initiatives in West Germany ranging from trade union activities to initiatives of the local authorities and autonomous organizations like the well-known 'Netzwerk Selbsthilfe'.

Calouste Gulbenkian Foundation (1981), *Whose Business is Business?* London: UK and Commonwealth Branch.

This booklet is based on the analysis of forty community enterprises in Britain. Eleven of them are examined in detail.

Commission of the European Communities (1981), *Prospects for Workers' Cooperatives in Europe*, Volume I, *Overview*; Volume II, *Country Reports, First Series: Denmark, Greece, Republic of Ireland, The Netherlands, Spain, United Kingdom*; Volume III, *Country Reports. Second Series: Belgium, France, Federal Republic of Germany, Italy*. Vols I and II London: Mutual Aid Centre; Vol. III Paris: TEN Cooperative Advisers.

• SOCIO-ECONOMIC ALTERNATIVES

Volume I identifies how co-operatives are disadvantaged and discusses the pros and cons of providing special help of various kinds.

Hennicke, Martin, and Tengler, Hermann (1986), *Industrie- und Gewerbeparks als Instrument der kommunalen Wirtschaftsförderung. Eine empirische Analyse in der BRD*, Stuttgart: C. E. Poeschl.

This study empirically examines local economic and employment policies in Germany. Of special importance are technology parks which proved to be a good instrument for promoting young and innovative entrepreneurs.

Hult, Juhani, and Koliseva, Esko (1981), 'Activating villages, some viewpoints and experiences', Publications of Social and Regional Sciences, Joensuu, Finland: University of Joensuu.

This article describes 'activity models' in Finland and examines the relationship between the village movement and administration and between the school and village.

International Foundation for Development Alternatives (1987), *Urban Self-Reliance Directory*, Sponsored by the Food–Energy Nexus Programme, Nyon, Switzerland: United Nations University.

• NON-EUROPEAN CASE-STUDIES

This booklet contains short descriptions of 214 projects of local self-reliance which range from small, grassroots organizations to urban projects of large, international agencies. These projects cover industrialized as well as developing countries.

Johannisson, Bengt (1985), 'Business and local community – Swedish experiences in bottom-up planning for local industrial development', paper presented at the Canadian Institute of Planners' National Conference 1985 on 'Sustainable Community: The Next Frontier', 23–6 June, Sudbury, Ontario, Canada.

This booklet offers a local perspective on economic activity for Sweden which considers social networks as crucial. This new approach is examined in examples about revitalization and bottom-up mobilization.

Laske, Stephan, and Schneider, Ursula (1985), '... und es funktioniert doch!', *Selbstverwaltung kann man lernen*, Vienna: Bundesministerium für Soziale Verwaltung.

• SOCIO-ECONOMIC ALTERNATIVES
This book contains a theoretical overview of self-management and gives practical advice for creating self-managed firms themselves. It also describes in detail a project of a furniture-producing firm in the Tyrol, Austria.

Mayer, Jean (ed.) (1988), *Bringing Jobs to the People. Employment Promotion at the Regional and Local Levels*, Geneva: International Labour Office.
• LOCAL DEVELOPMENT
This volume brings together a number of articles that have appeared in the *International Labour Review* in recent years. In his chapter on local employment initiatives in Western Europe Künstler gives an excellent overview on this subject.

Miglbauer, Ernst (1985), *Betriebliche Selbstverwaltung in Österreich. Einführung und Bestandsaufnahme*, Linz: Eigenverlag.
• SOCIO-ECONOMIC ALTERNATIVES
Especially in times of crises self-management is seen by trade unions as a possible alternative to capitalist development. This work examines the Austrian experiences with five years of self-management.

Moore, Chris, and Skinner, Vivienne (1983), *Community Business in the Clydeside Conurbation*, Working Paper No. 2, The Inner City in Context. Glasgow: University of Glasgow.
This study is based on interviews with thirteen local projects which clarify what a community business is, how it starts up, legal structures and key issues of management and organization.

Morison, Hugh (1987), *The Regeneration of Local Economy*, Oxford: Clarendon Press.
• LOCAL DEVELOPMENT
This study examines recent British initiatives to stimulate local economic regeneration and considers the institutional and political constraints on implementing the policies which economic analysis suggests should be adopted.

Österreichische Arbeitsgemeinschaft für eigenständige Regionalentwicklung – ÖAR (ed.) (1988), *Peripherie im Aufbruch – Eigenständige Regionalentwicklung in Europa*. Vienna: ÖAR Publikationsreihe, ÖAR.
• LOCAL DEVELOPMENT
This publication which includes the results of a seminar held in Austria offers a broad variety of activities of indigenous development. Its emphasis is laid on peripheral rural areas and a broad variety of case studies from all over Europe is given.

Österreichische Arbeitsgemeinschaft für eigenständige Regionalentwicklung - ÖAR (n.d.), *Advice for Self-Help Projects in Disadvantaged Regions*, Vienna: ÖAR.

This booklet is an informational brochure which gives an overview of the acitivites of the ÖAR and the diverse variety of self-help projects in Austria.

Sutton, Alan S. (ed.) (1987), *Local Initiatives: Alternative Path for Development?*, Maastricht: Presses Interuniversitaires Européennes.
• LOCAL DEVELOPMENT

This book poses important questions for the development of Europe: Is 'localism' compatible with 'Europeanism'? Marginal or mainstream - what is the future of local development? Instead of only giving a theoretical answer a broad variety of initiatives from all over Europe is presented ranging from the OECD ILE programme, the promotion of small-scale enterprises to support agencies of ENI in Italy and of the British Steel Corporation.

Todd, Graham (1984), *Creating Jobs in Europe: How Local Initiatives Work*, London: Economist Intelligence Unit.

The report examines national governments' attempts to create employment and also gives specific examples of local initiatives taken within certain European countries.

Ulbricht, Tilo L. V. (ed.) (1986), *Integrated Rural Development. Proceedings of a Symposium*, The Hague: National Council for Agricultural Research.

This volume analyses the problems of development at local level by means of a study of twelve regions in eight EEC countries. The study shows how great the interaction is between different factors which determine prosperity and welfare.

Verein zur Förderung der Eigenständigen Regionalentwicklung in Hessen e.V. - VER (eds.) (1987), *Ansätze einer eigenständigen Regionalentwicklung. Beiträge aus Österreich, der Schweiz, England, Schottland und Hessen*, Kassel: DDV Offsetdruckerei.

This book contains the written results of a conference held in Melsungen in 1986. It gives an overview of the indigenous development policies in Hessen and deals with political and practical obstacles and possibilities. Furthermore, it describes experiences from Austria, the United Kingdom and Switzerland. Finally, the results of the working groups on agriculture, ecology and economics, handicraft and business, tourism and village development are summarized.

Young, Ken, and Mason, Charlie (eds) (1983), *Urban Economic Development: New Roles and Relationships*, London: Macmillan.
• LOCAL DEVELOPMENT

This volume focuses on the new involvement of local government in economic policy. Empirical research analyses the impact of these employment initiatives; the editors emphasize the ambiguous effects of serving welfare as well as the market and the problems of actually reversing the trend towards deindustrialization.

2 CASE STUDIES OF LOCAL DEVELOPMENT INITIATIVES IN NON-EUROPEAN COUNTRIES

Alvarado, Manuel León (1980), *Genossenschaften als Träger der Diffusion von Innovationen. Die Bedeutung der Agrargenossenschaften für die Diffusion von Innovationen in Entwicklungsländern dargestellt am Beispiel Boliviens*, Tübingen: J. C. B. Mohr.

This book deals with the problem of diffusing innovations to rural areas. An empirical investigation of Bolivian agriculture shows that co-operatives could play a decisive role in diffusing innovations if they were offered a favourable framework.

Hirschman, Albert O. (1984), *Getting Ahead Collectively. Grassroots Experiences in Latin America*, New York: Pergamon.

This book, written with immediate personal experience, attempts to combine eyewitness reporting on some of the more interesting situations and project histories with the establishment of analytical categories through which the dynamics of these projects can be better understood. It takes grassroots experiences seriously and, at the same time, establishes analytic categories through which the experiences become more intelligible.

Hyden, Goran (1980), *Beyond* Ujamaa *in Tanzania. Underdevelopment and an Uncaptured Peasantry*, Berkeley and Los Angeles: University of California Press.

An ethnocentric bias in the production of knowledge is particularly apparent in Africa, where Western scholars still dominate what is being written in the academic domain. This book tries to overcome this bias and attempts to explain the structural anomaly of rural Africa, which allows small to be powerful, and tries to apply its thesis to Tanzania.

International Foundation for Development Alternatives (1987), *Urban Self-Reliance Directory*, Sponsored by the Food–Energy Nexus Programme, Nyon (Switzerland): United Nations University.

• EUROPEAN CASE STUDIES

This booklet contains short descriptions of 214 projects of local self-reliance which range from initiatives taken by small, grassroots organizations to urban projects of large, international agencies. These projects cover industrialized as well as developing countries.

Max-Neef, Manfred (1982), *From the Outside Looking In: Experiences in 'Barefoot Economics'*, Uppsala: Dag Hammarskjöld Foundation.
This book centres around two Latin American experiences in grassroots initiatives. The first is ECU-28, a project that promotes horizontal communication for peasants' participation and self-reliance. The second is a community revitalization project in Tiradentes, Brazil.

Mayer, Jean (ed.) (1988), *Bringing Jobs to the People. Employment Promotion at the Regional and Local Levels*, Geneva: International Labour Office.
• LOCAL DEVELOPMENT
This volume brings together a number of articles that have appeared in the *International Labour Review* in recent years. Rabevazaha analyses how in the case of Madagascar the control of development by the people can influence regional planning and basic needs. Jatobá delivers a case study of the labour market in the recession-hit Brazilian North-East. A further chapter deals with the effects of special public works programmes on local labour market and income distribution.

Mbithi, Philip M., and Rasmusson, Rasmus (1977), *Self Reliance in Kenya. The Case of Harambee*, Uppsala: Scandinavian Institute of African Studies.
The Kenyan Harambee self-help movement offers an interesting example of more than a decade of 'bottom-up' development, with little and often no governmental financial support. The movement appears to reflect pragmatic local priorities and offers an opportunity to test what local people 'really want'.

Myrdal, Jan (1986), *Indien bricht auf*, Freiburg/Breisgau: Mersch.
This study is an excellent analysis of social movements in India. It concentrates on the situation of the rural population, the class struggle and the growing strength of the marginalized part of the population. Furthermore, it analyses Indian history 'from below'.

Schneider, Bertrand (1986), *La Révolution aux Pieds Nues*, Paris: Fayad.
This book grew out of the conviction that traditional largescale projects have failed to improve the situation of the poor. It therefore promotes another strategy, which relies on self-help and self-organization of the poor themselves.

Scott, James (1985), *Weapons of the Poor. Everyday Forms of Peasant Resistance*, New Haven and London: Yale University Press.
This sensitive picture of local class relations and the constant and circumspect struggle waged by peasants materially and ideologically against their oppressors in general and the introduction of the 'green revolution' in particular shows that techniques of evasion and resistance may represent the

most significant and effective means of class struggle in the long run. By showing this it gives impressive insights into Malayan rural societies and their local organization.

Scott Fossler, R., and Berger, Renée A. (1982), *Public–Private Partnership in American Cities. Seven Case Studies*, Lexington: Lexington Books.

This book examines how urban communities successfully organize government, business and community resources to meet a wide range of needs, from economic development and downtown revitalization to environmental finality, improved services, development of minority enterprises, jobs and housing.

Scurrah, Martin, and Podestá, Bruno (1986), *Experiencias autogestionarias en América Latina*, Lima: GREDES.

While this book cannot offer the final word on the advisability of adopting self-management, it does provide useful lessons grounded in practical experiences and gives an overview of experiences in self-management in Latin America where it has attracted a variety of proponents.

Swantz, Marja-Liisa (1987), 'Development: from bottom to top or top to bottom? Grassroots dynamics and directed development: the case of Tanzania', *EADI Newsletter No. 3*, Participation and Needs Working Group, Helsinki: Valtion painatuskeskus.

This article examines why the top-down policy approach of village-centred development in Tanzania did not succeed as it did not take into account the acute needs of the people.

Zamosc, Leon (1986), *The Agrarian Question and the Peasant Movement in Colombia. Struggles of the National Peasant Association 1967–1981*, UNRISD/CINEP, Cambridge: Cambridge University Press.

Rural social movements are particularly relevant when their outcome focuses on the crucial alternative between an evolution based upon a peasant pattern of farming or one dominated by landlords and entrepreneurs. This book examines the Colombian situation: the agrarian contradictions, the peasant–state relationship and the ideology of the peasant movement.

3 SUBJECT AREAS RELATED TO LOCAL DEVELOPMENT INITIATIVES

3.1 LOCAL DEVELOPMENT

Arocena, José, Cavallier, Marcel Paul, Richard, Pierre, and Sartori, Dominique (1984), *Initiative locale et dévelopment*, Paris: Groupe de sociologie de la création institutionelle (GSCI), Centre d'études sociologiques (CNRS).

This study poses the question whether the taking in hand of local economic problems by those concerned contributes to a new orientation of development.

Bandman, M. K. (ed.) (1984), *Lokalnye komplexnyie programmy*. Novosibirsk: Izdatelstvo, 'Nauka', Sibirskoie Oddrielenye.
This publication describes methodological problems connected with the preparation of local development programmes and in this connection cites examples of experiences in the development of the province of Novosibirsk, the city of Moscow and some other places.

Bideleux, Robert (1987), *Communism and Development*, London and New York: Methuen.
This book, which mainly deals with problems concerning Communism, also gives strong arguments in support of village communism and market socialism, which gain increased importance in the politics of *glasnost* in the East.

Blakely, E., and Bowmann, K. (1985), *Taking Local Development Initiatives: The Local Development Planning Process*, Canberra: Australian Institute of Urban Studies.
This study examines economic development which is seen as a process and a product. The development process is analysed in its different stages until its implementation.

Boddy, Martin, and Fudge, Colin (1984), *Local Socialism?*, London: Macmillan.
This book attempts to apply a theoretical analysis of the character of the central and local state to the key issues of the 1980s. It shows the attempt of the British Left to develop alternatives to the Thatcher government, which is enforcing spatial and social polarization. Alternatives, however, must not concentrate on central planning but instead should promote local and regional potentials.

Bönisch, R., Mohs, G., and Ostwald, W. (eds) (1982), *Territorialplanung*, (Ost-) Berlin: Die Wirtschaft.
This book contains several references to experiences of local and regional development in East Germany.

Brodhead, P. D., Decter, Michael, and Svenson, Ken (1981), *Community-based Development: A Development System for the 1980s. Technical Study 3*, Canada: Ministry of Supply and Services.
This report analyses the past decade of direct job creation programmes and advocates a development system approach with a community base. It includes a review of existing community-based economic development.

Bullmann, Udo, Cooley, Mike, and Einemann, Edgar (eds) (1986), *Lokale Beschäftigungsinitiativen. Konzepte – Praxis – Probleme*, Marburg: Norbert Schüren.

• EUROPEAN CASE STUDIES

This book aims to deepen theoretical work on alternatives in local employment policy. Trade-union activities and programmes as well as alternative initiatives and conceptions are described, a decentralized policy is elaborated and its necessity and limitations are mentioned. Furthermore, a section deals with the conversion of technology and production.

Commission of the European Communities: Programme of Research and Actions on the Development of the Labour Market. Local Employment Initiatives: various publications. Brussels: Office for Official Publications of the European Communities.

The Commission of the European Communities has published various monographs on local employment initiative within this research and action programme. Of special interest are reports on a series of local consultations by the Centre for Employment Initiatives, London. Those interested in the work of the Commission are referred to the bibliography of Chapter 15.

Commission of the European Communities (1981), *Prospects for Workers' Cooperatives in Europe*, Volume I, *Overview*; Volume II, *Country Reports, First Series: Denmark, Greece, Republic of Ireland, The Netherlands, Spain, United Kingdom*; Volume III, *Country Reports. Second Series: Belgium, France, Federal Republic of Germany, Italy*. Vols I and II London: Mutual Aid Centre, Vol III Paris: TEN Cooperative Advisers.

• EUROPEAN CASE STUDIES

Volumes II and III focus on individual countries, discuss the background and development of co-operatives and examine some of the support available and some of the main barriers.

Davies, Tom, and Mason, Charlie (1984), *Government and Local Labour Market Policy Implementation*, Aldershot: Gower.

This study analyses labour-market policy responses to three problem areas: training for information technology industry; large-scale redundancies; and training women for non-traditional jobs. The response by both central and local government to these problems is examined and it is shown how the nature of the government involvement serves to support the market imperative at the expense of social policies.

Daviter, Jürgen, Gessner, Volkmar, and Höland, Armin (1985). *Rechtliche, steuerliche, administrative und soziale Hindernisse für örtliche Beschäftigungsinitiativen. Zwischenbericht*, Bremen: Zentrum für europäische Rechtspolitik an der Universität Bremen.

This discussion paper examines legal, taxational, administrative and social obstacles for local employment initiatives. It discusses the sustainability of a self-managed economy, and the disposition over capital, consultancy and legal contacts.

Floyd, Michael (1984), *Policymaking and Planning in Local Government. A Cybernetic Perspective*, Aldershot: Gower.

In contrast to planning from above, cybernetics offers a fundamentally different perspective, which emphasizes the importance of processes, whereby the plans and policies of different organizations are mutually adjusted to each other from the bottom up. This book examines the dynamics of this process and shows the insights cybernetics can offer.

Friedrich-Ebert-Stiftung (ed.) (1986), *Fachtagung: Örtliche Beschäftigungsinitiativen. Berichte und Arbeitspapiere von der Fachtagung: Welche Bedeutung haben lokale Beschäftigungsinitiativen bei der Arbeitsplatzbeschaffung und der Bewältigung des industriellen und sozialen Wandels?* Bonn: Friedrich-Ebert-Stiftung.

This meeting critically examined the relationship between trade unions and local employment initiatives. Trade unions now usually accept that local employment initiatives have a role to play, but they still insist on structural solutions to labour-market problems and do not accept local initiatives to undermine workers' bargaining power.

Friedmann, John, and Weaver, Clyde (1979), *Territory and Function. The Evolution of Regional Planning*, London: Edward Arnold.

This classic work about territorially integrated development analyses the American movement of regionalism and its impact on regional policy. Furthermore the mainstream theory of regional politics is criticized and in the final section a new paradigm based on the recovery of territorial life is introduced.

Goldsmith, Michael (ed.) (1986), *New Research in Central–Local Relations*, Aldershot: Gower.

This volume includes an assessment of theories of central–local relations, their macro-economic context and relationship to policy implementation. It examines the work of local actors in the United Kingdom and the crucial area of local government finance and, finally, looks at these relations in sectoral policy areas.

Hahne, Ulf (1985), *Regionalentwicklung durch Aktivierung intraregionaler Potentiale. Zu den Chancen 'endogener' Entwicklungsstrategien*, Munich: Schriften des Instituts für Regionalforschung der Universität Kiel.

This work criticizes traditional policies from above and examines the theory of indigenous development and its applicability as a new paradigm for regional development. It furthermore elaborates strategies for utilizing intraregional potentials in Germany.

Internationale Chronik zur Arbeitsmarktpolitik (1985), 'Örtliche Beschäftigungsinitiativen', *Internationale Erfahrungen mit einem neuen Ansatz auf lokaler Ebene*, no. 19, Berlin: Wissenschaftszentrum Berlin.

This article offers an excellent overview on local employment initiatives.

Maier, Hans E., and Wollmann, Hellmut (eds.) (1986), *Lokale Beschäftigungspolitik*, Basel/Boston/Stuttgart: Birkhäuser Verlag.

Owing to the lack of national solutions to unemployment, local actors are becoming increasingly involved in labour-market policy. Local employment policies include the regionalization of programmes of the nation state and their implementation by local authorities as well as autonomously initiated local or regional employment policy.

Mayer, Jean (ed.) (1988), *Bringing Jobs to the People. Employment Promotion at the Regional and Local Levels*, Geneva: International Labour Office.

• EUROPEAN/NON-EUROPEAN CASE STUDIES

This volume brings together a number of articles that have appeared in the *International Labour Review* in recent years. As well as specific case studies there are contributions on the theory and practice of regional employment development, urban and rural approaches to this problem and manpower information at local and branch levels.

Midgley, James, Hall, Anthony, Hardiman, Margaret, and Narin, Dhanpaul (1986), *Community Participation, Social Development and the State*, London and New York: Methuen.

This book examines the question of the role of the state in community participation and asks whether state involvement is beneficial or harmful to community participation and whether community and state resources can be harmonized to promote social development.

Morison, Hugh (1987), *The Regeneration of Local Economy*, Oxford: Clarendon Press.

• EUROPEAN CASE STUDIES

This study examines recent initiatives to stimulate local economic regeneration and considers the institutional and political constraints on implementing the policies which economic analysis suggests should be adopted.

Nassmacher, Hiltrud (1987), *Wirtschaftspolitik 'von unten'. Ansätze und Praxis der kommunalen Gewerbestandspflege und Wirtschaftsförderung*. Basel/Boston/Stuttgart: Birkhäuser Verlag.

This study analyses the variety of instruments available to influence the local economy directly or indirectly. It examines the range of action of local agents, the neglect of an employee-oriented policy and the tendency to aggravate the problems of technical change.

Niedersächsisches Institut für Wirtschaftsforschung e.V. (ed.) (1985), *Lokale Entwicklungsstrategien. Neue Perspektiven für die regionale Wirtschaftspolitik*, NIW Workshop 1985, Hanover: NIW.

This volume analyses whether local development strategies can serve as an alternative to regional economic policy. Innovation strategies in the industrial sector, rural development and integrated strategies are examined in detail.

Österreichische Arbeitsgemeinschaft für eigenständige Regionalentwicklung – ÖAR (ed.) (1988), *Peripherie im Aufbruch – Eigenständige Regionalentwicklung in Europa*, Vienna: ÖAR.

• EUROPEAN CASE STUDIES

This publication, which includes the results of a seminar held in Austria, offers a broad variety of activities of indigenous development. Its emphasis is on peripheral rural areas, and a broad variety of case studies from all over Europe is given.

Organisation for Economic Co-operation and Development: Cooperative Action Programme on Initiatives for Local Employment Creation. Paris: OECD.

Within this programme the OECD published various monographs on local employment initiatives. Their publications include *Feedback ILE*, which gives a short overview on special topics such as universities, trade unions or women. *ILE Notebooks* contain more detailed information and present a variety of case studies. Finally, there is a broad literature of monographs whose circulation is limited. In the bibliography of Chapter 15 the interested reader can find further OECD publications.

Organisation for Economic Co-operation and Development (1985), *Creating Jobs at the Local Level. Local Initiatives for Employment Creation*, Paris: OECD.

This report is the official statement of views of how the Directing Commitee of the ILE Programme sees the problems of local employment initiatives and the role public authorities might take. Furthermore, it presents selected case studies and, finally, a very helpful annotated bibliography of ILEs, ordered thematically. Of utmost importance is the bibliography covering the new roles for local authorities, co-operatives and self-managed firms and about support systems for ILEs.

Perry, Martin (1986), *Small Factories and Economic Development*, Aldershot: Gower.

Property initiatives for small business have been given high priority in national and local economic development policy. This book assesses the wisdom of this investment and evaluates the importance of property availability in fashioning the industrial landscape.

Ross, David P., and Usher, Peter J. (1986), *From the Roots Up: Economic Development as if Community Mattered*, Croton on Hudson, New York: The Bootstrap Press.

This work places strong emphasis on informal economic activity. Neglecting this part of the economy causes the importance of local and regional networks to be underestimated. The authors, therefore, propose a new 'politics of small scale'.

Sutton, Alan S. (ed.) (1987), *Local Initiatives: Alternative Path for Development?*, Maastricht: Presses Interuniversitaires Européennes.

• EUROPEAN CASE STUDIES

This book grew out of the conviction that many economic and social problems could be significantly eased by the creation of vigorous and autonomous processes of local development. Strong emphasis is given to the dichotomy between business and employment creation and between social and economic values. It poses important questions for the development of Europe: is 'localism' compatible with 'Europeanism'? Marginal or mainstream – what is the future of local development?

Weaver, Clyde (1984), *Regional Development and the Local Community: Planning, Politics and the Social Context*, Chichester: Wiley.

This book attempts to reinterpret the social and intellectual evolution of regional planning in France, Britain and the United States. It furthermore offers a brief sketch of an alternative, self-sustaining form of regional development, based upon the fulfilment of local needs by means of bottom-up use-value creation.

Williams, R. H., Cameron, S. J., Gillard, A. A., and Willis, K. G. (1986), *Promoting Local Economic Development. Policies and Evaluation*. Beckenham: Croom Helm.

All over the world promoting local economic development has recently become a function of local as well as of central government. Most local authorities now have their own economic development officers and their own strategies for promoting local economic growth; and they give incentives to induce large firms to locate in their areas and small firms to start up. This study analyses and assesses this recent boom in local economic development policies.

Young, Ken, and Mason, Charlie (eds) (1983), *Urban Economic Development. New Roles and Relationships*, London: Macmillan.

• EUROPEAN CASE STUDIES

This volume focuses on the new involvement of local government in economic policy. Empirical research analyses the impact of these employment initiatives; the editors emphasize the ambiguous effects of serving welfare as well as the market and the problems of actually reversing the trend towards deindustrialization.

3.2 SOCIO-ECONOMIC ALTERNATIVES

Badelt, Christoph (1980), *Sozioökonomie der Selbstorganisation. Beispiele zur Bürgerselbsthilfe und ihre wirtschaftliche Bedeutung*, Frankfurt: Campus.

This book examines self-organization as part of the emerging 'third sector'. It combines a political and economic analysis of this phenomenon with detailed descriptions of self-help organizations in various countries.

Centre d'études de l'emploi (1984), *Le créateur d'enterprise. Profil et besoins. Propositions d'action*, written by Félix S. Dossou, Dossier de recherche no. 9, Paris: CEE.

This study tries to improve the efficiency of the diverse social actors and the different institutions that promote job creation.

Clement, Werner (1985), *Die Fragmentierung des Arbeitsmarktes. Beispiele neuerer Formen der Beschäftigung und der Arbeitsmarktpolitik in Frankreich und den USA*, Nuremberg: Institut für Arbeitsmarkt- und Berufsforschung der Bundesanstalt für Arbeit.

This study presents a typology of new forms of employment and analyses public labour market policies, employee-owned firms, volunteer work and new work schedules in the USA and employment pacts, 'chomeurs créateur d'enterprises', 'emploi d'initiative locale' and 'contrats de solidarité' in France.

Einemann, Edgar, and Lübbing, Edo (1985), *Anders produzieren: Alternative Strategien in Betrieb und Region*, Marburg: SP-Verlag.

This study analyses the labour process from the perspective of those affected by the crisis. Furthermore, an alternative regional policy is elaborated that orients itself on experiences of the Greater London Council and is based on decentralized structural policy. Finally, alternatives in the German energy policy based on regionalization are presented.

European Centre for Social Welfare, Training and Research (ed.) (1987), International Expert Meeting on 'The Role of the Social Economy in the Creation of Local Employment: Between the Social and the Economic' organized by Eurosocial in cooperation with the Centre National de la Recherche Scientifique, Vienna.

This seminar analysed the new phenomenon of local initiatives from different angles to take account of their social as well as their economic dimension. The analysis centres around Austrian and French experiences.

Gijsel, Peter de, *et al.* (1985), *Ökonomie und Gesellschaft 4: Jahrbuch 3: Jenseits von Staat und Kapital*, Frankfurt/New York: Campus.

This yearbook proposes the third sector as an alternative to state and capital and examines the possibilities of an alternative economy. But the authors themselves are aware that the alternative economy need not contribute to the decline of capitalism. It might on the contrary help to renew profitability.

Hegner, Friedhart, and Schleglmilch, Cordia (1983), *Formen und Entwicklungschancen unkonventioneller Beschäftigungsinitiativen*, Berlin: International Institute of Management.

This discussion paper proposes analytical criteria by which to understand the different forms of unconventional employment creation and offers a typology relating to the way resources are obtained.

International Labour Organisation (1986), *Promoting Small and Medium-Sized Business*, Geneva: ILO.

Prepared for the 72nd session of the International Labour Conference (June 1986), this report puts the accent on policies and programmes destined to encourage the creation and development of small and medium-sized enterprises (in the widest interpretation of the concept).

Keating, Michael, and Hainsworth, Paul (1986), *Decentralisation and Change in Contemporary France*, Aldershot: Gower.

This book sets out to compare the original aims and promises with the reality of decentralization. While finding that the achievements have been real it concludes that decentralisation has brought a dispersal of power to existing élites rather than a 'new citizenship'.

Laske, Stephan, and Schneider, Ursula (1985), '... und es funktioniert doch!' *Selbstverwaltung kann man lernen*, Vienna: Bundesministerium für Soziale Verwaltung.

• EUROPEAN CASE STUDIES

This book contains a theoretical overview of self-management and gives practical advice for creating self-managed firms. It also describes in detail a project of a furniture-producing firm in the Tyrol, Austria.

Lopes, Marguerite T., and View, Jenice (1983), *Women, Welfare and Enterprise*, The American Enterprise Institute. Neighborhood Revitalization Project.

This paper explores the reasons for limited female entrepreneurship and suggests steps that can be taken to increase the rate of business formation by women, including low-income women and those on welfare.

McRobie, G. (1981), *Small Is Possible*, London: Jonathan Cape.

This work is the completion of a trilogy started by E. F. Schumacher with *Small Is Beautiful*. It describes local initiatives based on the use of intermediate technology in Europe and in other continents.

Miglbauer, Ernst (1985), *Betriebliche Selbstverwaltung in Österreich: Einführung und Bestandsaufnahme*, Linz: Eigenverlag.
• EUROPEAN CASE STUDIES
Especially in times of crises self-management is seen by trade unions as a possible alternative to capitalist development. This work examines the Austrian experiences with five years of self-management.

Ronen, Joshua (ed.) (1983), *Entrepreneurship*, Price Institute for Entrepreneurial Studies, Lexington: Lexington Books.
This volume reaffirms that the malaise of economic theory and policy lies in the neglect of the entrepreneur, who is seen as a person who identifies an opportunity and gives it an existence. Entrepreneurial behaviour and entrepreneurship in Yugoslavia and the Soviet Union are analysed from very different theoretical points of view.

Tavistock Institute of Human Relations (1986), *Worker Co-operatives in France, Italy and the United Kingdom. A Study of Organisation, Employment and Participation*, Vol. 1, *Main Report*, London: Tavistock Institute.
Among the issues given particular attention in this report are contemporary expectations of co-operatives and how far these are being fulfilled, the similarities and differences between co-operatives and small and medium-sized enterprises, the outcome of effective co-operative organization and the strategies of trade unions towards co-operatives.

3.3 SELF-RELIANCE

Bassand, Michael, Brugger, Ernst A., Bryden, John M., Friedmann, John, and Stuckey, Barbara (eds) (1986), *Self-reliant Development in Europe: Theory, Problems, Actions*, Aldershot: Gower.
This volume investigates how the practice of self-reliance affects the division of labour and vice versa. Special attention is given to the point where the division of labour becomes a hindrance rather than a help for individual and social development.

Evers, Hans-Dieter, Senghaas, Dieter, and Wienholtz (eds) (1983), *Auf dem Weg zu einer Neuen Weltwirtschaftsordnung? Bedingungen und Grenzen für eine eigenständige Entwicklung*, Nomos Verlagsgesellschaft, Baden-Baden: Nomos.
Inspired by Senghaas' *autozentrierter Entwicklung* (self-reliant development), this book examines the development of underdevelopment and looks for an alternative development strategy of dissociation which is more compatible with peace. The possibilities and limitations of a self-reliant development strategy are examined in national case studies.

Galtung, Johan, O'Brien, Peter, and Preiswerk, Roy (eds) (1980), *Self-Reliance: A Strategy for Development*, London: published for the Institute for Development Studies, Geneva, by Bogle-L'Overture.

This book offers self-reliance as an alternative to mainstream development policy. The roots of the theory of self-reliance are elaborated and mobilization, self-management, authenticity and an alternative lifestyle are seen as important components of self-reliance. Furthermore, sectoral self-reliance in energy, food, health, learning and technology is analysed and detailed national case studies are delivered.

Musto, Stefan (ed.) (1985), *Endogenous Development: A Myth or a Path? Problems of Economic Self-Reliance in the European Periphery*, Tilburg: EADI.

A common negative denominator of the individual contributions is provided by the confirmation that the overall functionalist growth model has failed to meet the expectations raised by it. The concept of indigenous development is analysed conceptually and theoretically. Furthermore, the challenges to development in southern Europe are analysed, special attention being given to old industrial regions.

Stöhr, Walter, B., and Taylor, Fraser D. R. (eds) (1981), *Development from Above or Below? The Dialectics of Regional Planning in Developing Countries*, Chichester: Wiley.

This volume analyses the theory of regional development. It critically assesses neoclassical 'trickle down' theory from above and offers 'bottom-up' planning as an alternative. In case studies of twelve developing countries these two approaches are compared and their respective successes and failures are described.

3.4 INNOVATION AND TECHNOLOGY

Allen, David N., and Levine, Victor (1986), *Nurturing Advanced Technology Enterprises: Emerging Issues in State and Local Economic Development*, New York: Praeger.

This work investigates the relationship between technological change and development. The nature of an advanced-technology workforce is characterized and training schemes are analysed. The role of universities is described. Finally, financial problems of start-ups are analysed and special attention is given to small business promotion.

Amin, Ash, and Goddard, John (eds) (1986), *Technological Change, Industrial Restructuring and Regional Development*, London: Allen & Unwin.

This book attempts an interdisciplinary approach to technological, industrial and regional problems. Giving special attention to the British situation it analyses regional implications of the internationalization of production due to technological changes. The role of small businesses in regional development is given special attention.

Aydalot, Phillippe (ed.) (1986), *Milieux innovateurs en Europe: Innovative Environments in Europe*, Paris: GREMI (Groupe de Recherche Européen sur les Milieux Innovateurs).

• EUROPEAN CASE STUDIES

This volume focuses on the environment, which is crucial for the innovation process. Individual firms have to be seen as part of an environment that is favourable or hostile to innovation. It examines this general framework in case studies about metropolitan and old industrial regions.

Boulianne, Louis-M., and Maillat, Denis (1983), *Technologie, enterprises et région: une étude régionale empirique*, Saint Saphorin: Georgi.

This book examines the regional impact of technological change and the principal determinants for understanding the differing capacity of diverse regions to adapt to external changes, which are the regional enterprise structure, integration in the global production system and other regional characteristics.

Brown, Lawrence A. (1981), *Innovation Diffusion: A New Perspective*, London and New York: Methuen.

Innovation plays an important role in the rise and fall of civilizations. This book focuses on the ways in which innovation and the conditions for adoption are made available to individuals, households or enterprises which constitute the population of potential adopters. It also elaborates a market and infrastructure perspective of the innovation process.

Castells, Manuel (ed.) (1985), *High Technology, Space and Society*, Beverley Hills: Sage.

This volume is based on the conviction of the importance of technological change for the evolution of spatial and social forms, together with an emphasis on the need to integrate technology in a broader framework of social relationships in order to understand the diversity of its effects on people's lives, on institutions and, ultimately, on spatial forms and processes.

Federwisch, Jacques, and Zoller, Henry G. (eds) (1986), *Technologie nouvelle et ruptures régionales*, Paris: Economica.

This volume studies non-European experiences in technology policy and tries to adapt these - if possible - to the European situation. It is shown empirically that the local conditions are of utmost importance for the localization of high-technology industries. Besides empirical work it also contains a theoretical and methodological criticism of mainstream regional theory.

Freeman, Christopher, Clark, John and Soete, Luc (1982), *Unemployment and Technical Innovation: A Study of Long Waves and Economic Development*. London: Frances Pinter.

Select annotated bibliography

Until recently economists treated technical change as a black box. It is the contention of this book that these simplifying assumptions about technical change can obscure the real processes of change rather than clarify them, if we are considering long periods such as half a century. Statistical analysis, therefore, must be complemented by economic, social and technological history.

Giaoutzi, Maria, and Nijkamp, Peter (eds) (1988), *Informatics and Regional Development*, Aldershot: Gower.

This volume analyses a broad variety of subject matter related to the relationship between regional development and informatics. Case studies from all over Europe are included; all affirm the importance of the growing informatics sector in the process of economic restructuring following the last recession.

Hakansson, Hakan (1987), *Industrial Technological Development: A Network Approach*, Beckenham: Croom Helm.

Innovations should not be regarded as the product of only one actor but as the result of an interplay between two or more actors, in other words as a product of a network of actors. This study elaborates this conclusion in relation to knowledge development, resource mobilization and coordination.

Hesp, Paul, and Stuckey, Barbara (1987), 'Local development in the global network – the role of individual creativity and social entrepreneurship', *Journal für Entwicklungspolitik*, Vienna: Mattersburger Kreis für Entwicklungspolitik.

This article examines the consequences of the 'Third Technological Revolution' on local development. It proposes innovation in the sense of individual creativity and moral judgement and social entrepreneurship as a means for regions once again to become history-makers.

Lundstedt, Sven B., and Colglazier, William E. Jr. (1982), *Managing Innovation: The Social Dimension of Creativity, Invention and Technology*, Published with the Aspen Institute for Humanistic Studies and the Ohio State University, Oxford: Pergamon Press.

The special focus of this book is on the synergism between social and technological innovation, or – stated more directly – how new arrangements in social structures or processes can facilitate the use and diffusion of new technological products or processes and thereby generate increased productivity and economic growth.

Nijkamp, Peter (ed.) (1986), *Technological Change, Employment and Spatial Dynamics*, Berlin: Springer-Verlag.

This volume examines the relationship between technological development and shifts in the labour market and the interaction between the urban or regional production environment and technology changes. These general reflections are deepened by empirical works on actual trends and econometric modelling on these topics.

Organisation for Economic Co-operation and Development (1979), *Trends in Industrial R&D in Selected OECD Member Countries: 1967–1975*, Paris: OECD.

The aim of the study is to examine recent trends in industrial R&D in the OECD area to see whether potentials are being maintained especially in the changed economic context since the oil crises of 1973.

Organisation for Economic Co-operation and Development (1984), *Industry and University: New Forms of Co-operation and Communication*, Paris: OECD.

This booklet aims to explore the current initiatives for closer interaction, the new motivations for collaboration, novel approaches for individual countries, and problems and prospects for strengthening relations between industry and universities.

Thwaites, A. T., and Oakey, R. P. (eds) (1985), *The Regional Economic Impact of Technological Change*, London: Frances Pinter.

This book reflects the recent work of a number of researchers concerned with the regional dimension to technological change within advanced industrial countries. Various aspects of technology are examined, ranging from research and development efforts to the diffusion of new technologies in the public and service sectors of the economy.

3.5 REGIONAL SCIENCE

Boddy, Martin, Lovering, John, and Bassett, Keith (1986), *Sunbelt City? A Study of Economic Change in Britain's M4 Growth Corridor*, Oxford: Clarendon Press.

This book deals with the processes of urban change. It contains case studies of UK cities with severe problems of economic decline, unemployment and social distress, on the one hand, and successful adapters, on the other. It tries to identify the reasons for these different developments.

Bradbury, J. H. (1985), 'Regional and industrial restructuring processes in the new international division of labour', *Progress in Human Geography*, vol. 9, no. 1, London: Edward Arnold.

This article analyses restructuring as a geographic phenomenon by stating that fluctuations in the 'global fabric' are the sum of the differences between and within regions. Furthermore, the influence of long-term economic swings on regional performance is taken into account.

Castells, Manuel, and Henderson, Jeffrey (eds) (1987), *Global Restructuring and Territorial Development*, London: Sage.
This book aims at a reconceptualization of the analysis of the restructuring process, which takes global, systemic changes as their starting point. In a second step in a variety of case studies some of the social and spatial consequences of global restructuring for particular economic sectors, regions and cities are analysed. These case studies range from the electronic and automobile industry to national case studies of Mexico and Malaysia.

Forbes, D. K., and Rimmer, P. J. (eds) (1984), *Uneven Development and the Geographical Transfer of Value*, Human Geography Monograph 16, Canberra: Research School of Pacific Studies, Australian National University.
The new categorization of the regional problem is important because it implies a switch to a structural analysis utilizing Marxist theoretical concepts. This book gives an overview of major debates in geography and argues that articulation research has reached an impasse.

Hamilton, Ian F. E. (ed.) (1986), *Industrialization in Developing and Peripheral Regions*, Beckenham: Croom Helm.
The broad message of the contributions in this book is one of caution and doubt about the achievements and future prospects for real industrial progress in peripheral regions. Their comparative advantages in lower technology and higher labour-intensive sectors do not enable them to close the economic gap with the core regions. The role of science and technology in economic development is discussed and examples of peripheralism and regional change in East European economies and 'sunbelt' countries in southern Europe are introduced.

Keeble, David, and Wever, Egbert (eds) (1986), *New Firms and Regional Development in Europe*, Beckenham: Croom Helm.
With the decline of the traditional heavy industries there has been much emphasis placed by governments in most EC countries on the importance of developing new, small manufacturing firms. This book reviews the extent and reasons for geographical variations in the number of new firms, examines the nature of such firms and assesses the regional impact and policy implications in various EC countries.

Marshall, Michael (1987), *Long Waves of Regional Development*, London: Macmillan.
This book applies the theories of long waves of Kondratieff, Schumpeter and Mandel to the analysis of the rise and decline of regions. Furthermore, a novel description of regional development in Great Britain is given, based on the contradiction of industrial and financial capital. Marshall's theory of long waves is no deterministic model but offers broad space for individual action and the struggle of classes and interests at the local level.

Massey, Doreen (1984), *Spatial Dimensions of Labour: Social Structures and the Geography of Production*, London: Macmillan.

The overall argument of this book is that behind major shifts between dominant spatial divisions of labour within a country lie changes in the spatial organization of capitalist relations of production. Such shifts in spatial structures are a response to changes in class relations and their development is a social and conflictual process.

Moulart, Frank, and Salinas, Patricia W. (eds) (1983), *Regional Analysis and the New International Division of Labor: Application of a Political Economy Approach*, Boston/The Hague/London: Kluwer-Nijhoff Publishing.

This book contrasts the theories, concepts and methodologies used in regional political economy and neoclassical regional science. Political economy models take account of change and movement, history and struggle. Therefore, the state and social movements are of utmost importance as space is viewed as the locus of a set of social relations.

Muegge, Hermann, and Stöhr, Walter B. (eds) (1987), *International Economic Restructuring and the Regional Community*, Aldershot: Gower.

This volume grew out of a new awareness of the firm as an individual entrepreneurial decision-making unit, caught up in the international process of restructuring. Bottom-up strategies are seen as a possibility for the region to become once again a history-maker. A variety of case studies deepens the theoretical insights of the book.

Musto, Stefan A., and Pinkele, Carl F. (eds) (1985), *Europe at the Crossroads. Agendas of the Crisis*, New York: Praeger.

This volume takes the European crisis as a starting point. It examines the phenomenon of the crisis in detail and poses the question whether dependency or autonomy is going to be the future of Europe. A representative sample of peripheral nations is analysed in detail. Finally a discussion is offered of whether self-reliance might be an alternative to mainstream regional development.

Quévit, Michel (1986), *Le pari de l'industrialisation rurale: la capacité d'entreprendre dans les régions rurales des pays industrialisés*, Paris: Editions Anthropos.

This work analyses the potential of industrialization in peripheral rural areas to promote indigenous development. It analyses the variety of initiatives and examines policy instruments to support them.

Scott, Allen J., and Storper, Michael (eds) (1986), *Production, Work, Territory: The Geographical Anatomy of Industrial Capitalism*, London: Allen & Unwin.

This book tries to elucidate the geographical underpinnings of modern industrial capitalism. In what amounts to a reformulation of human geography this book breaks new ground in building up a problematic of productive activity, labour markets and territorial organization in modern capitalism. Several new approaches are developed to location theory, local labour markets in cities and regions and to regional analysis in general.

Smith, Michael Peter, and Feagin, Joe R. (eds) (1987), *The Capitalist City: Global Restructuring and Community Politics*, Oxford: Basil Blackwell.

The world of modern capitalism is a global network both of corporations and of cities. This book analyses the intricate relationship among cities, state politics and urban politics at a time of economic restructing at global, national and local levels and deals with local responses of the community, households and urban politics to global restructuring.

Stöhr, Walter B. (1987), 'Regional economic development and the world economic crisis', *International Social Science Journal*, May, Oxford: Basil Blackwell.

This article examines the new regional problems arising from changes in external conditions which no longer allow the export of problems to other regions. Therefore, indigenous resource mobilization and innovation is presented as an alternative to external transfers.

Törnqvist, Gunnar, Gyllström, Björn, Nilsson, Jan-Evert, and Svensson, Lennart (eds) (1986), *Division of Labour, Specialisation, and Technical Change: Global, Regional, and Workplace Level*, Lund: Locus Liber.

This volume contains integrated studies that discuss division of labour, specialization and technical development, simultaneously considering the individual as well as the systemic perspective. The contributions are structured spatially in a global and international, a national and regional and a workplace level.

Wadley, David (1986), *Restructuring the Regions: Analysis, Policy Model and Prognosis*, Paris: OECD.

Sectoral failures created equally serious regional problems and severe employment repercussions. These new regional problems are examined in this report and alternative strategies for regional development as well as country studies and prospects are proposed.

4 JOURNALS

Development Dialogue. A journal of international development cooperation published by the Dag Hammarskjöld Foundation, Uppsala, Sweden.

The Dag Hammarskjöld Foundation has become famous by promoting 'Another Development'. *Development Dialogue*, therefore, offers diverse examples of another development in the Third World, local initiatives are described and alternatives to dominant policies from above are presented.

IFDA Dossier. International Foundation for Development Alternatives, Nyon, Switzerland (bimonthly).

IFDA has created the term 'Third System' to indicate people's power independent of political or economic power. In its dossiers it offers a wide range of short articles about development initiatives all over the world. Of special interest are articles about the 'local space' and 'news from the Third System'.

International Journal for Urban and Regional Research. Edward Arnold, London (quarterly).

Although local responses to global challenges are not at the centre of concern in this journal there are interesting papers on this subject, especially:

Feagin, Joe R., 'The social costs of Houston's growth: a sunbelt boomtown re-examined', vol. 9, no. 2, June 1985.

Trachte, Kent, and Ross, Robert, 'The crisis of Detroit and the emergence of global capitalism', vol. 9, no. 2, June 1985.

O'Dowd, Liam, and Rolston, Bill, 'Bringing Hong Kong to Belfast? The case of an enterprise zone', vol. 9, no. 2, June 1985.

Galster, George C., 'What is neighbourhood? An externality–space approach', vol. 10, no. 2, June 1986.

Clarke, Susan E., and Meyer, Margit, 'Responding to grassroots discontent: Germany and the United States', vol. 10, no. 3, September 1986.

O'Leary, Brendan, 'Why was the GLC abolished?' vol. 11, no. 2, June 1987.

Harris, Richard, 'A social movement in urban politics: a reinterpretation of urban reform in Canada', vol. 11, no. 3, September 1987.

Regional Development Dialogue. United Nations Centre for Regional Development, Nagoya, Japan (quarterly).

Of special interest is vol. 6, no. 1, Spring 1986. One form of decentralization is devolution of decision-making power to local governments. But without resource-mobilization powers of their own, local governments are in a difficult position. This analysis is supported by case studies from the Philippines, Thailand, Sudan and the Republic of Korea. Vol. 5, no. 1, Spring 1985 analyses the role of the village.

Regional Studies. Journal of the Regional Studies Association, Cambridge University Press, Cambridge (bimonthly). Of special interest are:

O'Farrell, Patrick N., 'Entrepreneurship and regional development, some conceptual issues', vol. 20, no. 6, October 1986.

Lewis, J. R., and Williams, A. M., 'Productive decentralisation or indigenous growth?: small manufacturing enterprises and regional development in central Portugal', vol. 21, no. 4, August 1987.

Index

Index

Index

Index